Systems Analysis for Water Technology

Willi Gujer

Systems Analysis
for Water Technology

 Springer

Prof. Dr. Willi Gujer
ETH Zurich
Institute of Environmental Engineering
Wolfgang-Pauli-Straße 15
8083 Zurich
Switzerland
gujer@ifu.baug.ethz.ch

ISBN 978-3-540-77277-4   e-ISBN 978-3-540-77287-1

DOI 10.1007/978-3-540-77287-1

Library of Congress Control Number: 2008924075

Cover design: Frido Steinen-Broo, eStudio Calamar, Spain

Printed on acid-free paper

9 8 7 6 5 4 3 2 1

springer.com

Preface

This book has a rather long history. It goes back to 1980, when environmental engineering evolved from sanitary engineering as a new and broader engineering discipline. I had the assignment to teach a course in *mathematical modeling of technical systems* as part of a new postgraduate course in Urban Water Management and Water Pollution Control at ETH in Zurich. I decided to teach this course on a rather abstract level, with the goal of introducing methods that are generally applicable across the different disciplines of what was then defined as environmental engineering. Now I teach a graduate course in *methods for urban water management*, which heavily relies on the material I compiled in the 1980s. This course is offered in the first semester of the master education in environmental engineering at ETH; it requires four hours of lecturing and problem sessions a week during one semester. The students earn six credit units (ECTS).

Of all the engineering disciplines, environmental engineering appears to be among those that maintain the most intimate contacts with the natural science disciplines. Only a detailed understanding of chemical, physical, and microbial processes will lead to engineered systems that fulfill the requirements of society and the environment and at the same time do not require excessive economic and natural resources. Mathematical models are a crucial base for engineering design – in environmental engineering they typically combine a quantitative description of chemical and microbial transformation processes with the description of the physical transport processes within the system of interest.

This book introduces methods and generic models that support the development of detailed system-specific mathematical models, primarily of technical water and wastewater treatment systems. It concentrates on methods which are required for the development of these models; it does not introduce a detailed discussion of specific processes or systems. In combination with an in-depth education in physical, chemical, and microbial processes for water and wastewater treatment these methods and models are of eminent value for the professional analysis of engineered systems.

Frequently mathematical modeling leads to coupled, nonlinear differential equations and thus requires the application of numeric integration. In addition the tools for systems analysis, parameter identification, sensitivity analysis, and error propagation are essential for responsible engineering work. A vast array of software products for this purpose are available on the market. A steep learning curve, ease of availability, economics, spectrum of tools, and efficiency led me to choose Berkeley Madonna (www.berkeleymadonna.com) as the simulation tool – the code is easy to read and the software, even in its free demo version, is sufficient for most student work. This book provides many examples of code for this software. Software and computer sessions are an essential part of learning to use the tools that are introduced in this book. Many different software systems can provide support for this; preference should be given to a general tool in the academic environment.

This book touches on many topics. Some are dealt with in depth (kinetics, stoichiometry, conservation of mass, reactor hydraulics, residence time distribution), while in others an introduction primarily based on case examples is provided (parameter identification, sensitivity analysis, error propagation, process control, time series analysis, design under uncertainty). Typically PhD students will subsequently follow more in-depth systems analysis and statistical courses whereas professional engineers should at least obtain the basis for their continued education.

In order to support the use of this text I will make some additional material, especially my lecture notes, available online at:

http://www.sww.ethz.ch

Finally I would like to thank my collaborators, assistants, PhD students, and students in general who have helped me to find errors and improve details of this book. I had the opportunity to translate and revise this book during a sabbatical leave that I spent during the summer of 2006 at DTU in Lyngby.

Zürich, summer 2007   Willi Gujer

Content

Chapter 1
Introduction

Urban water management is the engineering discipline with technical and professional responsibility for the design and operation of the extensive structures, installations, and institutions that are necessary for modern society to deal efficiently and comprehensively with water in urban areas. In the training of engineers, the first concern is to instill an understanding of the principles of the functioning of the entire technical system of urban water management. This is the primary goal of a first lecture series on urban water management (Gujer, 2002). The present text assumes that students have such a basic understanding and will introduce a set of methods that support productive and successful work in this engineering discipline.

Thus the focus of this text is on scientific methods that are useful in the analysis and prediction of the behavior of the systems, processes, and operations used in urban water management. Specific topics such as physicochemical or biological treatment of water are explicitly not the subject of this text. However, working efficiently with these topics requires the methods introduced herein.

1.1 Goal and Content of This Text

In urban water management we frequently concern ourselves with dynamic, i. e., time-dependent technical and natural systems, which we analyze with the aid of mathematical models. Causes of the dynamics are diurnal and seasonal variations as well as random events such as rain. Typically our problems require that we deal with many different materials and processes and analyze rather complex technical systems.

This text aims to introduce the methodological basis for the development and application of dynamic mathematical models. The goal is to introduce the basics such that the student will be able to develop and apply his or her own models and plan and interpret associated experiments and data. This will lay the groundwork for the future development of more complicated models and independent acquisition of his or her own preferred methods.

Figure 1.1 shows a simplified flow scheme for the treatment of problems, starting from a question and ending with an answer:

- We start from a task and an associated question that refers to a natural or technical system from the real world.
- First we delimit this system from its environment. This demarcation is an abstraction step which considers that the environment may affect the system under consideration. However, the system may not affect the environment such that there is feedback to the system again. Therefore effects of the system on the environment do not have to be considered. This demarcation refers to space, time, and the spectrum of state variables (state variables are time-dependent values that are important to the problem under consideration).
- In order to model the system mathematically, we need an understanding of the transport processes (here, primarily the topics of material flux, reactor hydraulics, mixing, advection, dispersion, and diffusion will be introduced). Transport processes do not change the materials; they affect only the concentration of the materials that are available at a particular location and time.

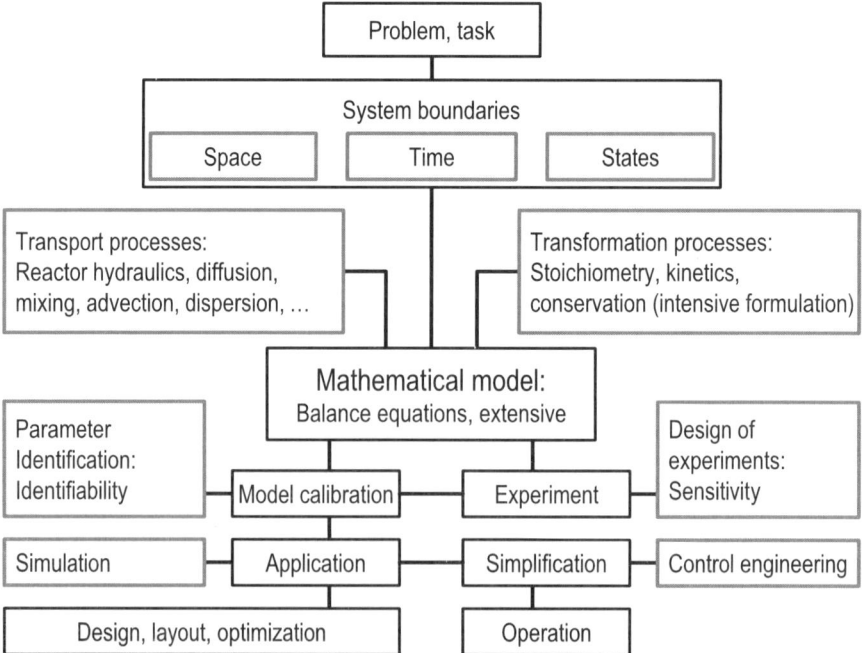

Fig. 1.1 From a question to a model to a solution of the problem. In *gray*: The different tools and themes of this text

- Chemical, biological, and to some extent physical processes convert educts into products. With the help of kinetics we characterize how the local environment affects the rate of these processes. Stoichiometry indicates the ratio in which educts are converted into products. Conservation laws provide us with a priori information, which is used to simplify our task.
- The mathematical model is now compiled in the form of a set of equations (with dynamic systems usually as ordinary and partial differential equations). For this we use material balance equations that combine the effects of the environment, transport, transformation, and changes in the state variables. The model should be simple, but describe the relevant phenomena with sufficient accuracy.
- Now, the parameters of the model must be determined, typically based on experiments or our professional experience. In addition, we will plan experiments which react sensitively to the parameters. For this we use the methods of sensitivity analysis.
- In order to calibrate the model (adapt it to experimental results) and to validate it (gain confidence in its validity), the parameters must be identifiable and identified. For this we make use of experimental observations and possibly nonlinear regression techniques.
- If our question concerns problems in the domain of planning, design, and optimization, the model may now be used with the help of simulation (usually based on the numerical solution of a system of differential equations).
- If the question concerns the operation of a system, the model equations are often simplified, and methods from control engineering will be applied.

The scheme, introduced here in the form of an abstract analysis will be developed systematically in the text and explained with many examples. Sophisticated concepts cannot be evaluated without a detailed understanding of the methods and tools for urban water management that are introduced here.

While the problems that we deal with in urban water management change continuously, the methods that we use are fairly stable. Classical methodical textbooks, in particular from chemical and process engineering, partly go back to the 1950s and 1960s.

Chapter 2
Modeling and Simulation

It is a characteristic of environmental engineering that we must frequently predict the behavior of extensive, complicated systems with highly variable boundary conditions. Often, important data and information are missing. With models that allow us to transfer experience from one system to another, we can partially compensate this lack of information.

Thus, modeling is an important aspect of our profession — simulation makes use of these models and permits us to make statements about the expected behavior of rather complicated systems. To simulate means to predict the behavior of a system of interest with the help of typically numeric solutions of model equations. We answer questions such as "What would be, if ...?"

2.1 System, Model, Simulation

We speak of a *system*, if some objects and their interactions are separated by a plausible demarcation from their environment (i. e., from the complex reality). The objects and interactions that are of importance relative to the question posed must be part of the system. All other objects and interactions are to lie outside of the system boundaries. We then describe such a system with the aid of mathematical *models*, which we can analyze instead of the real system. Finally, we will apply to the real system what we learnt from the behavior of the models.

We differentiate between physical models (a model railway, a geographical map, a pilot plant, etc.) and abstract, usually mathematical, models. In this text abstract models are of interest.

Modeling always starts with a process of abstraction: we reduce system complexity in view of the question posed. Only important processes, state variables, parameters, and interactions are maintained. We will first develop qualitative, verbal descriptions (models), which can then be transformed into quantitative, mathematical models used to describe system behavior.

Simulation means to experiment with abstract models in order to answer questions like *"what would be, if ...?"*. With the aid of mathematical methods we analyze the possible behavior of a system. We will then use what we learn from model predictions to design, optimize, and operate real-world systems.

2.2 Models in Natural and Engineering Sciences

Research in natural sciences is the attempt to fathom the mechanisms that lead to the observed behavior of a natural system. Engineering has the task to plan, design, realize, and operate systems based on scientific knowledge, and thereby to achieve a set goal with the least possible effort.

It is common to both disciplines that they are successful only by abstractions or models of reality. The task of the natural scientist is to develop models that provide an ever-improving understanding of reality. The method of the engineering sciences is to compile models which, at small expense, provide valid statements about the problem at hand. Thus, in environmental engineering the focus is not on modeling per se, but rather on developing models that serve a purpose. In the environmental natural sciences, the development of models that uncover interactions may well be the actual goal of the work.

Apart from analysis (research) and prediction (project engineering, consulting), we also use models in teaching of concepts and communication of relationships (Fig. 2.1). Here the models are encapsulated knowledge that we can discuss in an internationally standardized and uniform language (mathematics).

2.3 Types of Mathematical Models

In the context of this text, we are interested primarily in the following groups of models:

- *Deterministic models* assume that the behavior of a system is clearly determined by its present condition and the future external factors of influence. Given the same initial and boundary conditions, a deterministic model will always lead to the same result. This is in contrast to *stochastic models*, which consider

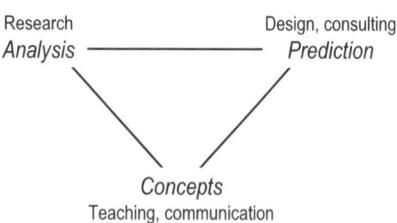

Fig. 2.1 Different tasks and goals of modeling

also uncertainties in the model parameters and possible random fluctuations of external variables and thus, even with defined initial conditions, lead to uncertain future behavior.

- *Dynamic models* make statements about the time-dependent development of a system, i. e., the future conditions of a system can be assigned to the time axis. *Static systems* do not make a statement about temporal development. They may well consider changes of external factors, but they do not allow prediction of the temporal transition from one condition to the next. The concept of *steady state* describes the situation of a static system.
- *Spatially discrete models* make statements about subsystems that are considered to be well mixed, and where gradients of the state variables are not modeled; in dynamic models they lead to ordinary differential equations. *Spatially continuous models* also describe gradients of state variables in space and accordingly lead to partial differential equations. If the transport processes are dominated by one or two directions (the anisotropic case), then we can often limit the models to fewer than three spatial dimensions.
- *Temporally continuous models* lead to differential equations; they regard time as a continuum and make statements about the time course of a system. *Temporally discrete models* lead to difference equations; they make statements only for certain times and do not permit the description of the transition between such times in detail.

Example 2.1: The bank account is a temporally discrete system

Interest is credited at regular intervals to a bank account. After the annual credit of the interest, the capital and its increase grow with compound interest: a temporally discrete system. If the bank account were a temporally continuous system, then the interest would be constantly credited (each fraction of a second), thus the interest would rise, because we would receive compounded interest much earlier. The banks naturally noticed this – accordingly, we must typically pay interest for our debt every month, whereas we obtain interest only once a year.

Similarly a small population of organisms that grows in discrete steps, e. g., by cell division and duplication, could be described as a temporally discrete system. However, once the number of organisms is so large that the birth of an individual new organism is no longer noticeable, we usually describe the behavior of the system as temporally continuous.

Example 2.2: Models used for dimensioning are usually static models

Historically biological wastewater treatment plants were dimensioned based on a sludge loading rate B_{TS} (frequently called F/M, the food-to-microorganism ratio). Here, only time-independent values are used and it is not possible to forecast the treatment efficiency on an hourly basis. These design models are static.

Example 2.3: The cascade of stirred tank reactors is a spatially discrete model

If we model a plug-flow type reactor with the aid of a cascade of stirred tank reactors, we select a locally discrete model. We can then only make statements about conditions in the individual reactors of the cascade. The model does not provide intermediate values that might be observed in the real reactor. The resulting model consists of a system of coupled ordinary differential equations. If alternatively we were to use a model with dispersion and advection, then a spatially continuous model would result, and accordingly we would obtain a system of partial differential equations.

Example 2.4: Box models are spatially discrete

In the environmental sciences, we frequently model natural systems with so-called box models, i. e., models that consider a part of the system as a completely mixed, homogeneous box (without internal gradients) and transport processes only at the surface of this box (simple examples of box models are batch reactors or stirred tank reactors).

Example 2.5: Rain events are stochastic events

In the dimensioning of sewers, we use information about the intensity, duration, and frequency (IDF) of rain events, which we compile by using statistical methods. A model rain event is computed with the help of a statistical model and will thus never occur in the form used for the design. However, the characteristics of the rain event that is used have a certain probability of occurring.
The statistical characteristics of the real observed rain event are extracted in the form of a constant, average intensity which afterwards leads in the context of the rational method for design (a deterministic model) to a unique result.

2.4 Systems Analysis

The tasks of systems analysis are (i) to identify a suitable structure of a mathematical model for the description of the behavior of a system of interest, (ii) to identify the associated parameters of the model, including their uncertainty, (iii) to analyze the mathematical behavior of the models, (iv) to evaluate the quality of the model, (v) to analyze and estimate the uncertainties of the model predictions, and (vi) to plan and design experiments with the best yield of information.

Figure 2.2 shows an abstract representation of a real-world system. We cannot observe all the external factors that influence the behavior of the system, and those which we can observe are subject to measuring errors. We track the behavior of the real system again with observed conditions and measured output variables,

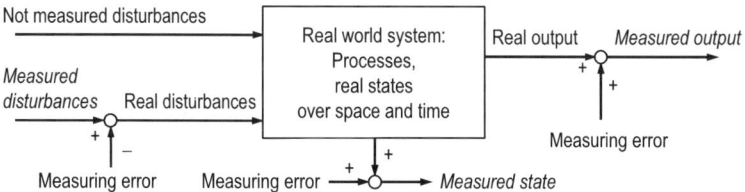

Fig. 2.2 Observation of reality and the effect or measuring and observation error

which are also subject to measuring errors. The system itself is subject to processes, which we characterize with abstract mathematical models. The structure of the model that we compile corresponds to our perceptions and concepts. The picture that we create for ourselves for a system is therefore different to a greater or less extent from the unknown reality.

With the aid of systems analysis, we try to grasp the difference between the model prediction and the real-world behavior of the system. Scientific methods are used to keep these differences small and under control. For this we use the following methods:

- The choice of a *model structure* that can cope with system behavior and be identified with the aid of experimental evidence.
- The *estimation of parameters* based on experimental observations given a particular model structure. This is also called system identification. Identified parameters values are burdened with uncertainty.
- *Consideration of sensitivity* will tell us whether the parameters are identifiable and which experiments would yield most information.
- *Error propagation* provides us with estimates of model prediction uncertainties.

Example 2.6: Unobservable properties

The distribution of the rain intensity over an urban catchment area can be observed only at great expenditure with the spatial and temporal resolution necessary to describe the discharge behavior of a drainage system reliably. Only a very fine observation grid would yield a reasonably exact picture.

The various organic pollutants in wastewater cannot be described individually, although the detailed composition of the wastewater determines the behavior of the wastewater treatment plant. We must rely on sum or group parameters such as COD and additional experience in similar situations.

On a statistical basis, the water requirement of individuals is well known. We cannot, however, assign an accurate instantaneous value to the requirement of an individual person or a single house.

2.5 Calibration, Validation, and Verification

No model illustrates the whole of reality. Calibration and validation give us confidence that we can make meaningful use of mathematical models in a limited state and time space. Environmental sciences cannot typically verify their models.

We *calibrate* a model with a given mathematical structure by achieving an optimum agreement between experimental observations and associated model prediction by adaptation of the model parameters. We *validate* a model by testing its validity in the state and time space required to answer the question posed.

The terms *verification* and *validation* are frequently used as synonyms. However, here we will define verification much more narrowly, using the example of the way in which the law of gravitation has been verified in a very broad sense, such that we can make reliable predictions even for situations where the law has not yet been tested and measurements are not possible. Large and complex models such as those developed and used in the environmental sciences cannot be verified in this sense; we have to neglect too many details that may dominate the behavior of the system in new situations. Typically, we will only validate our models and thereby we are restricted in the possibilities of reliable extrapolation.

Summary: We calibrate based on a limited set of experiments. We validate for a limited space of application (a range of temperatures or concentrations, a time period). We verify for a broad field of application without restriction of the validity.

Example 2.7: Calibration, validation, and verification

For apples falling from a tree, we can find a simple model that allows us to predict the time t required to reach the ground as a function of their initial height h: $t = k \cdot h^{1/2}$. Based on many experiments, we can determine the model constant k (approximately $0.22 \, \text{s} \, \text{m}^{-1/2}$). We have calibrated the model. We can easily use this result in order to interpolate, but not yet in order to extrapolate very far. We could definitely not accurately determine the height of an airplane based on dropping an apple, as at high velocities air resistance becomes important.

If we now conduct our experiments in a vacuum and find the same value for k when we instead use a feather or an apple, our confidence increases. The model is now validated for a broad range of application and we would even risk extrapolating somewhat. However, we would certainly not yet board a space ship that is expected to fly us to the moon based on our limited experience.

Only with the gravitation law (and of course a few more details) which can describe many different phenomena and which could be verified from the very small up to the astronomic scale are we willing to risk our life in the space ship. We can then speak of verification and are ready to rely on very broad validity.

In physics the goal is typically to obtain verified models.

Figure 2.3 introduces a general procedure for the calibration of a mathematical model. We compare measured values from reality, which suffer from measuring

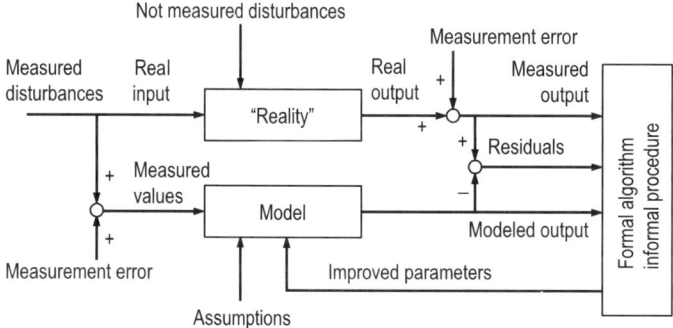

Fig. 2.3 Calibration of a model. When validating, the improvement of the parameters is omitted

errors, with the predicted values from the model, which are computed based on disturbances (external influences) that likewise suffer from observation errors. With the help of a formal or informal (trial-and-error) procedure, we improve the values of the model parameters until we reach sufficiently good agreement between reality and the prediction of the model.

In validating a model, we frequently use observations of the real system that were not used for the parameter estimation (calibration). If these observations cover the space in which we want to use the model, and the comparison between the prediction and observation is satisfactory, we regard the model as validated for the application. In order to improve the reliability of parameter estimation, it is advisable to include now the data that was used for validation and re-estimate the parameters.

Verification is not an important procedure in environmental engineering sciences.

2.6 Model Structure

If we do not succeed in calibrating a model with sufficient security (accuracy), then we must improve the mathematical structure of the model. We may have chosen too simple a mathematical model, we may have neglected important processes, or we may have chosen mathematical expressions that cannot capture the properties of the individual processes. In addition, problems may arise from parameters that are not identifiable based on the experimental evidence at hand.

No scientific procedure guides us in the development of a suitable model structure; rather we have to rely on the application of our professional knowledge, our technical expertise, and the application of scientific findings. Later (Sect. 2.8 , and Chap. 12), we will get to know suitable tools and methods that will support us and facilitate this task.

Example 2.8: Air resistance in the law of falling objects

If we want to apply the law that we developed for the description of the falling apple in Example 2.7 to a feather in the wind, we will obviously not succeed. Neglecting the resistance of air is in this case of such importance that we must adapt the model structure in order to obtain a model that can reasonably be calibrated.

2.7 Simulation

The use of mathematical models to answer questions such as "What would result, if ...?" is called simulation. In a simulation we typically proceed from a validated model.

Frequently, it is too expensive, illegal, or impossible, to run experiments with the plants or systems that we plan, design, build, and operate because:

- Physical models (pilot plants) are expensive, time consuming, and difficult to operate.
- Experiments can endanger humans (drinking water) or the environment (wastewater), if we leave our field of expertise.
- Natural systems (running waters, lakes, groundwaters) cannot be endangered.
- Rain cannot be imitated in real systems.

With the aid of simulation, we can investigate the behavior of the real world or analyze results from pilot plants based on mathematical models. We can even make a prediction for the circumstances under which a certain danger will arise, etc.

Example 2.9: A definition of the term "simulation"

Many definitions exist. The VDI guideline 3633 provides the following definition (translated from German):
"Simulation is emulation of a system with its dynamic processes in an experimentable model in order to arrive at results which are transferable to reality".

Example 2.10: Real-time simulation

In large cities the behavior and control of sewer systems are increasingly based on "real-time" simulations of rain events. With the help of weather radar and many signals from the sewer system, the behavior is simulated and the underlying model is calibrated. With a calibrated model it is then possible to simulate different control strategies. The results of such scenarios are then the basis for operators to take decisions for manual interference.
Here calibration with the latest available data is sufficient, because the extrapolation extends only over a short period of time.

Example 2.11: Simulation in hydrology

With hydrological models, prognoses are made for the development of flood conditions in the lowland areas of rivers. This provides early warnings of upcoming dangers.
After an accidental pollutant discharge into a river, simulation programs are used to predict the propagation of the pollutant. This provides early warnings that allow water withdrawals for water supply purposes to be suspended.

Example 2.12: Simulation in process engineering

Today, the performance of entire wastewater treatment plants under variable hydraulic and pollutant loads is simulated in the context of plant design. In addition simulation is used to develop new control concepts for such plants.

Example 2.13: Flight simulators

Pilots are trained in simulators, because it is cheaper than the real thing, and because dangerous situations that in reality would be too risky to consider are available.
Sitting in front of a personal computer (PC) even laypeople can obtain a realistic picture of the task of a pilot from a simulation program.

2.8 Components of a Model

Today, technical and natural systems in engineering or natural sciences are frequently modeled with systems of algebraic equations as well as ordinary and partial differential equations. These model systems consist of structural components whose characteristics are discussed here. We must strictly differentiate between the real-world system and its mathematical model (a picture). Here we concern ourselves only with the models; however, the individual structural components have of course their parallels in the real world.

2.8.1 Structural Components of a Mathematical Model

Figure 2.4 shows the structural components of a mathematical model, as they are usually applied in the description of aquatic systems (and many other systems). The arrows suggest the order in which these structural components are defined when a new model is compiled. In detail these elements include:

Fig. 2.4 The most important structural components of a mathematical model. The *arrows* suggest the order of the definition of the individual elements (extended from Reichert, 1998)

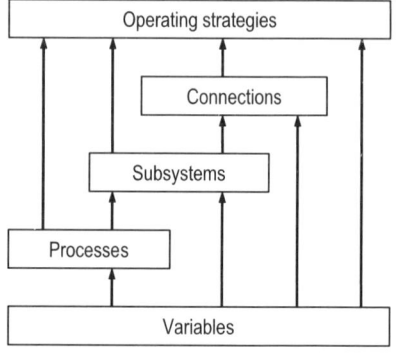

Variables

With the variables, we connect numeric values that may be constant or may change with time and/or over space. It is the numeric values of the variables which we must enter and adapt for a specific system, or which we examine, analyze, or display in detail. We distinguish the following types of variables:

- *Intensive properties* are variables that are defined for a specific location and are independent of the size of the system, e. g., concentrations, temperature, flow velocity, and viscosity. *Extensive properties* are proportional to the size or mass of the system and therefore always refer to a certain system, e. g., flow, volume, diameter, length, mass, and energy.
- *System variables:* Statements about the condition of a system always refer to a certain time and a certain location in a specific subsystem. We call the variables that allow us to localize a specific condition in time and space system variables. Typical system variables are time, local coordinates, and the designation of the subsystems. Depending upon the type of the model (stationary, homogeneous, discrete, etc.) we may neglect some of these variables.
- *State variables* are the dependent variables whose prediction as a function of the system variables is the actual goal of modeling. Examples are concentrations, temperature, flow (not influent), flow velocity, and time-dependent volumes.
- *Disturbances* are factors or properties that affect a system from the outside and which we (in the context of our investigation) cannot influence. Examples are influents and their pollutant content or measuring error.
- *Initial and boundary conditions:* In order to be able to solve ordinary and partial differential equations, we must provide absolute values and possibly derivatives for state variables at specific times and locations. These can result from data (possibly as model parameters which must be identified) or may be chosen as the steady (not time-dependent) state of the system under mean operating conditions. A steady state may be obtained with the help of numeric integration (relaxation) of the relevant model equations.

- *Parameters* are variables whose absolute value must be known before model equations can be solved. Examples are fixed volumes, kinetic rate constants (e. g., a growth rate), stoichiometric ratios (e. g., a yield coefficient), hydraulic variables (e. g., wall roughness or viscosity), and material properties (e. g., Henry or diffusion coefficients). It may be necessary to identify kinetic or stoichiometric parameters in the course of model identification, but this requires us to provide strategies to choose the absolute values before we solve model equations. System-dependent parameters (volume, flow, diameter, mass, etc.) can typically be specified a priori based on the problem to be solved.
- *Data* are values from observations or measurements in the real system that is to be simulated. These are measured values for which absolute values (subject to measuring error) as well as system variables (time, location, also subject to error) are available. Frequently it is assumed that time and location but not the data are observed free of error.

Example 2.14: System variables

In a batch reactor (a completely mixed model system with neither influent nor effluent), we track the concentration of pollutants (state variables). Complete mixing leads to the fact that we do not expect concentration gradients over the volume. Therefore the space coordinates are omitted as system variables and only time remains.

Models of rivers frequently neglect the region of transverse mixing of pollutants below a sewage outlet. The assumption is that the river is completely mixed in a cross section, but that gradients of state variables remain in the longitudinal direction. Only the longitudinal coordinate and time remain as system variables, and the remaining model is called one dimensional (1D). If mixing across the river is of interest, we have to choose a two-dimensional (2D) model, and only a three-dimensional (3D) model could provide the full details of the mixing process.

Example 2.15: Intensive and extensive variables

The load of pollutants $[M_S\ T^{-1}]$ entering a wastewater treatment plant is proportional to the size of the city that produces the wastewater, and is thus an extensive variable. We cannot transfer the experiences with this specific load to another city of different size. The specific production of pollutants per inhabitant (60 $gBOD_5$ per person per day) is an intensive property that can be used for the design of different plants.

The sludge loading rate $B_{TS} = Q \cdot BOD_5/(V_{AT} \cdot X_{TSS})$ is an intensive property. By dividing the flow Q by the volume V_{AT} of the aeration tank it becomes independent of the size of the plant. Experience with a certain sludge loading rate is to a large extent independent of the system and can therefore be transferred to another treatment plant.

The consumption of drinking water in a city (an extensive property) is a characteristic of this city. The water consumption per inhabitant of this city (an intensive property) may, however, be a point of reference for planning in another city.

Processes, Sources, and Sinks

Processes result in the physical, chemical, or biological transformation of the properties and location of materials of interest. They determine and change the state of the system. In our models they affect the state variables. We distinguish between transformation and transport processes.

Transformation processes produce or consume materials and energy. They characterize sources and sinks that we do not seize with transport processes. We characterize transformation processes as intensive properties, only dependent on time, local state variables, and parameters.

- *Dynamic processes* are material and energy transformation processes that we characterize by stoichiometry and kinetics. The rate of transformation is comparable with the time constants of interest in the analysis of the system.
- *Equilibrium processes* are transformation processes for which the rates of forward and backward reaction are very large in comparison with the time constants of interest in the analysis of the system. Equilibrium is reached when the production and consumption in these processes is very much larger than the sum of all other processes (transport and transformation) that affect these materials. If equilibrium is reached, we do not have to pursue the time dependence of these transformations. Given that time is the only system variable, the state variables involved may be predicted from coupled algebraic equations.

Example 2.16: Equilibrium process

Many acid–base reactions are very fast. We typically characterize them as an acid–base equilibrium. We could also capture this behavior as a combination of a fast forward and backward reaction.

For the equilibrium: $CO_2 + H_2O \leftrightarrow HCO_3^- + H^+$ the equation

$K_S = \dfrac{[HCO_3^-] \cdot [H^+]}{[CO_2]}$ provides us with a relationship between the concentration of

the acid $[CO_2]$ and the base $\left[HCO_3^- \right]$. In place of this relationship we could also introduce two fast reactions:

Forward reaction: $CO_2 + H_2O \rightarrow HCO_3^- + H^+$ with $r_{forward} = k_f \cdot [CO_2]$.

Backward reaction: $CO_2 + H_2O \leftarrow HCO_3^- + H^+$ with $r_{backward} = k_b \cdot [HCO_3^-] \cdot [H^+]$.

In equilibrium the two reaction rates are equal ($r_{forward} = r_{backward}$). This results in:
$K_S = \dfrac{k_f}{k_b}$. If we want to use the two reaction equations in place of the equilibrium,
we can, e. g., select k_f very large and afterwards derive the required value of k_b from this relationship. The Berkeley Madonna software provides a special module to enter equilibrium reactions in the form of fast forward and backward reactions.

Transport processes bring materials and energy from one location to another, neither changing their quantity or mass nor their kind. Transport processes are directional and are quantified as vector variables. We distinguish between (Fig. 2.5):

- *Advection*, which is a directional transport that affects all materials as well as the surrounding water in an equal way. On the average there is no relative velocity of the materials transported and the surrounding water. We characterize advection by a flow velocity.
- *Sedimentation and flotation*, which characterize the falling or rising of particles due to gravity, buoyancy or a *centrifugal* or *magnetic* force. Unlike advection, the particles move relative to the surrounding water.
- *Dispersion:* If we model a three-dimensional flow field with only one- or two-dimensional models, we must neglect some induced flows and velocity gradients. We can compensate this simplification by the introduction of an additional, locally averaged transport process called *dispersion:* we introduce an additional surrogate transport process that averages transport at a particular time over a certain area (e. g., a cross-sectional area).
- *Turbulence:* As a consequence of the dissipation of mechanical energy eddies arise in the water which we hardly can grasp in all detail. Parcels of water together with their content of materials are exchanged, which results in a transport process. We quantify turbulence for a specific location, averaged over time.
- *Diffusion:* The transport of material resulting from the random movement of individual molecules and small particles (thermal motion) is called diffusion. Diffusion is a characteristic of the transported material or particle size and the surrounding water. We quantify the transport of a material, at a specific location, averaged over time. Diffusion makes a statement about the probability that, within one time period, a particle moves in a certain direction.

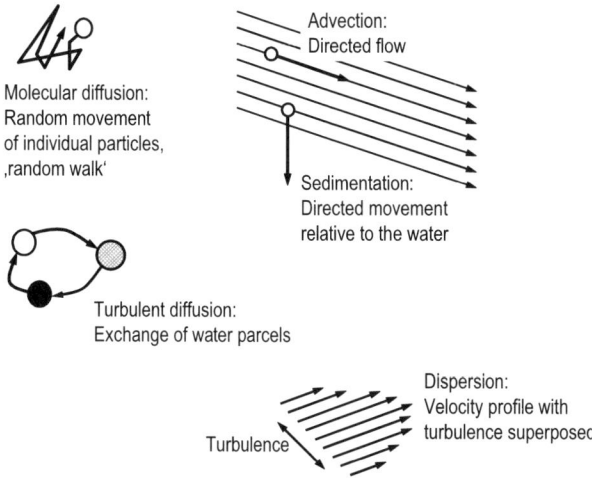

Fig. 2.5 Characterization of the transport processes

Technical processes and *controlled elements or members* are procedures that we control purposefully, based on predefined mainly mathematical rules. Examples are aeration equipment, dosing of chemicals, operation of heat exchangers, flow control, and valve positions.

Subsystems

Subsystems are defined parts of a system for which we can compile independent models. Frequently models for subsystems are based on *idealized, typed elements*, for which we can describe the transport processes and the boundary conditions mathematically exactly. Therefore we can write the balance equations for the state variables of these subsystems. The model of the entire system can then be developed based on the subsystems. Subsystems can affect the states of the system such that feedback to the subsystems may exist. As a consequence we cannot analyze the subsystem autonomously, but only as a part of the entire system.

- *Ideal reactors*: When modeling technical systems, we use ideal reactors as templates for the models of subsystems. Ideal reactors are defined such that we have an exact mathematical description of their behavior. They are discussed in detail in Chap. 6.
- For the description of natural systems we can use predefined subsystems or compile them, depending upon the problem to be solved. Environmental scientists have a suite of models and programs available which offer such predefined model structures for lakes, 1D, 2D or 3D models for rivers and groundwater, soil columns, activated sludge plants, wastewater treatment processes, etc. Often these models provide predefined transport and mixing processes, whereas the transformation processes must be defined by the user.

Connections

Connections (links) of subsystems are extensive resources; they *transport* materials, energy, etc. (quantified by state variables) from one subsystem to another. Usually we neglect the volume of these connecting elements; accordingly no transformation processes must be considered in these elements. Influents and effluents are links to the environment for which one end is not considered when we analyze a system.

- *Simple connections* lead mass and energy unchanged from one subsystem to the next. In the case of a material flow they are defined, e. g., by the flow rate and the material concentrations contained in it.
- *Recirculations* lead mass flows against the general direction of flow from a later to an earlier subsystem. They typically lead to rather intensive coupling of the mass balance equations of different subsystems.
- *Unions* combine and mix the mass flows from several subsystems and feed them united into a new subsystem.

- *Bifurcations* divide the mass flows in a connection and feed them into several subsystems. It is possible that a bifurcation distributes the different materials contained in the flow in different fractions to the effluents of the bifurcation. An example of a bifurcation is a simple model of a sedimentation tank: the influent is split into sediment (with elevated concentration of solids) and treated effluent (with a reduced concentration of solids).
- *Diffusive connections* describe the mass transport over boundary layers, e. g., as a consequence of turbulence, molecular diffusion, or solubility at surfaces. They contain a forward and an appropriate backward reaction and are typically characterized by *mass transfer coefficients*.
- *Disturbances* are inputs into a system. They cannot be controlled at short notice and are not affected by the system itself.

Operating Strategies

In technical systems we use the existing *degrees of freedom* based on operating strategies in order to reach a desired performance.

- The decisions that operators of technical systems make, are subject to *default operating procedures*; e. g., in the operation of an activated sludge plant we will require a minimum sludge age (solids retention time) to be maintained; in the operation of a sewer system we will require that a certain amount of combined wastewater is directed towards the wastewater treatment plant; or in water treatment we will demand that a minimum concentration of disinfection chemical is maintained over a specified time, etc.
- With the help of *control strategies* and *control engineering,* we can automate the application of *operating procedures*. This requires us to build additional mathematical relations into the systems that make use of control possibilities but decrease the number of remaining degrees of freedom in the operation.

2.8.2 Case Study

On the basis of a simple case study, the various structural components of a model (a system) are introduced. Fig. 2.6 shows a schematic for a simple activated sludge system, as could be used to develop a mathematical model. This figure does not want to depict reality but rather the abstraction captured in the model.

The model system is divided into two subsystems (activated sludge tanks), in which the biological transformation processes are active. In these subsystems we must grasp the internal mixing processes. The secondary clarifier, which in reality is a rather large structure, is modeled with zero volume and based on two bifurcations: in the upper layer the clear water that contains soluble materials is separated from the solids, which settle out; in the lower layer the concentrated activated sludge is split into return sludge and excess sludge.

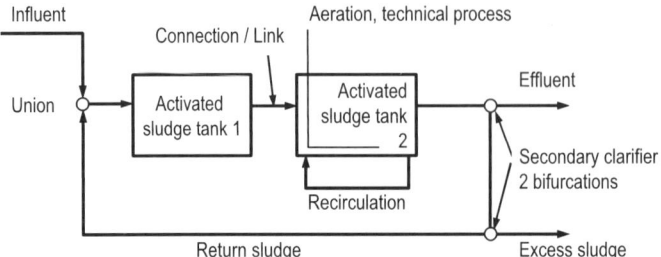

Fig. 2.6 Flow scheme of a simple activated sludge system that could be the basis for model development. See the text for explanations

Sedimentation and thickening in the secondary clarifier are thus not modeled as transport processes, but captured with a bifurcation, thereby neglecting the time necessary for sedimentation and thickening of the sludge as well as the rather large volume of the clarifier. A total of four links to the environment are included: influent, discharge, excess sludge, and air flow. The material flows in these links do not change the environment in a way that must be considered in the analysis of our system. There is a simple connection between the two subsystems (an aperture), and a technical process (aeration) supplies the required oxygen to the second activated sludge tank. The recirculation inside the second activated sludge tank only has an influence if this subsystem is not modeled as a stirred tank reactor, and thus gradients of state variables in this tank are captured by the model. The return sludge is combined with the inlet and afterwards led into the first reactor.

In this system a set of state variables must be captured to make a comprehensive statement about its performance: oxygen, organic materials, microorganisms, nutrients, temperature, etc. In addition, models for transformation processes (e. g., growth of microorganisms) and associated model parameters such as growth rates (kinetics) and yield coefficients (stoichiometry) are necessary. Moreover, we must define initial and boundary conditions for all the state variables from which our computations proceed. Disturbances (or forcing functions) that affect the system from the outside, such as the influent and its content of pollutants, must be given (we cannot influence these functions at short notice). From the real plant we may have data at our disposal that allow us to calibrate the compiled model. Our goal is to describe the course of pollutant concentrations (state variables) in different locations over time (system variables). This is possible only if we also include the operating strategies, which might be to maintain a constant activated sludge concentration, a steady oxygen concentration, and a fixed return sludge flow rate (degrees of freedom).

2.9 Dimensions and Units

The development, calibration, and validation of models require observations and measurements of state variables. Measurement compares the dimension of variables with a set of fixed standards for mass, time, length, etc. such as the kilogram, second, and meter, respectively, and expresses the result in units.

Herein variables are defined with their dimension given in straight brackets: [M] for mass (including moles), [L] for length, and [T] for time. Specific values of variables are given in units of grams (g), kilograms (kg), and sometimes moles (mol), meters (m), centimeters (cm), seconds (s), hours (h), and days (d). In a few instances the temperature will have to be expressed either in degrees Celsius (°C) or Kelvin (K).

In systems analysis and mathematical modeling it is good practice *always* to check the homogeneity of the dimensions after a model equation has been derived and the homogeneity of the units once a model is applied (see Example 2.18).

Example 2.17: Dimension and units of a concentration and a diffusion coefficient

The concentration C_{NH} of ammonium in water may be defined as the mass of nitrogen in the form of ammonium per unit volume. Herein it will be introduced as:

C_{NH} = concentration of ammonium nitrogen $[M_N \, L^{-3}]$

A specific measured value is given with units such as $C_{NH} = 2.4$ gN m^{-3}.
The diffusion coefficient of oxygen in water is introduced as:

D_{O2} = molecular diffusion coefficient of oxygen in water $[L^2 \, T^{-1}]$

Table 4.1 provides specific values for this parameter in pure water at 25°C in two different units:

$D_{O2} = 2.42 \cdot 10^{-5}$ cm^2 s^{-1} and $D_{O2} = 2.04 \cdot 10^{-4}$ m^2 d^{-1}.

Both values express (nearly) the same absolute value of D_{O2}.

Example 2.18: Homogeneity of dimensions and units in a mass balance equation

Equation (9.8) describes the concentration profile of a pollutant across a spherical activated sludge flock in the format of a differential equation derived from a mass balance:

$$\frac{d^2 C_A}{dr^2} = -\frac{2}{r} \cdot \frac{dC_A}{dr} - \frac{R_A}{D_A}.$$

The definitions of the symbols are as follows:

C_A = concentration of the material A $[M_A \, L^{-3}]$ measured in the units of gA m^{-3}
r   = radius coordinate [L] measured in units of m

R_A = transformation rate of material A [$M_A L^{-3} T^{-1}$] measured in units
    of $gA\,m^{-3}\,d^{-1}$

D_A = molecular diffusion coefficient of compound A [$L^2 T^{-1}$] measured in units of
    $cm^2\,s^{-1}$

An analysis of the dimensions results in:

$$\left[\frac{d^2C_A}{dr^2}\right] = \frac{M_A}{L^3} \cdot \frac{1}{L^2} = \frac{M_A}{L^5}$$

$$\left[\frac{2}{r}\frac{dC_A}{dr}\right] = \frac{1}{L} \cdot \frac{M_A}{L^3} \cdot \frac{1}{L} = \frac{M_A}{L^5}.$$

$$\left[\frac{R_A}{D_A}\right] = \frac{M_A}{L^3 \cdot T} \cdot \frac{T}{L^2} = \frac{M_A}{L^5}$$

Obviously the dimensions of the three terms of Eq. (9.8) are equal.
An analysis of the units results in:

$$\left[\frac{d^2C_A}{dr^2}\right] = \frac{g_A}{m^3} \cdot \frac{1}{m^2} = \frac{g_A}{m^5}$$

$$\left[\frac{2}{r}\frac{dC_A}{dr}\right] = \frac{1}{m} \cdot \frac{g_A}{m^3} \cdot \frac{1}{m} = \frac{g_A}{m^5}.$$

$$\left[\frac{R_A}{D_A}\right] = \frac{g_A}{m^3 \cdot d} \cdot \frac{s}{cm^2} = \frac{g_A \cdot s}{m^3 \cdot cm^2 \cdot d}$$

The three terms of Eq. (9.8) are not homogeneous relative to their units. The units of the molecular diffusion coefficient must be changed by multiplying its value by the (dimensionless) factor

$86{,}400\,s\,d^{-1}/10{,}000\,cm^2\,m^{-2}$.

The check of units identifies possible problems in the prediction of absolute values. The check of dimensions identifies possible problems in the derivation of the model equation. However, neither homogeneous dimensions nor units are a guarantee of correct model derivation.

Chapter 3
System Boundaries and Material Balances

The material, energy, and momentum balance equations are the fundamental equations for the derivation of mathematical models for the description of technical and natural systems in the environmental sciences. Here we deal only with the material balance equation, which connects the accumulation of a material in a system with transport and transformation of this material and permits us to make predictions for the change of state variables (material concentrations) with time. Balance equations always refer to a system, which can be infinitely small, or finite and large.

3.1 System Definition

We define a system as a separated part of the environment that covers the conditions and processes which we want to include in our analysis as we work out the solution of a problem. The defined system may be affected by the environment; however, its effect on the environment must not be relevant for the analysis of our system, i. e., there must be no feedback from the system to the environment and back to the system again.

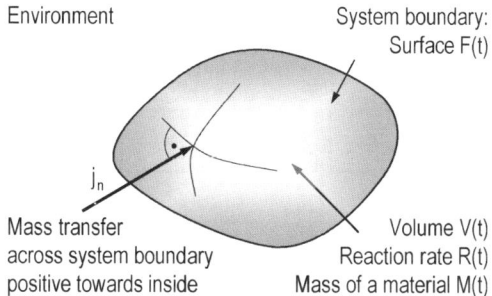

Fig. 3.1 Schematic representation of a system: demarcation from the environment, surface, volume, transport, material inventory, and reaction

Environment

System boundary: Surface F(t)

j_n

Mass transfer across system boundary positive towards inside

Volume V(t)
Reaction rate R(t)
Mass of a material M(t)

In Fig. 3.1 a system is shown schematically as separated from the environment. For all the materials and things that are of interest in view of our question, we must know at any time whether they are within or crossing the system boundaries. In addition we must know whether they are being produced or consumed within the system. A system may be a specific technical construction or reactor or a natural phenomenon to which we apply models; in addition, an abstract concept that we use to develop and deduce models can be regarded as a system.

We differentiate between the following kinds of systems:

- An *open system* can exchange materials (mass) and energy with the environment (the most general definition of a system).
- A *closed system* cannot exchange materials (mass) with the environment; but it may exchange energy (a hot water bottle).
- An *isolated system* exchanges neither mass nor energy with the environment (a thermos flask).
- An *autonomous system* is not affected by the environment; it can, however, deliver mass and energy to the environment (a reactor without influent, a leaking tank, etc.)
- A *subsystem* may influence the overarching system, which again may create a feedback loop to the subsystem. Thus, the subsystem and the comprehensive system are interlinked, which may result in the coupling of the balance equations (returning sludge in an activated sludge system affects the performance of an activated sludge tank and this again affects the composition of the activated sludge, as shown in Fig. 3.2).

Example 3.1: Open and closed systems

A river is an open system; it exchanges materials (e. g., water) and energy (heat) with the environment.

As a first approximation the planet Earth can be regarded as a closed system. The sun provides energy. However, for many questions the exchange of mass with the environment is negligible.

Many questions on the solar system can be worked on based upon the assumption that the solar system is an isolated system which takes up neither energy nor mass from the outside.

The sun is to a large extent an autonomous system that delivers mass and energy to the environment, but (to a first approximation) does not pick up anything.

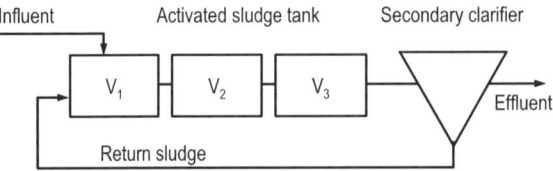

Fig. 3.2 An activated sludge process, modeled with four subsystems that affect each other mutually

3.2 System Boundaries

System boundaries can be selected freely. We select them such that we can answer the question posed with the analysis of the system as simply and comprehensively as possible. Thus, the demarcation of a system from the environment is more an art than a science. With increasing experience we succeed in defining systems in such a way that the questions posed can be worked on quite simply. There exists no single correct demarcation, but there are infinitely many unsuitable or even wrong system demarcations.

Example 3.2: Suitable and unsuitable system boundaries

To quantify the performance of a wastewater treatment plant, we must determine as exactly as possible how large the load of the plant is. To do this it is advantageous to measure the water flow and the pollutant content of the wastewater. This is definitely much simpler than determining the pollutant load at the source, i. e., from each discharge into the sewer system. An appropriate suitable system boundary is shown in Fig. 3.3.

The separation between system and environment can be chosen freely, as long as the system does not affect the disturbances (inputs from the outside). In view of answering questions, systems and, if necessary, its subsystems should be selected as simply and appropriately as possible. The application of material balance equations (as introduced below) requires the reaction rate to be integrated over the entire volume and the material transport over the whole surface of the system. Therefore we define the system in such a way that these integrals can be solved as easily as possible. Frequently we define first a subsystem in which no local gradients of state variables must be considered; this may be:

- A small volume element ΔV or even an infinitesimally small element dV
- a cross section of a reactor with a small or infinitesimally small height Δz or dz, respectively
- an ideal completely mixed reactor, in which the mixing eliminates the gradients
- an individual air bubble or particle within a reactor

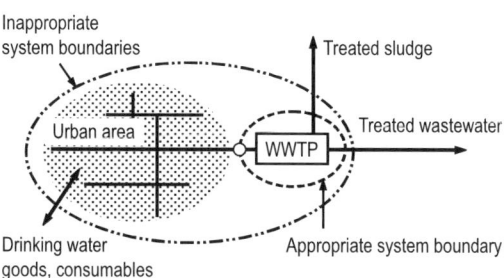

Fig. 3.3 Comparison of a suitable and an unsuitable system boundary for the evaluation of the performance of a wastewater treatment plant (WWTP)

The results of this microscopic view are then used in a macroscopic view and inserted into balance equations for the entire system. If the subsystem consists of an infinitesimally small element, then the microscopic result must first be integrated over a larger subsystem or an entire system. If the subsystem already has a finite extension (e. g., a completely mixed compartment or reactor), then the result can frequently be introduced directly into the overarching balance equations. Examples for this procedure will be introduced in Chap. 9.

The mass of a material within a system is affected by the *transport* of material across the system boundaries as well as its *transformation* within the system (production and consumption); see Fig. 3.1. These two groups of processes are discussed in the next sections.

3.3 General Balance Equation

The starting point in the modeling, dimensioning, and design of a reactor system in drinking water processing and wastewater treatment is the material balance equation. It links the different processes within a single equation.

A material balance in verbal formulation has the following form:

Rate of   Influent through   Effluent through   Net
accumulation in = boundaries of   −   boundaries of   + production in
a control volume   control volume   control volume   control volume

Transport processes   (3.1)

In Eq. (3.1) the accumulation characterizes the change of the state variables in a system. This is the consequence of the transport processes and the net production processes in the system. Any control volume can be used to define the system boundaries as long as the restrictions discussed above are considered.

Equation (3.1) is based on a well-defined rule of signs: influent, storage, and production of a material are positive, whereas effluent, depletion, and consumption are negative. The terms influent and effluent must be interpreted comprehensively as transport into the system (input) or evacuation from the system (output).

In the following sections the individual terms of Eq. (3.1) will be discussed and will then be put together in a mathematical context.

3.3.1 Inventory and Accumulation

The change of the quantity of a material within a system is the consequence of a change of the state of the system in space and time and is a cause for the change of the output or effluent concentrations. Their prediction is a central element of the modeling effort.

The mass M_A of the material A in the entire system results from the integral of the local concentration C_A over the whole volume of the system:

$$M_A(t) = \int_{V(t)} C_A(t,x,y,z) \cdot dV' \ [M_A].\tag{3.2}$$

The change of M_A with time corresponds to the rate at which material A accumulates in the system. The accumulation or storage rate is computed from the partial derivative of M_A with respect to time:

$$\frac{\partial M_A}{\partial t} = \frac{\partial}{\partial t} \int_{V(t)} C_A \cdot dV' \ [M_i \ T^{-1}].\tag{3.3}$$

Example 3.3: Amount of material and accumulation with ideal, complete mixing

What is the mass M_A of the material A in a system, if as a consequence of the intensive mixing the concentration of the material C_A is constant across the entire system?
With $C_A \neq f(x,y,z)$, Eq. (3.2) results in

$$M_A(t) = \int_{V(t)} C_A \cdot dV' = C_A(t) \cdot \int_{V(t)} dV' = C_A(t) \cdot V(t).$$

How large is the rate of accumulation of material A?
From Eq. (3.3) we obtain $\dfrac{\partial M_A}{\partial t} = \dfrac{\partial(C_A \cdot V)}{\partial t} = V \cdot \dfrac{\partial C_A}{\partial t} + C_A \cdot \dfrac{\partial V}{\partial t}$.

With constant volume (a frequent case) this results in:

$$\frac{\partial M_A}{\partial t} = \frac{\partial(C_A \cdot V)}{\partial t} = V \cdot \frac{\partial C_A}{\partial t} = V \cdot \frac{dC_A}{dt}.$$

3.3.2 Transport Processes

Transport of material is the result of advection (transport with the flowing water), diffusion (as a consequence of turbulence or molecular motion) or sedimentation/flotation (as a consequence of the force of gravity, or centrifugal or magnetic force). Transport processes bring materials from the environment into the system (input) or from the system back into the environment (output).

Material A can be transported over the system boundary into or out of a system. Its transport processes are characterized by j_A, the specific mass flux of material A, which is an intensive, directed, vector property that indicates how much of the material A is transported per time per unit area (see Fig. 3.1). The total transport

J_A of material A into a system is defined here as a scalar, extensive property obtained from:

$$J_A(t) = \oint_{F(t)} j_{n,A}(t) \cdot dF', \qquad (3.4)$$

J_A = total transport of material A into the system, a scalar, extensive property $[M_A\ T^{-1}]$

$j_{n,A}$ = specific mass flux of material A normal (perpendicular) to the surface F of the system, a scalar property $[M_A\ L^{-2}\ T^{-1}]$

$F(t)$ = surface area of the system as a function of time $[L^2]$

$j_{n,A}$ is positive if locally the material is transported into the system and is negative if it is transported out of the system. From this we obtain J_A positive, if overall more material A is transported into the system than is lost.

Formally the specific flux $j_{n,A}$ can be obtained from the dot product:

$$j_{n,A} = -\vec{j}_A \cdot \vec{n} \qquad (3.5)$$

$\vec{j}_A$ = specific flux of material A, a vector property $[M_A\ L^{-2}\ T^{-1}]$

$\vec{n}$ = unit vector, normal to the surface of the system, directed outwards $[-]$

Frequently the specific flux of material is constant over some parts of the surface (e. g., over the cross section of inlet and outlet pipes) and outside of these surfaces (e. g., at reactor walls) it is zero. Here Eq. (3.4) with some active areas F_i, which are assumed to be independent of time, results in:

$$J_A = \sum_i \int_{F_i} j_{n,A,i} \cdot dF' = \sum_i j_{n,A,i} \cdot F_i . \qquad (3.6)$$

If the flow of material is dominated by advection with the flow velocity u, then the following applies:

$$j_{n,A,i} = u_i \cdot C_{A,i} \quad \text{or with} \quad Q_i = u_i \cdot F_i \quad J_{A,i} = Q_i \cdot C_{A,i} . \qquad (3.7)$$

Thus, if all input and output occurs only via influent and effluent pipe Eq. (3.6) becomes (see Example 3.5):

$$J_A = \sum_i Q_i \cdot C_{A,i} . \qquad (3.8)$$

Chapter 4 is devoted to the systematic introduction of transport processes which are used to characterize $j_{n,A}$.

Example 3.4: Extensive flow of material J_A with constant, intensive (specific) flux $j_{n,A}$

What is flux of the material A (J_A) through the surface F, if the specific flow of material ($j_{n,A}$) is constant (not depending on location)?

According to Eq. (3.4) we obtain for constant $j_{n,A}$:

$$J_A = \oint_{F(t)} j_{n,A} \cdot dF' = j_{n,A} \cdot \oint_{F(t)} dF' = j_{n,A} \cdot F(t).$$

An example of such a system is a rising gas bubble that exchanges oxygen with its environment.

Example 3.5: Flow of material in a sewer

What is the mass flux of COD in the influent of a wastewater treatment plant?
The average flow velocity is $u = 1\,m\,s^{-1}$ and the material concentration is $C_{COD} = 250\,gCOD\,m^{-3}$. The inlet channel is rectangular, 0.4 m wide, and flows 0.3 m deep.
The specific mass flux $j_{n,COD}$ in the longitudinal direction of the sewer is:

$$j_{n,COD} = u \cdot C_{COD}.$$

From Eqs. (3.6), (3.7), and (3.8) we obtain for J_{COD}:

$$J_{COD} = \int_F u \cdot C_{COD} \cdot dF' = C_{COD} \cdot \int_F u \cdot dF' = C_{CSB} \cdot Q = 30\ gCOD\ s^{-1},$$

where $F = 0.12\,m^2 =$ cross-sectional area and $Q = F \cdot u = 0.12\,m^3\ s^{-1}$ is the flow of wastewater.

3.3.3 Reaction, Production, and Consumption

Chemical and biological processes transform materials such that these may both be produced or consumed (degraded). Thus, the material concentrations and the amount of material in a system are subject to change. Consumption corresponds to negative production.

The mass of material A within a system can increase due to production or decrease due to consumption. The reaction rate r_A $[M_A\ L^{-3}\ T^{-1}]$ is an intensive variable that indicates how much of material A is locally produced (positive) or degraded (negative). The total production of the material R_A results from the integral of the reaction rate r_A over the entire volume V of the system:

$$R_A(t) = \int_{V(t)} r_A(t, x, y, z) \cdot dV' \tag{3.9}$$

R_A = total production of material A in the entire system, an extensive variable $[M_A\ T^{-1}]$
r_A = specific production rate of the material A per unit volume, an intensive variable in space and time $[M_A\ L^{-3}\ T^{-1}]$
V = volume of the entire system $[L^3]$

Chapter 5 is devoted to the systematic introduction of kinetic and stoichiometric information, which is used to characterize the production rates r_A of materials.

Example 3.6: Production in a completely mixed system

How much material A is transformed in the volume V, if the reaction rate r_A is the same over the entire volume?

Equation (3.9) with constant r_A yields:

$$R_A = \int_{V(t)} r_A \cdot dV' = r_A \cdot \int_{V(t)} dV' = r_A \cdot V .$$

3.3.4 Mathematical Form of the Balance Equation

The balance equation can be written for each system and subsystem as well as for each material of interest. This corresponds to a bookkeeping process that connects accumulation, transport, and reaction.

If we introduce accumulation, transport, and transformation (reaction) in mathematical format into Eq. (3.1) we obtain:

$$\frac{\partial M_A}{\partial t} = J_A + R_A , \tag{3.10}$$

and with Eqs. (3.3), (3.4), and (3.9):

$$\frac{\partial M_A}{\partial t} = \frac{\partial}{\partial t} \int_{V(t)} C_A \cdot dV' = \oint_{F(t)} j_{n,A} \cdot dF' + \int_{V(t)} r_A \cdot dV' . \tag{3.11}$$

Equation (3.11) follows our rule of sign: storage, influent, and production are positive, whereas depletion, discharge, and consumption are negative. Equation (3.11) is written here in a very general form and serves as basis for the derivation of balance equations for very different systems. It will usually be simplified with consideration of the special characteristics of the modeled system.

In this text we will always in a first version write Eq. (3.11) in the structure introduced here: to the left we will write the accumulation term $\partial M_A / \partial t$, which indicates how the mass of material A or the state of the system changes with time; to the right we will first include the transport processes, which indicate the mass of material transported over the system boundaries, which will lead to concentration gradients in space; and to the far right will stand the reaction term, which indicates how much of the material is gained or lost by reactions.

Example 3.7: Simple mass balance for water

A reactor contains a constant volume of water. Influent and effluent are only possible via appropriate pipes and amount to Q_{in} and Q_{out}. *What can we derive from this description?*

The concentration of the water corresponds to its constant density ρ_W. In reactions water is only produced or consumed in smallest amounts, so that r_W can be neglected. Equation (3.11) results in:

$$\frac{\partial M_W}{\partial t} = V \cdot \frac{\partial \rho_W}{\partial t} + \rho_W \cdot \frac{\partial V}{\partial t} = Q_{in} \cdot \rho_W - Q_{out} \cdot \rho_W + r_W \cdot V \quad .$$

With constant volume V, constant density ρ_W, and $r_W = 0$ it remains:

$$0 = Q_{in} - Q_{out} \quad \text{or} \quad Q_{in} = Q_{out} = Q .$$

We obtain the statement, which we can also derive intuitively, that with constant water volume the influent must correspond to the discharge.

Example 3.8: A simplified form of the macroscopic material balance equation

Making the following, frequently valid, assumptions we can simplify the general macroscopic mass balance Eq. (3.11):

1. The volume V of the system (reactor) is constant.
2. The reactor is well mixed such that the concentrations of all substances and all other intensive properties do not vary over the entire volume.
3. Influent and effluent are through pipes which have a cross section A_{in} and A_{out} and a flow velocity u_{in} and u_{out}, respectively.

With assumption 3, Eq. (3.4) becomes:

$$\int_{F(t)} j_n \cdot dF' = \int_{Fin} C_{in} \cdot u_{in} \cdot dF + \int_{Fout} C_{out} \cdot u_{out} \cdot dF = Q_{in} \cdot C_{in} - Q_{out} \cdot C_{out} .$$

Assumption 1 leads to $Q_{in} = Q_{out} = Q$ (the same amount of water flows into the system as gets lost from the system, see Example 3.7). In addition Eq. (3.3) applies:

$$\frac{\partial}{\partial t} \int_V C \cdot dV' = \frac{\partial}{\partial t} C \cdot \int_V dV' = V \cdot \frac{dC}{dt} .$$

From assumption 2 it follows that the reaction rate r is constant over the entire system, and therefore Eq. (3.9) becomes:

$$\int_{V(t)} r \cdot dV' = r \cdot V .$$

After substitution into Eq. (3.11) we obtain:

$$V \cdot \frac{dC}{dt} = Q \cdot C_{in} - Q \cdot C_{out} + r \cdot V .$$

Application of this equation is simple, if we can specify how the reaction rate r can be computed. It is equivalent to the mass balance for a continuous flow stirred tank reactor (CSTR, see Sect. 6.3)

The balance equation (3.11) relates extensive variables. Accordingly it can only be applied to systems of finite size; we speak of a *macroscopic material balance*.

If inside the balanced system gradients of the intensive variables (concentrations, reaction rates, etc.) arise, an analytic solution of this equation is available for special cases only. A balance equation for a differentially small system does justice to this situation, in the sense that such gradients appear as partial derivatives. We then speak of a *microscopic balance*, which takes the form of a partial differential equation and must be integrated over the entire macroscopic system. Thereby the influents and effluents become boundary conditions and the microscopic transport processes describe internal mixing.

The integral theorem of Gauss (Bronstein et al., 1993) states:

$$\oint_F \vec{j} \cdot \vec{n} \cdot dF' = \int_V \text{div } \vec{j} \, dV' .\tag{3.12}$$

With Eq. (3.5) this becomes:

$$-\oint_F j_n \cdot dF' = \int_V \text{div } \vec{j} \, dV'\tag{3.13}$$

$\vec{j}$ = vector field, here the directed field of the specific mass flux $[M\,L^{-2}\,T^{-1}]$
$\vec{n}$ = unit vector, normal to the system surface, positive towards the outside $[-]$
F = surface of the system $[L^2]$
V = volume of the system $[L^3]$

After the substitution of Eq. (3.12) into the macroscopic balance Eq. (3.11) and derivation with respect to V we obtain the material balance in its differential, microscopic form:

$$\frac{\partial C_i}{\partial t} = -\text{div}(\vec{j}_i) + r_i = -\left(\frac{\partial \vec{j}_i}{\partial x} + \frac{\partial \vec{j}_i}{\partial y} + \frac{\partial \vec{j}_i}{\partial z} \right) + r_i .\tag{3.14}$$

Written in one-dimensional form, as is frequently applied to a river, a sewer, or a plug-flow reactor, there remains:

$$\frac{\partial C_i}{\partial t} = -\frac{\partial j_i}{\partial x} + r_i .\tag{3.15}$$

Equation (3.14) is as general as Eq. (3.11). It is, together with Eq. (3.15), the basis for the derivation of balance equations for systems in which internal mixing is not sufficient to eliminate the gradients of intensive properties. The term $\partial j/\partial x$ characterizes the internal mixing processes.

Example 3.9: One-dimensional form of the differential material balance equation

In rivers and many plug-flow type technical systems one flow direction dominates the behavior of the system. Accordingly we can use Eq. (3.14) in its one-dimensional form (3.15).

With the assumption that only the material flux in the x-direction is of importance and that this flux is composed of advection (u·C, for directed flow velocity u and concentration C) and turbulent diffusion (D_T = turbulent diffusion coefficient [L^2 T^{-1}]), we obtain for the total mass flux:

$$j = u \cdot C - D_T \cdot \frac{dC}{dx}.$$

Substituted into Eq. (3.15) yields for constant u and D_T:

$$\frac{\partial C}{\partial t} = -u \cdot \frac{\partial C}{\partial x} + D_T \cdot \frac{\partial^2 C}{\partial x^2} + r.$$

In this form we will apply Eq. (3.14) to turbulent plug-flow reactors (Sect. 6.6).

Example 3.10: Bookkeeping as an example of the balance equation

An example of a balance is our private banking account (system). The amount deposited (state variable) changes (change of state), starting from an empty account (initial condition) as a consequence of incoming and outgoing payments (transport processes) as well as the deduction of taxes and fees (consumption) and the addition of interest (production).

3.4 Special Cases of the Material Balance Equation

In the following three special cases we can set one of the three terms of the material balance equation to zero. This simplifies the application of balance equations.

3.4.1 Stationary Balance or the Steady State

In the steady state the state of a system does not change with time, i. e., there is no storage or accumulation of material. The according term of the balance equation does not have any importance and may therefore be dropped. The result is a stationary balance which cannot make any statements about the time-dependent transition from one state to another one.

The stationary balance equation has the form:

$$0 = \int_F j_{n,A} \cdot dF' + \int_V r_A \cdot dV' \qquad (3.16)$$

Equation (3.16) describes a system in which over a long time period all disturbances (influent, temperature, etc.) as well as all system properties (F, V) remain constant. In addition, neither the mixing nor the operating conditions may be changed in the system. The resulting condition is called a *steady state*, which is an important starting point in the analysis of systems. In comparison to the dynamic,

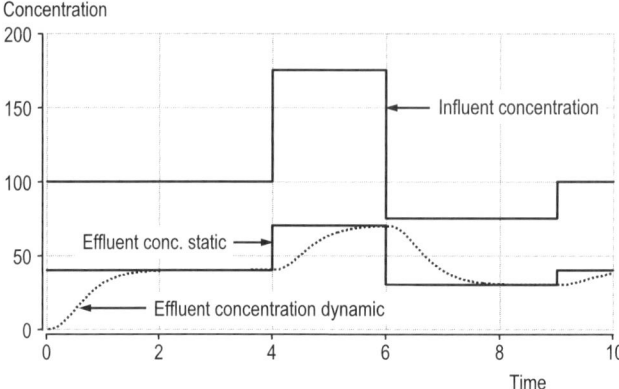

Fig. 3.4 Comparison of the static and the dynamic response of a system to changes in load

time-dependent balance equation (3.11) the steady-state equation (3.16) is a powerful simplification: partial differential equations frequently become ordinary differential equations, and ordinary differential equations may become algebraic equations.

With Eq. (3.16) we cannot predict any dependence of the states C_i of the time. If a transport variable $(j_{n,i})$ changes with time, a direct change in the state variables C_i results. The temporal transition from one steady state to the next cannot be described, because the attenuation obtained from the accumulation process is omitted. We speak of stationary (or static) models in which the disturbances (inputs) do not have to be independent of time. However, the fact that the static equation neglects the storage term in Eq. (3.11) results in an immediate approach to a new equilibrium. Figure 3.4 compares the immediate response of a static model with the attenuated response of the dynamic model. Clearly the static model cannot capture the full time dependency of the system behavior.

Example 3.11: A stationary model

In Example 3.7 a stationary model is derived: in a system a constant amount of material M_W (here a constant volume of water) is given; any storage of water is excluded. The effluent reacts immediately to a change of the influent, $Q_{in} = Q_{out}$.
In reality this immediate response is not possible. Either the volume of the system must change to adjust the water height in the overfall weir to accommodate the required effluent or the water must be compressed (the density ρ_W is thereby changed), so that the correct pressure is established. These adjustments are typically so small and occur so fast that we can assume steady state in most situations.

Example 3.12: Balance of an intensively mixed reactor in steady state

A reactor with the constant volume V is fed with a constant influent Q that contains the material A with constant concentration $C_{A,in}$. The reactor is intensively

Fig. 3.5 The Roman fountain

mixed, so that no gradients of material concentrations can be formed. The material A is degraded with the rate $r_A = -k \cdot C_A$.

What is the concentration of the material C_A in the effluent?

Equation (3.16) together with Eq. (3.8) and considering that the concentration C_A is the same throughout the reactor and therefore also in the effluent results in:

$$0 = Q \cdot C_{A,in} - Q \cdot C_A - k \cdot C_A \cdot V,$$

from which we obtain: $C_A = \dfrac{C_{A,in}}{1 + k \cdot V / Q}$.

Example 3.13: The steady state in lyrics

A Swiss poet, C. F. Meyer (1825–1898) described the steady state in a poem as follows (original in German language):

Der römische Brunnen

Aufsteigt der Strahl, und fallend giesst
Er voll der Marmorschale Rund.
Die, sich verschleiernd überfliesst
In einer zweiten Schale Grund;
Die zweite gibt, sie wird zu reich,
Der dritten wallend ihre Flut.
Und jede nimmt und gibt zugleich
Und strömt und ruht.

Roman Fountain
(Translation by C.H. Séquin)

High soars the jet, – then falls and fills
a bowl of marble to its bound,
which yields; – the rippled surface
spills into a second stony round.
This second overflowing urn
then fills a third, – its surface sways;
and each one takes and gives in turn,
– and flows – and stays.

C.F. Meyer was particularly impressed by the steady state which results despite the active transport processes and the continuing motion in this Roman fountain.

3.4.2 Closed Systems

A closed system does not exchange materials with the environment; therefore no transport processes at the system boundaries exist and Eq. (3.11) becomes:

$$\frac{\partial M_i}{\partial t} = \int_{V_{(t)}} r_i \cdot dV' \tag{3.17}$$

The application of Eq. (3.17) to a system with variable volume V(t) is introduced in Example 3.16. In urban water management the reactor volume is usually referred to as the volume of the water that with neither influent nor effluent remains constant. If in addition intensive mixing eliminates all gradients of intensive properties, Eq. (3.17) can be simplified to (see Example 3.14):

$$V \cdot \frac{dC_i}{dt} = r_i \cdot V \quad \text{oder} \quad \frac{dC_i}{dt} = r_i. \tag{3.18}$$

Example 3.14: The test tube as a closed system

Chemical reactions are frequently followed in test tubes. Thereby the test tube (the reactor, the system) is shaken such that its content is completely mixed. In the course of the observation of the reaction there is neither influent nor effluent; the system is closed.

Since the volume V remains constant and the reaction rate r does not vary across the reactor, we can simplify Eq. (3.17) to:

$$\frac{\partial M}{\partial t} = V \cdot \frac{\partial C}{\partial t} + C \cdot \frac{\partial V}{\partial t} = r \cdot \int_V dV' = r \cdot V \quad \text{or} \quad \frac{dC}{dt} = r.$$

Example 3.15: Reaction rate r in a closed system

In former times it was customary to write a change of concentration with time in order to express the reaction rate r_A. Thus, e. g., for a first-order reaction the rate was written as:

$$\frac{dC_A}{dt} = -k \cdot C_A.$$

This form is not compatible with the representation selected in this text. For a closed, well-mixed system with constant volume the balance equation has the form of Eq. (3.18):

$$\frac{dC_A}{dt} = r_A.$$

If we insert the rate of a first-order reaction ($r_A = -k \cdot C_A$) into this simplified balance equation we obtain the representation given above.

In open systems it is possible that a reaction proceeds without change of the concentration of the reaction partners ($dC_A/dt = 0$), because the transport processes compensate the transformation immediately (see Example 3.12). Therefore it is mathematically not correct to write the reaction rate in the format of the differential equation given above. This early way of writing goes back to the time when chemists typically examined reactions in the laboratory in closed systems (in the test tube).

Example 3.16: Radioactive decay in a closed system

Tritium ^3H has a radioactive half-life of $\tau_{1/2} = 12.4$ yr, i.e., a certain initial mass of tritium disintegrates within 12.4 years to half its initial amount. How large is the decay constant of tritium, if the decay follows a first-order reaction?

$$r_{3H} = -k \cdot C_{3H}.$$

We define the system such that it always contains all initially existing tritium isotopes independently of any further distribution of the tritium. In addition the system is closed, so that no new tritium can be added. Here, the system boundaries are not necessarily defined geographically or geometrically, but correspond rather to an abstract concept, since individual isotopes may travel quite far over 12.4 yr. The material balance for ^3H in a closed system ($j_{3H} = 0$) results in:

$$\frac{\partial M_{3H}}{\partial t} = \int_{F(t)} j_{n,3H} \cdot dF' + \int_{V(t)} r_{3H} \cdot dV' = 0 - k \cdot \int_{V(t)} C_{3H} \cdot dV' = -k \cdot M_{3H} = \frac{dM_{3H}}{dt}.$$

With the initial condition: $t - 0$ $M_{3H}(t-0) = M_{3H,0}$
and the information about the half-life $t = \tau_{1/2}$ $M_{3H}(\tau_{1/2}) = M_{3H,0}/2$
we obtain after integration:

$$\frac{M_{3H}(\tau_{1/2})}{M_{3H,0}} = \frac{1}{2} = \exp(-k \cdot \tau_{1/2}) \text{ or } \ln(0.5) = -0.69 = -k \cdot \tau_{1/2},$$

$$k = 0.69/\tau_{1/2} = 0.69/12.4 \text{ yr} = 0.056 \text{ yr}^{-1}.$$

Because radioactive decay is a linear process, the decay of a certain initial amount of material can be described without knowing the local material concentrations. The only conditions are that the system contains the entire initial amount of material of interest and that the system is closed. The linear process manifests itself mathematically in the fact that the integral over the reaction ($r_{3H} \cdot dV'$) can be solved independently of the local concentration C_{3H} to yield $-k \cdot M_{3H}$.

3.4.3 Conservative Material

A conservation law applies to so-called conservative materials (see Sect. 5.7), which states that material A can neither be produced nor consumed, i.e., $r_A = 0$. For conservative materials Eqs. (3.11) and (3.14) become:

$$\frac{\partial M_A}{\partial t} = \oint_{F(t)} j_{n,i} \cdot dF' \text{ and} \tag{3.19}$$

$$\frac{\partial C_A}{\partial t} = -\text{div}(\vec{j}_A). \tag{3.20}$$

Typical conservative materials are chemical elements, electrical charge, and the mass of all materials and water. In addition, we will introduce other conservative properties such as the theoretical oxygen demand (TOD).

Summary: Each of the three special cases introduced here refers to one term of the balance equation: the accumulation term refers to the steady state, the transport term to closed systems, and the reaction term to conservative materials. In addition, combinations of these special cases are possible. A glass of water standing on a table is a closed system containing a conservative material and must therefore be at steady state.

Example 3.17: Water as conservative material

Reactions are only very small sources and sinks for water flowing through a system. Therefore for most questions in urban water management water is treated as a conservative material and its concentration remains constant, according to its density ρ_w. The use of Eq. (3.20) then results in:

$$\frac{\partial \rho_w}{\partial t} = 0 = -\mathrm{div}(\vec{j}_w) = -\mathrm{div}(\vec{v} \cdot \rho_w) \text{ and } \mathrm{div}(\vec{v}) = 0 .$$

This equation corresponds to the so-called *continuity equation* in technical hydraulics. It describes a restriction which must be obeyed in the three-dimensional flow field of a current.

Example 3.18: Conservative material in a closed system

For a conservative material (reaction term $=0$) in a closed system (transport term $=0$) with constant volume ($dV/dt=0$) a steady state will result (accumulation term $=$ transport term $+$ reaction term $=0$).

If the volume is subject to change or we have to consider internal mixing, then the local concentrations are also subject to change, which results in a dynamic, time-dependent state. The balance equation for a conservative material in a well-mixed but closed system without gradients of intensive variables has the form:

$$\frac{\partial M}{\partial t} = C \cdot \frac{\partial V}{\partial t} + V \cdot \frac{\partial C}{\partial t} = 0 \text{ or } C \cdot \frac{\partial V}{\partial t} = -V \cdot \frac{\partial C}{\partial t} .$$

Pressure surges (water hammer) in rapidly closed water pipelines are closed, dynamic systems in which a conservative material (water) is subject to local concentration variations (density) due to pressure change, and thus the volume of the system must respond (the lines expand).

Example 3.19: Steady state for conservative materials

For a conservative material (reaction term $= 0$) the balance equation yields for steady state that the input must correspond to the output of the system (net transport $= 0$ or inputs $=$ outputs).

3.5 Summary

The verbal (Eq. (3.1)), extensive (Eq. (3.11)), and intensive (Eq. (3.14)) forms of the material balance equations summarize the influence of different processes on the state of a system. Various simplifications are used, the most important ones being characterized in Table 3.1.

Table 3.1 Compilation of the simplifications of the balance equation for different systems, models or state variables (materials and energy)

	Storage	Transport across system boundary	Reaction	Energy ex-change
Open system	+	+	+	+
Closed system	+	none	+	+
Isolated system	+	none	+	none
Autonomous system	+	only outputs	+	only output
Subsystem	+	+ may affect input	+	+
Stationary system	none	+	+	+
Conservative material	+	+	none	+

Chapter 4
Transport Processes

There are two types of transport processes: (i) deterministic processes following, e. g., the streamlines of flowing water or force lines of force fields, which can be predicted with the help of technical hydromechanics and physics, and (ii) stochastic processes, which are caused by many random events, e. g., Brownian motion and turbulence.

Example 4.1: A drop of dye in water

If we introduce a drop of a colored aqueous solution into a container with standing water, then locally large gradients of the concentration of the coloring material develop. With time, the color will be distributed evenly throughout the entire container. We observe the consequence of transport processes: coloring molecules move and the local concentration of the coloring material is changed. We can follow this process by measuring the color intensity locally. In contrast, the concentration of the water changes only slightly. Even though the individual water molecules move too, we cannot capture this movement with simple analysis.

The microscopic transport processes apparently affect the materials contained in the water at low concentration differently to the way in which they affect the water that fills the space.

In this experiment we observe an important concept: the color moves from locations with high concentration to locations with small concentration, until finally the entire container is filled with the same concentration of dye and apparently no more transport takes place (even though individual molecules of color still move around).

4.1 Characterization of Transport Processes

In both technical and natural systems transport processes are of central import-
ance. We differentiate between deterministic, advective processes, and stochastic,
diffusive processes.

Figure 4.1 shows the movements of water packages and individual particles on
different scales. On a large scale we may follow the macroscopically directed
current of flowing water, which is easily observed. On a smaller scale energy is
dissipated and turbulence becomes important. Finally on the molecular scale thermal motion dominates the relative movement of the particles.

These transport processes have very different characteristics. Here we will differentiate between five types of processes:

- *Advection* describes a deterministic transport process in which external forces
  cause a flow field that carries the water and the materials and particles contained in the water along the streamlines. The characterization of advection
  (flow velocity) is a central task of technical hydraulics.
- *Diffusion* is a stochastic process in which transport at a specific location is
  caused by random movements of the interesting particles. We differentiate between *molecular diffusion*, which is a consequence of the material-specific
  thermal motion, and *turbulent diffusion*, which is caused by turbulence (dissipation of kinetic energy) and is therefore not specific to a material.
- *Dispersion* describes the transport of materials which results from the deviations of the real flow field from the one predicted from hydromechanics or
  technical hydraulics. Our hydraulic models are simplifications; they frequently
  yield only one-dimensional flow and thereby provide a flow velocity which is
  averaged over the cross section. This deviates from reality, as individual water

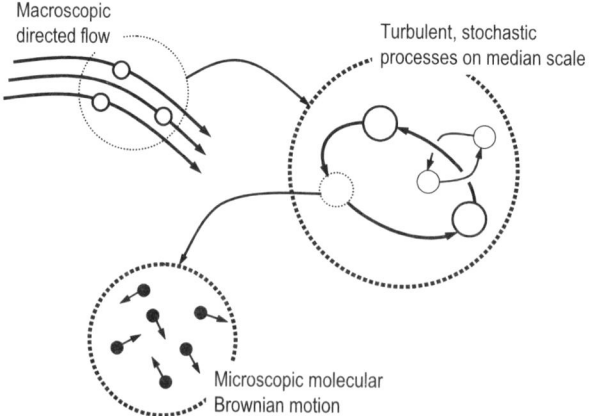

Fig. 4.1 Properties of transport processes depend on their scale, from large, directed, advective
processes to progressively smaller ones with increasingly stochastic elements

packages travel on different flow paths and with differing flow velocities. The process of dispersion corrects this deviation for the transport of materials; it is typically based on empirical information.

- *Sedimentation/flotation/centrifugation* are deterministic processes that are caused by external forces such as gravitation, buoyancy or inertia and cause particles to divert from the streamlines. Similar effects may be caused by magnetic fields.
- *Convection* describes the advective currents that are caused by gradients in the density of the water. Warm water with low specific density ascends, whereas dense (cold) water sinks. The process has both deterministic and stochastic aspects.

4.2 Modeling of Transport Processes

Simple models and empirical parameter values for the modeling of transport processes are introduced.

4.2.1 Advection

In advection we regard the temporally averaged, directed, local flow velocity v which we obtain from hydromechanics. Materials and particles are transported as part of the flowing water.

Forces that affect a system from the outside cause a directed current of the water (Fig. 4.2). The field of the temporally averaged flow velocities changes only slowly in comparison to the time constants of interest. Small particles and dissolved materials are carried by this flow; their transport is proportional to the local flow velocity and their concentration:

$$j_A = v \cdot C_A \tag{4.1}$$

j_A = specific flow of material i $[M_A\, L^{-2}\, T^{-1}]$, a directed variable (vector)
v = local flow velocity $[L\, T^{-1}]$, a vector variable
C_A = local concentration of material A $[M_A\, L^{-3}]$, a scalar variable

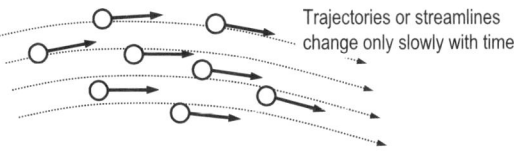
Trajectories or streamlines change only slowly with time

Fig. 4.2 In the context of advection the flow field is due to outside forces and changes, compared to the time constants of interest, only slowly and in a predictable way. Water packages and particles follow the trajectories (streamlines)

In analyzing the material transport in a pipe, we frequently accept as a first approximation that the velocity of flow v is evenly distributed over the cross section (Fig. 4.3a). With this assumption a concentration profile remains stable and shifts laterally with the flow velocity. In reality, however, friction along the walls induces profiles of varying flow velocities, which are particularly pronounced in the case of laminar flow (parallel streamlines, Fig. 4.3b). From differing flow velocities a distortion of the concentration profile results. If the flow is turbulent, this distortion is smaller, but not negligible. Since we typically use flow velocities averaged over a cross section (evenly distributed), we must combine advection with an overlaying process which describes the distortion of the concentration profiles; this process is called dispersion (see Sect. 4.2.6).

Example 4.2: Mass transfer in a water pipe

A pipeline with a diameter of 200 mm has a roughness factor after Strickler of $k_{St} = 85 \, m^{1/3} \, s^{-1}$ (Manning $n = 1/k_{St}$). The line is 2000 m long and the acceptable hydrostatic pressure loss along this line amounts to 6 m water gauge. The water carries a concentration of $C_{Ca} = 60 \, gCa^{2+} \, m^{-3}$.

How much Ca^{2+} is transported in this line?

From the equation of Manning/Strickler $v = k_{St} \cdot R_h^{2/3} \cdot I_E^{1/2}$ or

$$Q = 0.312 \cdot k_{St} \cdot D^{8/3} \cdot I_E^{1/2} \text{ we obtain: } Q = 0.020 \, m^3 \, s^{-1}.$$

The mass transfer is $J_{Ca} = C_{Ca} \cdot Q = 1.2 \, g \, s^{-1}$.

This problem is of course formulated very simple. We solved it based on the rules of technical hydraulics which allow us to compute an average flow velocity v. Alternatively we could have developed a velocity profile over the cross section (logarithmic distribution in turbulent flow) and then integrated over the cross section. The result would be the same. Only when concentration gradients in cross sections and length profile arise will the two results differ substantially.

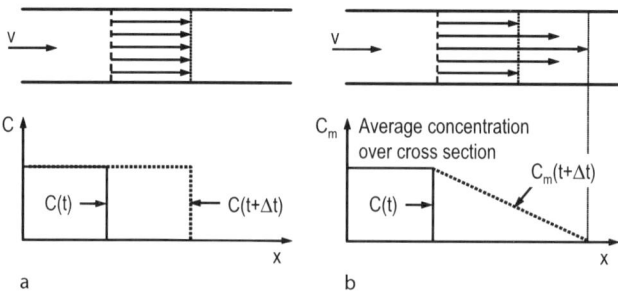

a   b

Fig. 4.3 (a) Homogenous flow velocity v over the entire cross section leads to a lateral shift of the concentration profile of a tracer. (b) Parabolic velocity profiles, as obtained in laminar flow, lead to a linear deformation of the front of the concentration profile averaged over the cross section. This parabolic flow field also only considers advection, but is now based on a more differentiated, two-dimensional model

For the intensive balance equation (3.15) with advection as a transport process, we obtain in one dimension:

$$\frac{\partial C}{\partial t} = -v \cdot \frac{\partial C}{\partial x} + r \,. \tag{4.2}$$

For the steady-state, Eq. (4.2) results in:

$$v \cdot \frac{dC}{dx} = r \quad \text{resp.} \quad \frac{dC}{d\tau} = r \quad \text{with } \tau = x/v = \text{space time} \tag{4.3}$$

4.2.2 Sedimentation

Particles sediment in water if additional outside forces affect the particles (gravitation, buoyancy, centrifugal forces, inertia, magnetic forces). Thus, particles are diverted from the flowing water (Fig. 4.4).

Stoke (1851) deduced what later became Stoke's law, which describes the interaction of small particles with the surrounding fluid. It applies to spherical particles within the laminar sedimentation range ($Re_P < 1$, small sedimentation velocities):

$$v_S = \frac{1}{18} \cdot \frac{\rho_P - \rho_W}{\rho_W} \cdot \frac{g}{v_W} \cdot d_P^2 \qquad \text{if} \qquad Re_P < 1, \tag{4.4}$$

$$Re_P = \frac{v_S \cdot d_P}{v_W} = \text{Reynolds number for sedimentation} \tag{4.5}$$

v_S = sedimentation velocity [L T^{-1}]
ρ_S, ρ_W = density of the particle and the water [M$_i$ L^{-3}]
g   = acceleration due to gravity g = 9.81 m s^{-2}
v_W = kinematic viscosity of the water, 1.0034 mm^2 s^{-1} at 20°C
d_P = ball diameter of the particle [L]

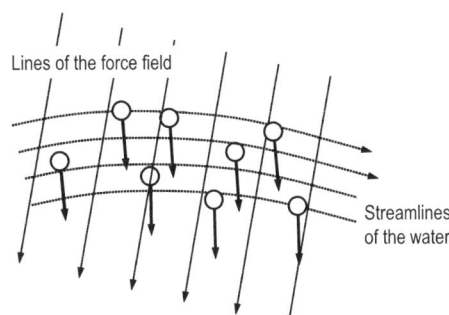

Fig. 4.4 If outside forces affect particles, they are diverted from the flow field by sedimentation

For sedimentation in the turbulent range ($Re_P > 2000$) one finds empirically approximately:

$$v_S \approx \sqrt{\frac{8}{3} \cdot \frac{\rho_P - \rho_W}{\rho_W} \cdot g \cdot d_P} \quad \text{based on a drag coefficient of } C_W = 0.5. \qquad (4.6)$$

Equations (4.4) and (4.6) are plotted in Fig. 4.5. The sedimentation velocities of bacteria ($d_P \approx 1\ \mu m$) up to stones with a diameter of 1 cm vary by a factor 10^7.

If sedimentation and advection overlay, then the two flows of the material add:

$$\vec{j}_{tot} = \vec{j}_A + \vec{j}_S = (\vec{v} + \vec{v}_S) \cdot C. \qquad (4.7)$$

The intensive mass balance equation (3.14) in one-dimensional form becomes:

$$\frac{\partial C}{\partial t} = -(v + v_S) \cdot \frac{\partial C}{\partial x} + r. \qquad (4.8)$$

Since sedimentation and advection frequently occur in different directions, a one-dimensional model may not be sufficient to capture the details.

Example 4.3: Sedimentation in a centrifuge

You have been offered a centrifuge with a radius of $r_Z = 0.1$ m and $n = 3000$ rpm (revolutions per minute), with the statement that bacteria may be centrifuged out of water in order to generate drinking water. *How do you judge the statement?*

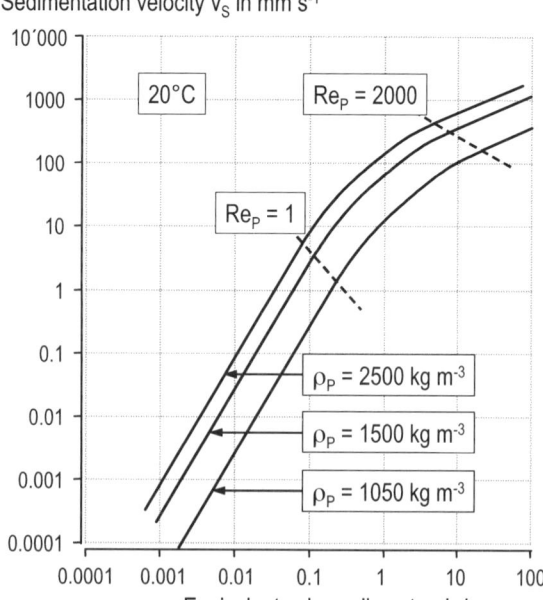

Fig. 4.5 Sedimentation velocity of spherical particles under gravity in pure water at 20°C

The maximum acceleration amounts to $a_\omega = r_Z \cdot \omega^2 = r_Z \cdot (2 \cdot \pi \cdot n)^2 = 10000 \, \text{m s}^{-2}$. With $d_P = 1 \, \mu m$ and $\rho_P = 1.05 \, \text{g cm}^{-3}$ and this acceleration a maximum sedimentation velocity of $v_S = 0.027 \, \text{mm s}^{-1}$ results, using Eq. (4.4). If the centrifuge is constructed in a way which limits the maximum required sedimentation path to 10 mm, this would require a residence time of more than 6 min. The apparatus is probably not very efficient.

Example 4.4: Sedimentation of bacteria and algae

A spherical bacterium with a diameter of $d_P = 1.2 \, \mu m$ and a density of $\rho_B = 1.2 \, \text{g cm}^{-3}$ is suspended in standing water. *How fast does it settle at 20°C?*
Equation (4.4) yields: $v_S = 0.00016 \, \text{mm s}^{-1} \approx 1.3 \, \text{cm d}^{-1}$ ($Re_P \approx 2 \cdot 10^{-7}$).
What is the sedimentation velocity of a spherical alga with 30 μm diameter under otherwise identical conditions?

$v_S = 0.098 \, \text{mm s}^{-1} \approx 8.4 \, \text{m d}^{-1}$ ($Re_P \approx 0.003$).

By using star-shaped structures at the surface, alga can reduce their apparent density. *What is the sedimentation velocity, if the density is reduced to $\rho_A = 1.02 \, \text{g cm}^{-3}$?*

$v_S = 0.0098 \, \text{mm s}^{-1} \approx 0.8 \, \text{m d}^{-1}$.

Example 4.5: Temperature and sedimentation

The grit chamber of a wastewater treatment plant is designed to remove sand grains with a diameter of $d_P \geq 0.1 \, \text{mm}$ and a density of $\rho_P = 2.65 \, \text{g cm}^{-3}$.
What is the sedimentation velocity of these sand grains at 5°C, 15°C, and 25°C?
The kinematic viscosity of water v_W amounts to (Linde, 1999):
5°C	$v_W =$	1.51 $mm^2 \, s^{-1}$
15°C		1.14
25°C		0.89

which yields from Eq. (4.4): $v_S =$ 6.0 mm s^{-1} with Re = 0.4
                                          7.9   0.7
                                         10.1   1.1
Sedimentation is just at the limit of the laminar range, so Stoke's law applies. The temperature effect is observable with $0.026°C^{-1}$, but is rather small. Nevertheless, differences in sedimentation between winter and summer are to be expected. We expect better separation of the sand in the summer.

Example 4.6: Gravitation and sedimentation

If you were to design a grit chamber to treat wastewater on the Moon, where the gravitational constant is only 1/6 of the 9.81 m s^{-2} observed on Earth, what would be the sedimentation velocity of a sand grain with $d_P = 0.1 \, \text{mm}$ and $\rho_P = 2.65 \, \text{g cm}^{-3}$ at 15°C (see also Example 4.5)?
Equation (4.4) yields: $v_S = 1.3 \, \text{mm s}^{-1}$, substantially smaller than on the Earth. This would require a much larger volume of the grit chamber.

Example 4.7: Dissolved air flotation

In a dissolved air flotation process small gas bubbles are formed from water super-saturated with air. These bubbles rise and thereby catch particulate pollutants which are carried to the surface. The bubbles have a diameter of approx. $d_P = 0.15$ mm. *How fast will they rise at 20°C?*
Assumptions: $\rho_{Air} = 0$, $v_{W,20°C} = 1$ mm^2 s^{-1}.
Equation (4.4) yields $v_S = -12$ mm s^{-1}. $v_S < 0$, thus the bubbles rise. $Re_P = 1.8$, just above the laminar range, nevertheless Eq. (4.4) is still quite accurate.
As soon as a rising bubble catches a flock, flotation is slowed.

Example 4.8: Thickening in a secondary clarifier

In the lower part of a secondary clarifier of an activated sludge system both advection (a consequence of the removal of return sludge) and sedimentation take place, whereby both processes point in the same direction. *How large is the total specific flux of activated sludge downward, if the advection velocity amounts to $v = 0.5$ m h^{-1}, the sedimentation velocity $v_S = 5$ m h^{-1}, and the local activated sludge concentration $X_{AS} = 0.6$ kg m^{-3}?*

$$j_{Advection} = v \cdot X_{AS} = 0.5 \cdot 0.6 = 0.3 \text{ kg}_{AS} \text{ m}^{-2} \text{h}^{-1},$$

$$j_{Sedimentation} = v_S \cdot X_{AS} = 5.0 \cdot 0.6 = 3.0 \text{ kg}_{AS} \text{ m}^{-2} \text{h}^{-1},$$

$$j_{total} = j_{Advection} + j_{Sedimentation} = 3.3 \text{ kg}_{AS} \text{ m}^{-2} \text{h}^{-1} \text{ (see also Eq. (4.7))}.$$

In the return sludge there is only advection, no sedimentation. This requires that the activated sludge is thickened sufficiently that the sludge inflowing from above can be removed. *What concentration must the activated sludge reach in the return line?*

$$j_{Advection} = j_{total} = v \cdot X_{ReturnSludge} \text{ and thus } X_{ReturnSludge} = 3.3 \text{ kg m}^2 \text{h}^{-1}/0.5 \text{ m s}^{-1}$$
$$= 6.6 \text{ kg}_{AS} \text{ m}^{-3}.$$

4.2.3 Random Walk

Usually we model transport processes in the continuum without following the movement of individual particles. The random walk is an alternative model that follows the fate of individual particles. This is possible with stochastic models that permit only a statistical interpretation of the results.

The random walk is introduced here to demonstrate the principle of diffusion. The model is not applied very frequently but rather has didactic merit.

The *random walk* describes a random movement of a particle in one- or multi-dimensional space; it is a stochastic model for the diffusion of particles in water. A collision of the particle with its environment is assumed to occur in each time step Δt. Subsequently, the particle will move in a new random direction with a new random velocity, before it collides again with the environment. The particle will thus move on a random path and thereby on average move further and further away from its place of origin.

The steps in local coordinates (in two dimensions) are given by:

$$\Delta x = \sigma_{\Delta x} \cdot N(0,1) = \sqrt{2 \cdot D_x \cdot \Delta t} \cdot N(0,1) \quad \text{and} \quad \Delta y = \sqrt{2 \cdot D_y \cdot \Delta t} \cdot N(0,1) \quad (4.9)$$

$N(0,1)$ = normal (Gaussian) random number with expected value of 0 and unit standard deviation

$\sigma_{\Delta x}, \sigma_{\Delta y}$ = standard deviation of the individual local step in the x- and y-directions [L]

D_x, D_y = molecular or turbulent diffusion coefficient in the x- or y-directions $[L^2 T^{-1}]$

Δt = Time step of the simulation [T]

The coordinates in one or multidimensional space are computed stepwise in a difference equation, whereby for each time step a new value of Δx and Δy is obtained:

$$x(t + \Delta t) = x(t) + \Delta x \quad \text{and} \quad y(t + \Delta t) = y(t) + \Delta y . \qquad (4.10)$$

Figure 4.6 provides an example of a two-dimensional random walk of two particles. Starting at the origin of the coordinate system, the resulting coordinates $x(t)$ and $y(t)$ follow a normal distribution (central limit theorem) with:

$$\mu_x = 0 \quad \text{and} \quad \sigma_x = \sigma_{\Delta x} \cdot \sqrt{\frac{t}{\Delta t}} = \sqrt{2 \cdot D_x \cdot t} \quad \text{as well as}$$

$$\qquad (4.11)$$

$$\mu_y = 0 \quad \text{and} \quad \sigma_y = \sigma_{\Delta y} \cdot \sqrt{\frac{t}{\Delta t}} = \sqrt{2 \cdot D_y \cdot t} .$$

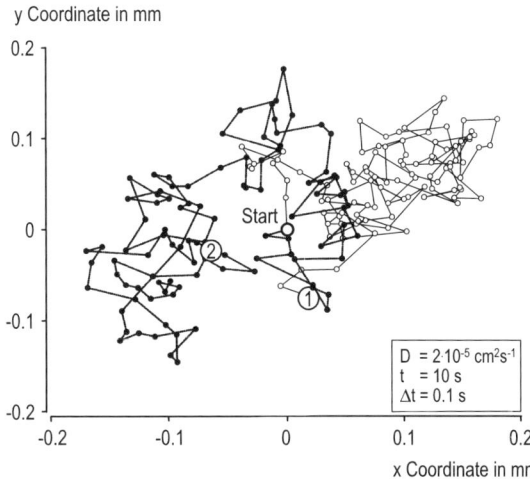

Fig. 4.6 Random walk of two particles, starting at the origin. One hundred time steps of $\Delta t = 0.1$ s over 10 s

If we let a large number of particles move based on a random walk and then analyze their local distribution, this will coincide with the distribution predicted from the deterministic Fick's law of diffusion (Eq. (4.13), Sect. 4.2.4). With increasing number of particles the computing time for the random walk model increases as well (whereas the solution of Fick's law is independent of concentration). Thus, we must balance the desired accuracy and the number of particles. Here the central limit theorem of statistics is of help. It says that the sum of n independent variables (here Δx and Δy) with expected value $\mu = 0$ and variance σ^2 tends for large n towards a normal distribution with $\mu_{Sum} = 0$ and $\sigma^2_{Sum} = n \cdot \sigma^2$ (here the sum is equivalent to x or y). The expected error of the sum μ_{Sum} is likewise normally distributed with:

$$\mu_{Error} = 0 \text{ and } \sigma^2_{Error} = \frac{\sigma^2_{Sum}}{n} = \frac{\sigma^2}{n^2}. \tag{4.12}$$

Example 4.9: Parameters of a random walk

Water contains $2 \, gO_2 \, m^{-3}$. We simulate the distribution of the oxygen molecules in one liter of water with the help of a random walk. We select a time step of 0.1 seconds and 10,000 particles.
How many oxygen molecules does a single particle represent?

$32 \, gO_2 = 1 \text{ mole } O_2 = 6 \cdot 10^{23}$ molecules.

One particle represents $\dfrac{2g}{32g} \cdot \dfrac{6 \cdot 10^{23}}{10'000} \cdot 0.001 \, m^3 l^{-1} = 3.7 \cdot 10^{15}$ molecules (that's a lot of molecules).

The diffusion coefficient of oxygen amounts to $D_{O2,25°C} = 2.42 \cdot 10^{-5} \, cm^2 \, s^{-1}$ (see Table 4.1).
According to Eq. (4.9) the step in space is normally distributed with:

$$\Delta x \cdot N(0,1) = \sqrt{2 \cdot D \cdot \Delta t} \cdot N(0,1) = 0.022 \, mm \cdot N(0,1).$$

This distance stands for a large number of individual movements of an individual molecule.

Example 4.10: Two-dimensional random walk in Berkeley Madonna

A two-dimensional random walk of a particle that is added at time $t = 0$ at the origin ($x = 0$, $y = 0$) and diffuses over 10 s, may have the following form in Berkeley Madonna:

```
STARTTIME = 0            ; Beginning of the simulation, sec
STOPTIME = 10            ; End of the simulation, sec
DT = 0.1                 ; Duration of a time step, sec
D = 2E-3                 ; Diffusion coefficient in mm² sec⁻¹
Dx = sqrt(2*D*DT)        ; Standard deviation of the local step in x, mm
init x = 0               ; Local coordinate x, mm
```

Fig. 4.7 Comparison of the standard deviation of the coordinate x of 1000 particles which diffuse in a random walk with the standard deviation of the concentration which diffuses based on Fick's law and the same diffusion coefficient

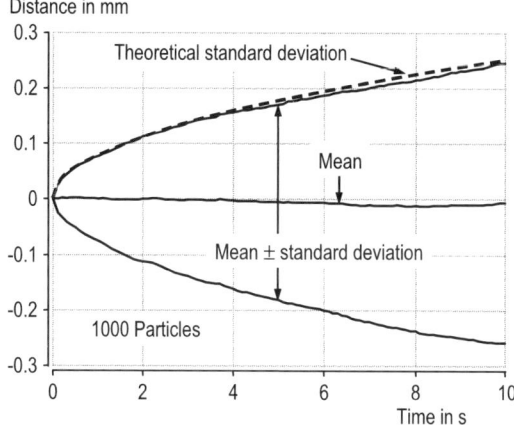

next x = x + normal(0,1)*Dx   ; Stochastic local step x, mm
Dy = Dx   ; Standard deviation of the local step in y, mm
init y = 0   ; Local coordinate y, mm
next y = y + normal(0,1)*Dy   ; Stochastic local step y, mm

Examples of the application are presented in Fig. 4.6.

The standard deviation of the local coordinate x, computed from the migration of 1000 particles, is compared in Fig. 4.7 with the standard deviation of the concentration that results from the application of the deterministic Fick's law (Eq. (4.13)). The deviation between the analytical solution and the stochastically computed solution (random walk) is:

$$\sigma_{\bar{x}} = \frac{\Delta x \cdot \sqrt{\dfrac{t}{\Delta t}}}{\sqrt{n-1}} = \frac{\sigma_x}{\sqrt{n-1}} = 0.03 \cdot \sigma_x .$$

The expected value of the local coordinate x is $\mu_x = 0$. The standard deviation of the average value μ_x is 3% of the standard deviation of the individual values of x and increases proportional to the root of time t.

Example 4.11: Random walk as a method for integration

The random walk is an alternative model to Fick's law for the description of diffusion processes which we can use to solve diffusion problems with difficult boundary conditions. If we want, e. g., to know how a volatile pollutant evaporates from an open bottle, we must solve a three-dimensional diffusion problem which we can describe only with rather involved methods based on Fick's law. With the help of a random walk, this problem becomes rather easy to solve. We distribute initially, e. g., 10,000 particles randomly over the volume of the bottle. Subsequently, we follow these particles with the help of their random walk. After each time step we count how many of the particles are still in the bottle.

The following code in Berkeley Madonna follows this strategy (in two rather than three dimensions to provide an overview) (tested, with only count as output):

```
STARTTIME = 0                           ; Start of the simulation
STOPTIME = 1000                         ; End of the simulation, d
DT = 0.1                                ; Time step, d
DTout = 10                              ; Output only every 10 days
xmin = 1      xmax = 2                  ; Bottle neck reaches horizontally from
                                          xmin to xmax, cm
dB = 3                                  ; Diameter of the bottle, cm
hB = 15                                 ; Height of the bottle, cm
hF = 20                                 ; Height of the bottle including neck, cm
D = 1                                   ; Diffusion coefficient cm² d⁻¹
Dx = sqrt(2*DT*D)                       ; Stand. dev. of the single step in the
                                          random walk, cm
n = 10000                               ; Number of particles
nF = n*dB*hB/(dB*hB+(xmax-xmin)*(hF-hB))  ; Number of particles in the
                                          bottle without neck
init x[1..nF] = random(0,dB)            ; Particles in the bottle without neck
init x[nF+1..n] = random(xmin,xmax)     ; Particles in the bottle neck
init y[1..nF] = random(0,hB)            ; Height of the particles in the bottle
init y[nF+1..n] = random(hB,hF)         ; Height of the particles in the bottle
                                          neck
Delx[1..n] = Dx*normal(0,1)+x[i]        ; New x coordinate after random step
Dely[1..n] = Dx*normal(0,1)+y[i]        ; New y coordinate after random step;
                                          The next two statements test that the
                                          particles do not move outside of the
                                          bottle (boundary condition)
next x[1..n] = if Dely[i] < hB then if Delx[i] > 0 and Delx[i] < dB then Delx[i]
else x[i] else if Delx[i] > xmin and Delx[i] < xmax then Delx[i] else x[i]
next y[1..n] = if y[i] > hF then y[i] else if Dely[i] < 0 then y[i]
else if Dely[i] < hB then if Delx[i] > 0 and Delx[i] < dB then Dely[i] else y[i]
else if Delx[i] > xmin and Delx[i] < xmax then Dely[i] else y[i]
InBottle[1..n] = if y[i] < hF then 1 else 0   ; Marks all particles which are still in
                                          the bottle
count = arraysum(InBottle[*])/n         ; Counts the particles which are still in
                                          the bottle
```

Figure 4.8 shows the result. It takes years until the material escapes. The evaporation of the water would have a greater impact. Every other way to solve this problem would require larger expenditure and possibly specialized software. Here the programming expenditure is minimized, but the cost of computation is high (about 1 min for a single run with 10,000 particles).

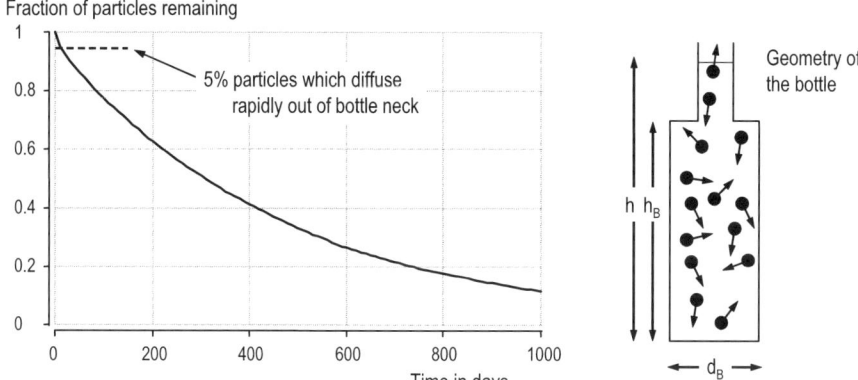

Fig. 4.8 Diffusion of a volatile material out of bottle with a narrow neck. Result obtained based on a random walk of 10,000 particles (see Example 4.11)

4.2.4 Molecular Diffusion

Diffusion is a transport process in which molecules, particles, or water packages at a certain location are moving around in a random direction and by a random amount. We characterize such a process by temporally averaging the net transport for a specific location, which leads to Fick's first law.

We can link the random walk to *molecular diffusion*: individual molecules collide with neighboring water molecules and change their direction and velocity as a consequence of the transferred momentum. In the gas phase the distance between two collisions is of the order of the magnitude of the mean free path or approximately the distance between two molecules. The average velocity is given by the temperature through the kinetic energy of the particles. Since the velocity of the molecules depends on their molecular weight, a different expected value for the covered distance results for different materials – we will find this again in diffusion coefficients, which depend on the specific material.

Fick's First and Second Law

Adolph Fick (1829–1901), a German scientist, published in 1855 (at the age of 26) the paper which gave the name to the law of diffusion.

Figure 4.9a shows the movements of individual particles in a system with different material concentrations. The probability that a particle from the area with high concentration changes to the low concentration range is proportional to the high concentration, and conversely only a few molecules from the region with low concentration reach the region with high concentration. There results a material

Fig 4.9 Examples of stochastic transport processes: (a) molecular diffusion, random direction of the thermal motion of individual particles. (b) turbulent diffusion, exchange of water packages by eddies in the water. Both processes lead to an observed transport of material from the left with a high concentration to the right with a low concentration

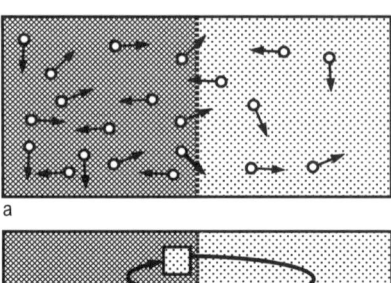

a

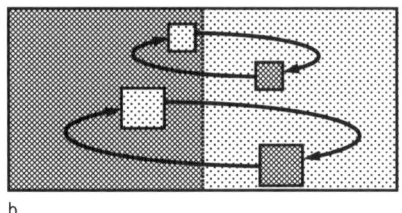

b

transfer that is proportional to the difference between the concentrations in the two system compartments. Fick's first law describes this situation based on a differential view in the following form:

$$j_A = -D_A \cdot \frac{\partial C_A}{\partial x} \tag{4.13}$$

j_A = specific flux of the material A $[M_A\ L^{-2}\ T^{-1}]$
D_A = diffusion coefficient of the material A $[L^2\ T^{-1}]$
C_A = local concentration of the material A $[M_A\ L^{-3}]$
x = space coordinate, the direction of the density gradient $[L]$

The diffusion coefficient D_A is tabulated in many standard works for many different materials in different solvents. Examples are given in Table 4.1. Since the kinetic energy of individual particles increases with temperature, also transport and therefore the diffusion coefficient increases:

$$D_T = D_{25°C} \cdot \exp\left(k_T \cdot (T - 25°C)\right), \tag{4.14}$$

where k_T depends on the material (Table 4.1) and varies between $0.018°C^{-1}$ and $0.04°C^{-1}$. A temperature change of 10°C increases D_T by $q_{10°C} = 1.2$–1.5 (see Example 5.5).

If we regard molecular diffusion as the only transport process, we can substitute Eq. (4.13) into the intensive material balance Eq. (3.15) and obtain:

$$\frac{\partial C_A}{\partial t} = D_A \frac{\partial^2 C_A}{\partial x^2}. \tag{4.15}$$

Table 4.1 Examples of diffusion coefficients in pure water at 25°C (Linde, 1999)

Material, molecule	Diffusion coefficient D_A		$q_{10°C}$
	$[10^{-5}\,cm^2\,s^{-1}]$	$[10^{-4}\,m^2\,d^{-1}]$	
Oxygen, O_2	2.42	2.09	1.45
Nitrogen, N_2	2.0	1.73	
Carbon dioxide, CO_2	1.91	1.65	1.31
Hydrogen, H_2	5.11	4.42	1.25
Hydrogen sulfide, H_2S	1.36	1.18	
Methane, CH_4	1.84	1.59	1.29
Glucose, $C_6H_{12}O_6$	0.69	0.60	
Acetic acid, CH_3COOH	1.25	1.08	
Acetate, CH_3COO^-	1.09	0.94	
Sodium, Na^+	1.33	1.15	
Potassium, K^+	1.96	1.69	
Ammonium, NH_4^+	1.96	1.69	
Calcium, Ca_2^+	0.40	0.35	
Magnesium, Mg^{2+}	0.71	0.61	
Chloride, Cl^-	2.03	1.75	
Bromide, Br^-	2.08	1.80	
Nitrate, NO_3^-	1.90	1.64	
Nitrite, NO_2^-	1.91	1.65	
Bicarbonate, HCO_3^-	1.19	1.03	
Carbonate, CO_3^{2-}	0.92	0.80	
$H_2PO_4^-$	0.88	0.76	
HPO_4^{2-}	0.88	0.76	

Equation (4.15) is called Fick's second law and presupposes that $D_A \neq f(x)$. It can be solved analytically for various boundary and initial conditions. An example is shown in Fig. 4.10. Here a material (dissolved oxygen) in an infinite, one-dimensional space diffuses from left to right. The solution of Eq. (4.15) for this problem has the form:

Initial conditions: $C(t=0, x \leq 0) = C_0$ and $C(t=0, x>0) = 0$ and
Boundary condition: $C(t, -\infty) = C_0 = 2$, $C(t, +\infty) = 0$

$$C(t,x) = \frac{C_0}{2} \cdot \left(1 - erf\left(\frac{x}{\sqrt{4 \cdot D \cdot t}}\right)\right). \tag{4.16}$$

The derivative of Eq. (4.16) with respect to x results in a normal distribution $N(0, \sigma_x)$ with standard deviation:

$$\sigma_x = \sqrt{2 \cdot D \cdot t}. \tag{4.17}$$

Fig. 4.10 Development of the concentration profile following molecular diffusion. Starting from a concentration step at $t=0$, the effective distance of diffusion is shown over many days (diffusion coefficient $2 \cdot 10^{-4}$ m^2 d^{-1}). It is assumed that the containers on the left and right extend to infinity

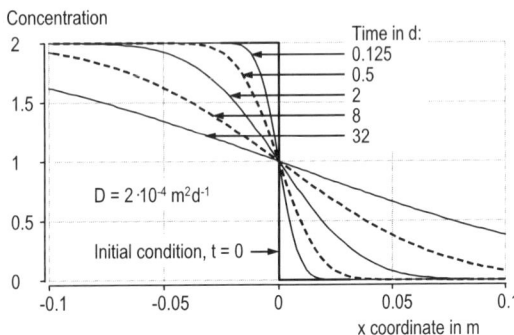

Table 4.2 Characteristic effective distance of molecular diffusion, computed for $D_A = 2 \cdot 10^{-4}$ m^2 d^{-1} (e. g., oxygen at 20°C)

Duration t	Effective distance (σ)	Duration t	Effective distance (σ)
1 s	0.07 mm	1 d	2 cm
1 min	0.5 mm	1 month	11 cm
1 h	4 mm	1 yr	40 cm

Comparing Eq. (4.17) with Eq. (4.11), σ_x is a measure for the distance over which a material diffuses in time t. In view of Fig. 4.10 the values in Table 4.2 result. The range of molecular diffusion is rather small, it has importance primarily in laminar boundary layers in the vicinity of surfaces and of microorganisms.

Example 4.12: Sugar in the coffee

You are leisurely sitting at a bar in Italy and order yourself an espresso. As is typical, you add a lot of sugar. In order not to cool the coffee you want to avoid unnecessary stirring.
How long will it take until the dissolved sugar distributes itself over the entire coffee?
In the unstirred coffee only molecular diffusion will be active at first.
Assumptions: The coffee stands 3 cm deep in the cup. The temperature of the coffee is 80°C. The molecular diffusion coefficient for sugar (molecular weight 360) is $D_{25°C} = 0.5 \cdot 10^{-5}$ cm^2 s^{-1} at 25°C. $q_{10°C} = 1.35$ (corresponding to $k_T = 0.03°C^{-1}$).

$$D_{80°C} \approx D_{25°C} \exp(0.03 \cdot (80-25)) \approx 2.6 \cdot 10^{-5} \, cm^2 \, s^{-1}.$$

From Eq. (4.17) you obtain the relevant time constant: $t_h = \dfrac{h^2}{2 \cdot D} = 2 \, d$.

It would take approximately 2 days to sweeten the coffee based on molecular diffusion. Convection currents would result from the cooling of the coffee (cold coffee would sink along the cooling walls of the cup), and would accelerate the mixing. However, if you like your coffee hot and sweet, you must stir it.

Example 4.13: Substrate variation in a rotary biological contactor (RBC)

In an RBC with a diameter of 3 m the discs will rotate at approximately 2 rpm (30 cm s^{-1} peripheral speed). Thus, the biofilm will dip into the polluted water for about 15 s and will afterwards be exposed to the air for 15 s.

Up to what depth of the biofilm must we expect strong variations of the pollutant concentrations?

If we take glucose as a typical substrate and 15°C as a typical temperature, then a typical diffusion coefficient is $D_{Gl} \approx 0.69 \cdot 10^{-5} \, cm^2 \, s^{-1}/1.3 = 0.53 \cdot 10^{-5} \, cm^2 \, s^{-1}$ (Table 4.1). The concentration will increase during 15 s and will then be reduced during another 15 s. A typical effective distance over the duration of 15 s becomes with Eq. (4.17):

$L_{Gl} = 0.13$ mm.

The active biofilm thickness is in the range 0.1–0.25 mm. Thus, we must expect that the rotation speed has an effect on the performance of the RBC. The faster the rotation, the better the performance will be.

Example 4.14: Algal mat in a river

In a river the sediment is overgrown with a green algal mat which sets oxygen free as a consequence of photosynthesis.

How deep does the oxygen penetrate into the sediment during a sunny summer day?

Assumptions: $t_{Sun} = 14$ h, temperature $= 25°C$. The diffusion coefficient for the oxygen amounts to:

$D_{O2,25°C} = 2.42 \cdot 10^{-5} \, cm^2 \, s^{-1}$. With Eq. (4.17),

this results in $L_{O2} = (2 \cdot 2.42 \cdot 10^{-5} \cdot 14 \cdot 3600)^{0.5} = 1.56$ cm.

Thus, the algae can supply a significant depth of sediment with oxygen, if the oxygen consumption underneath the algae is not too large.

Example 4.15: Laminar boundary layer at a gas bubble

The laminar boundary layer on the surface of a gas bubble is 50 μm thick. The oxygen concentration at the phase boundary (surface water/air) amounts to 10 gO_2 m^{-3} (corresponding to saturation). In the turbulent water phase 2 gO_2 m^{-3} are measured.

What is the specific mass flux of oxygen over the laminar boundary layer at 10°C?

In the laminar boundary layer the transport of oxygen is determined by molecular diffusion. The diffusion coefficient for dissolved oxygen at 10°C is (Table 4.1):

$D_{O2,10°C} = 2.09 \cdot 10^{-4} \, m^2 \, d^{-1} \cdot \exp(0.037 \cdot (10-25)) = 1.2 \cdot 10^{-4} \, m^2 \, d^{-1}$.

Assuming a linear decrease of the oxygen concentration over the boundary layer, the concentration gradient becomes:

$$\frac{\partial S_{O2}}{\partial x} = \frac{\Delta S_{O2}}{\Delta x} = \frac{10 - 2 \, gm^3}{50 \mu m} = 160'000 \, gm^{-4} \, .$$

From Fick's first law (Eq. (4.13)) this yields:

$$j_{O2} = -1.2 \cdot 10^{-4} \cdot 16 \cdot 10^4 = 19.2 \; g_{O2} m^{-2} d^{-1}$$

in the direction of decreasing concentration, thus from the gas to the water.

4.2.5 Turbulent Diffusion

Turbulent diffusion is not specific to a material. It is driven by the stochastic movements of turbulence. In a turbulent flow regime its transport is orders of magnitude larger than molecular diffusion.

In *turbulent diffusion* we follow individual water packages that move in a turbulent field with randomly distributed velocities (Fig. 4.9b). Colliding water packages are mixed again and again and continue to be transported with the next eddy. For individual molecules of a material a random walk results from this movement, whereby the individual distances are larger by orders of magnitude than with molecular diffusion. Since transport is provided by water packages, all materials move simultaneously; there is a characteristic turbulent diffusion coefficient D_T that characterizes turbulent eddy diffusion in a system. It is possible that the turbulence field is not isotropic, so that vertically and horizontally different coefficients $D_{T,x}$, $D_{T,y}$, and $D_{T,z}$ may result.

Figure 4.9b shows that with turbulent diffusion material is transported from regions with high concentration to regions with lower concentration. Thus, Fick's law can be used to characterize turbulent diffusion. We compute the temporally averaged transport of a material at a certain location:

$$j_{T,A} = -D_T \cdot \frac{\partial C_A}{\partial x} \qquad (4.18)$$

$j_{T,A}$ = flux of the material A due to turbulence $[M_A \, L^{-2} \, T^{-1}]$
D_T = turbulent diffusion coefficient $[L^2 \, T^{-1}]$

Table 4.3 Estimated values for turbulent diffusion coefficients D_T in rivers (Fischer et al., 1979). These estimates are subject to large uncertainty. Real values can deviate by a factor of 2–4 from estimated values

Crosswise and in direction of flow [a]:	
Technical, straight channel:	$D_{T,y} \approx D_{T,x} \approx 0.15 \cdot h \cdot u^*$
River, slowly meandering:	$D_{T,y} \approx D_{T,x} \approx 0.60 \cdot h \cdot u^*$
River, channel, vertical:	$D_{T,z} \approx 0.067 \cdot h \cdot u^*$

[a] In the direction of flow, dispersion outweighs turbulent diffusion. Crosswise and in the vertical direction turbulent diffusion is of importance, in particular in straight channels without secondary flows.

Fischer et al. (1979) provide empiric equations for the estimation of turbulent diffusion coefficients, especially for rivers (Table 4.3). These values can be transferred to other systems.

$$u^* = \sqrt{\frac{\tau_0}{\rho}} = \sqrt{g \cdot R \cdot I_E} \qquad (4.19)$$

$$\tau_0 = g \cdot \rho \cdot R \cdot I_E \qquad (4.20)$$

$$I_E = \frac{v^2}{R^{3/4} \cdot k_{St}^2} \qquad (4.21)$$

$$u^* = v \cdot \sqrt{\frac{g}{R^{1/3} \cdot k_{St}^2}} \approx 0.05 \quad \text{to} \quad 0.1 \cdot v \qquad (4.22)$$

h   = mean depth of the river [L]
u^* = shear velocity [L T^{-1}]
τ_0 = shear stress [M T^{-2} L^{-1}]
g   = acceleration of gravity, 9.81 m s^{-2}
ρ_W = density of water [M L^{-3}]
R   = hydraulic radius $\approx$ h = mean depth of water [L]
I_E = energy gradient [–]
v   = flow velocity [L T^{-1}]
k_{St} = roughness factor according to Strickler [m$^{1/3}$ s^{-1}] 1/n according to Manning

---

Example 4.16: Transport processes in an activated sludge tank

In a plug-flow type aeration tank the turbulent diffusion coefficient D_T is determined to be 2400 m^2 d^{-1}. The mean flow velocity v amounts to 250 m d^{-1}. The molecular diffusion coefficient for pollutant A is estimated to be $D_A = 0.5 \cdot 10^{-4}$ m^2 d^{-1}. At a sampling point in the basin the concentration $C_A = 20$ g$_A$m^{-3} and the gradient of the pollutant concentration is determined at $\Delta C_A / \Delta x = -2$ g$_A$m^{-4}.
What is the relationship between the three contributions to mass transfer (advection, turbulent diffusion, and molecular diffusion)?

$j_{Advection}$ = $v \cdot C_A = 250 \cdot 20$ =   5000 g$_A$m^{-2} d^{-1}
$j_{Turbulence}$ = $-D_T \cdot \Delta C_A / \Delta x$ =   4800 g$_A$m^{-2} d^{-1}
$j_{Diffusion}$ = $-D_A \cdot \Delta C_A / \Delta x$ =   0.0001 g$_A$m^{-2} d^{-1}

In order to capture the total mass transport at the sampling point, advection and turbulence must be considered, whereas molecular diffusion is negligibly small.

Example 4.17: Vertical and cross mixing of a pollutant in a river

Case 1: An industrial company discharges its effluent at the sole of a river, distributed over the entire width. *After which flow distance can we expect mixing over the entire depth of the water?*

After Eq. (4.17) the mixing length for diffusion amounts to $h = \sqrt{2 \cdot D_{T,z} \cdot t_h}$.
In a river $u^* = 0.1 \cdot v$ (Eq. (4.22)) is approximately valid.
The mixing length for vertical mixing becomes:

$$L_h = v \cdot t_h = v \cdot \frac{h^2}{2 \cdot D_{T,z}} = v \cdot \frac{h^2}{2 \cdot 0.067 \cdot h \cdot u^*} \approx 75 \cdot h .$$

Thus, after a flow distance of approximately 75 times the depth of the water the river is vertically mixed.

Case 2: An industrial company introduces its wastewater laterally into a river. *After which flow distance can we expect a mixing over the width of the river?*
With $B = \sqrt{2 \cdot D_{T,y} \cdot t_B}$ the mixing length for the entire river width B becomes:

$$L_B = v \cdot t_B = v \cdot \frac{B^2}{2 \cdot D_{T,y}} = v \cdot \frac{B^2}{2 \cdot 0.6 \cdot h \cdot u^*} \approx 8 \cdot \frac{B^2}{h} .$$

Characteristic examples are provided in the following table.
How long is the flow distance in the river Rhine and in a small brook, to achieve mixing over the depth and across the entire river?

Parameter	Rhine	Brook	Unit
Width, B	200	10	m
Mean depth, h	3	0.3	m
Mean flow velocity, v	2.5	1	$m\,s^{-1}$
Diffusion coefficient vertical, $D_{T,z}$	0.053	0.002	$m^2\,s^{-1}$
Mixing over depth, L_H	250	25	m
Diffusion coefficient lateral, $D_{T,y}$	0.45	0.018	$m^2\,s^{-1}$
Mixing over the entire width, L_B	100,000	2700	m

Whereas the mixing over the depth is rather rapid, lateral mixing requires very long flow distances. In the Rhine river approximately 10 hrs are necessary for flowing 100 km. However, usually secondary currents arise over long flow distances which lead to transverse dispersion and accelerate the mixing. Figure 4.11 shows that mixing effluent from a point source across a river is a very slow process.

Example 4.18: Sedimentation against turbulent diffusion

In a straight channel the water flows with a velocity of $v = 1$ m s^{-1} and a depth of $h = 1$ m. In the influent the water contains $C_{P,m} = 10$ g m^{-3} of a suspension of particles with a density of $\rho_P = 2500$ kg m^{-3} and a diameter of $d_P = 0.1$ mm. Vertical turbulent diffusion works against the sedimentation of these particles.
Which concentration profile of the particles over the depth of the channel will result, if all sediments are whirled up again once they reach the bed of the channel?

Fig. 4.11 Mixing of an industrial effluent into a receiving water

Fig. 4.12 Concentration profile of suspended particles in a turbulent channel

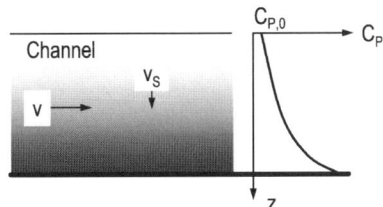

Assumptions: $u^* = 0.05 \cdot v = 0.05$ m s^{-1}. $D_{T,vertical} = 0.067 \cdot h \cdot u^* = 0.0034$ m^2 s^{-1}
(Table 4.3).

Sedimentation: $v_S = 9$ mm s^{-1} (Fig. 4.5).

The transport of the particles as a consequence of sedimentation will in the equilibrium just be compensated by turbulent transport against the concentration gradient.

$$j_S = v_S \cdot C_P(z) = -j_T = D_T \cdot \frac{\partial C_P}{\partial z}.$$

After separation of the variables C_P and z integration results in:

$$C_P = C_{P,0} \cdot \exp\left(\frac{v_S \cdot z}{D_T}\right),$$

where $C_{P,0}$ is the particle concentration at the surface (see Fig. 4.12).

If we assume that the sediments arc whirled up again from the sole, the mean
concentration of $C_{P,m}$ must correspond to the initial concentration in the influent:

$$C_{P,m} = \frac{\int_0^h C_P \cdot dz}{h} = \frac{C_{P,0} \cdot D_T}{h \cdot v_S} \cdot \left(\exp\left(\frac{v_S \cdot h}{D_T}\right) - 1\right),$$

$$\frac{v_S \cdot h}{D_T} = \frac{v_S}{0.067 \cdot u^*} \approx \frac{v_S}{0.0034 \cdot v},$$

$$C_{P,0} = C_{P,m} \cdot \frac{v_S \cdot h}{D_T} \cdot \frac{1}{\left(\exp\left(\dfrac{v_S \cdot h}{D_T} \right) - 1 \right)}$$

$$= C_{P,m} \cdot \frac{v_S}{0.0034 \cdot v} \cdot \frac{1}{\left(\exp\left(\dfrac{v_S}{0.0034 \cdot v} \right) - 1 \right)},$$

$$C_P(h) = C_{P,0} \cdot \exp\left(\frac{v_S \cdot h}{D_T} \right) = C_{P,0} \cdot \exp\left(\frac{v_S}{0.0034 \cdot v} \right),$$

resulting in: $v_S / (0.0034 \cdot v) = 2.64$, $C_{P,0} = 2.0 \, \text{g m}^{-3}$, $C_P(h) = 28.4 \, \text{g m}^{-3}$.

We expect an exponential increase of the concentration from top to bottom (Fig. 4.12): the larger v_S, the more pronounced the concentration profile becomes. Such concentration profiles are established, e. g., in sewers, where towards the bottom higher concentrations of suspended solids than on the surface are frequently observed. Sampling wastewater in sewers must consider this fact.

4.2.6 Dispersion

Dispersion describes the consequence of overlaying processes in a multidimensional flow field. Its cause is the flow velocity profile which establishes itself as a consequence of friction (and possibly other forces) in any advective current. We cannot capture these profiles with our one-dimensional models of technical hydraulics. This leads to the fact that water packages move with velocities of flow

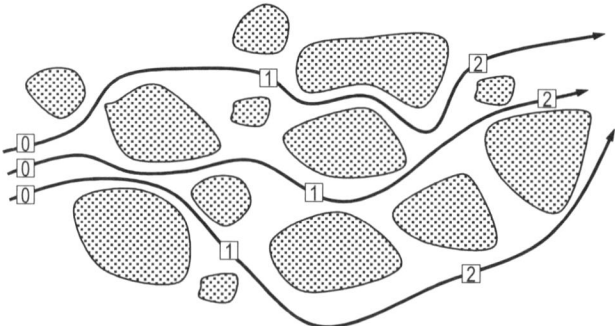

Fig. 4.13 Dispersion in porous media. Water packages or particles that enter the medium initially at the same time move on different streamlines through space and reach the control point at the discharge at different times

that deviate from the computed mean values. This is of particular importance for the transport of the materials contained in these water packages. Lateral and vertical turbulent transport (diffusion) mixes water packages with different ages.

Figures 4.13 and 4.14 show the profiles of flow velocities for different systems. They are reproducible and can be computed (possibly with significant cost). If we are interested in the mean local concentration of a material, we will have to average the local concentrations over the cross section of a current, with the result that we will average the concentration of water packages with different ages. This is equivalent to an apparent mixing in the direction of flow. If the flow is turbulent, water packages will not move along the streamlines, but they will be mixed by the turbulence (and to a much smaller degree by molecular diffusion) with their environment. Over some time period a molecule may occupy any position in the cross section of the flow and is therefore subjected to many randomly distributed flow velocities. The sum of many random variables of any distribution asymptotically strives towards a normal distribution (central limit theorem).

Overlaying the velocity profiles with vertical and horizontal turbulence leads after a long flow distance to a result that cannot be differentiated from longitudinal (turbulent) diffusion but is caused by different processes.

The positions of 100 particles are simulated in Fig. 4.15. The advection is subject to a parabolic distribution of flow velocity, and the overlaying turbulence is modeled with a random walk transverse to the direction of flow. The distribution of the particles over the cross section is initially skewed. After some time the particles have occupied many randomly distributed positions over the entire cross section, the distribution over the cross section is even, and a normal distribution results in the longitudinal direction. This overlay of velocity profile and turbulence is called dispersion. It has the consequence that water packages are mixed similarly as in turbulent diffusion. Thus, after some time the result of dispersion cannot be distinguished from turbulent diffusion.

After the complete transverse mixing of the regarded system (see Example 4.17 and Fig. 4.15), we can describe the effect of dispersion with overlaying turbulent diffusion with Fick's law. The model represents a one-dimensional analysis and averages the concentration and the flow of material over the cross section. Thus, dispersion is always bound to a three-dimensional system with its specific dimensions and velocity profile. It describes transport processes averaged over the cross section and is thus not truly an intensive process.

Fig. 4.14 Dispersion in laminar flow. As a consequence of the viscosity of the water a parabolic distribution of the flow velocity is established over the cross section. Water packages that start at the same time reach a later location at different times, cf. Fig. 4.3

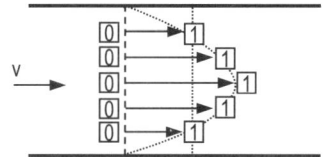

Fig. 4.15 Representation of the position of 100 particles in the simulation of a random walk. All particles start at x = 0, y = 0.4. The velocity profile is parabolic; the random walk applies only to the y-direction

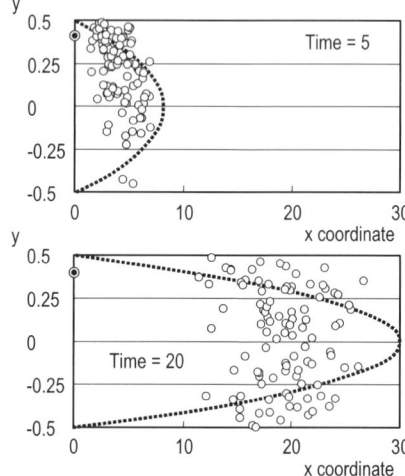

Fick's first law, integrated over the cross section, becomes:

$$J_{D,A} = -D_D \cdot \frac{\partial \overline{C_A}}{\partial x} \cdot F \qquad (4.23)$$

$J_{D,A}$ = flux of material A as a consequence of dispersion $[M_A \, T^{-1}]$, an extensive variable
$\underline{D_D}$ = dispersion coefficient $[L^2 \, T^{-1}]$
$\overline{C_A}$ = weighted mean concentration of material A in the cross section. The weight is according to the local flow velocity $[M_A \, L^{-3}]$
x = longitudinal coordinate, the model is one dimensional $[L]$
F = cross-sectional area of the system $[L^{-2}]$

A simplified (although less accurate) form of Eq. (4.23) is also frequently written in intensive representation:

$$j_{D,A} = -D_D \cdot \frac{\partial C_A}{\partial x} \cdot \qquad (4.24)$$

The dispersion coefficient D_D in turbulent flow is independent of the material, and relates only to flow conditions. In laminar flow molecular diffusion causes the transverse mixing, so that D_D then becomes specific to individual materials. For some systems the literature provides information to estimate dispersion coefficients; these estimates are, however, subject to large errors (a factor of 2–4). Examples are given in Table 4.4.

Combining the intensive mass balance equation (3.15) with advection and dispersion results in:

$$\frac{\partial C}{\partial t} = -v \cdot \frac{\partial C}{\partial x} + D_D \cdot \frac{\partial^2 C}{\partial x^2} + r \, . \qquad (4.25)$$

Table 4.4 Longitudinal dispersion coefficient

River, slowly meandering (Fischer, 1975)	$D_{D,x} = 0.011 \cdot \dfrac{v^2 \cdot B^2}{h \cdot u^*}$
Sewer, partially filled (Huisman, 2000)	$D_{D,x} = 0.003 \cdot \dfrac{v^2 \cdot B^2}{h \cdot u^*}$
Circular pipe, turbulent, full (Elder, 1959)	$D_{D,x} = 5.93 \cdot d \cdot u^*$
Circular pipe, laminar (Taylor, 1953)	$D_{D,x} = \dfrac{d^2 \cdot v^2}{192 \cdot D_A}$

B	=	width of the river [L]
h	=	mean depth
u^*	=	shear velocity [L T^{-1}], see Eq. (4.19)
v	=	mean flow velocity [L T^{-1}]
d	=	diameter of the circular pipe [L]
D_A	=	molecular diffusion coefficient of material A [L^2 T^{-1}]

If we add to the cross section of a prismatic channel at $x=0$ at $t=0$ the amount M of nonreactive tracer (Dirac pulse, $t=0$), then the following partial differential equation describes the mass transport:

$$\frac{\partial C}{\partial t} = -v \cdot \frac{\partial C}{\partial x} + D_{D,x} \cdot \frac{\partial^2 C}{\partial x^2} .$$

With initial conditions $C(x,0) = M \cdot \delta(0,0)$

$\delta(0,0) =$ Dirac pulse at $x=0$ and $t=0$

and the boundary condition $C(\pm\infty, t) = 0$
the result is:

$$C(x,t) = \frac{M}{\sqrt{4 \cdot \pi \cdot D_D \cdot t}} \cdot \exp\left(\frac{-(x - v \cdot t)^2}{4 \cdot D_D \cdot t} \right). \tag{4.26}$$

Equation (4.26) describes for a specific time along the flow direction x a normal distribution with the expected value and standard deviation according to:

$$\mu_x = v \cdot t \text{ and } \sigma = \sqrt{2 \cdot D_D \cdot t} . \tag{4.27}$$

For a specific location x Eq. (4.26) describes over time a skewed distribution (Fig. 4.16). The skewness results from the originally symmetrical normal distribution that continues to be deformed during the time it is observed at location x.

Example 4.19: Advection and dispersion in a pipe

The material balance in a long, turbulent flowing pipe has the form of Eq. (4.25)

$$\frac{\partial C}{\partial t} = -v \cdot \frac{\partial C}{\partial x} + D_D \cdot \frac{\partial^2 C}{\partial x^2} + r .$$

Fig. 4.16 Concentration profile at a fixed location x over time t after the addition of a Dirac pulse at $x = 0$ and $t = 0$

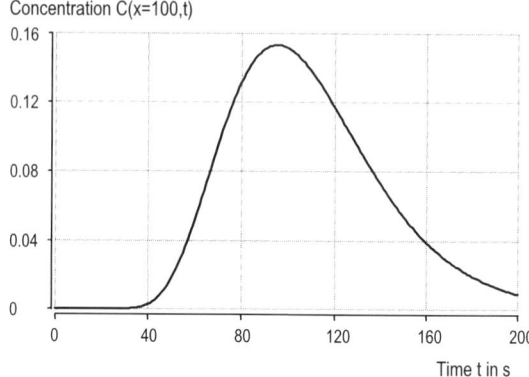

Concentration C(x=100,t)

Do we have to consider dispersion in a pipe which feeds a reactor?
Assumption: the pipe is in steady state or concentrations change only slowly. $u^* = 0.05 \cdot v$ (Eq. (4.22)).
The mass flux as a consequence of advection amounts to $j_A = v \cdot C$ or integrated $J_A = Q \cdot C$.
The mass flux as a consequence of dispersion amounts to $j_D = -D_D \cdot dC/dx$.
According to Table 4.4 (Elder, 1959), for circular pipes:
$D_D \approx 5.93 \cdot u^* \cdot d \approx 5.93 \cdot 0.05 \cdot v \cdot d \approx 0.3 \cdot v \cdot d$.
The balance equation in steady state with a first-order reaction ($r = -k \cdot C$) has the form:

$$\frac{\partial C}{\partial x} = \frac{D_D}{v} \cdot \frac{\partial^2 C}{\partial x^2} - \frac{k \cdot C}{v} \quad \text{and since} \quad \frac{\partial^2 C}{\partial x^2} > 0 \text{ is valid } 0 > \frac{\partial C}{\partial x} > -\frac{k \cdot C}{v}$$

$$j_D = -D_D \cdot \frac{\partial C}{\partial x} = -0.3 \cdot v \cdot d \cdot \frac{\partial C}{\partial x} \qquad \frac{j_D}{j_A} = \frac{0.3 \cdot d \cdot \left(-\dfrac{dC}{dx}\right)}{v \cdot C} < 0.3 \cdot \frac{d \cdot k \cdot C}{v \cdot C} = 0.3 \cdot d \cdot \frac{k}{v}.$$

Typical conditions in a water pipeline are:

$v = 0.5 - 2 \text{ m s}^{-1}$, $d = 0.1 - 2$ m, $k = 0 - 5000 \text{ d}^{-1} = 0 - 0.05 \text{ s}^{-1}$.

Thus: $\dfrac{j_D}{j_A} < 0.05$.

In transport pipes in the steady state we can estimate mass transport with a deviation of less than 5%, if we only consider advection and neglect dispersion. This considers fairly fast reactions that might take place in the pipe. If the reaction is much faster, the longitudinal gradient becomes large and we have to consider dispersion. For slow reactions dispersion is negligible.

Example 4.20: Dispersion in a wide river

Secondary flow is induced in meandering rivers. This causes the water to rotate in the longitudinal direction. These secondary flows induce dispersion also in the transverse direction, which accelerates transverse mixing. Since there is only a small resistance to such rotations, cross-flows once induced affect the river over a long flow distance.

Example 4.21: Dispersion in a sewer

A sewer has a diameter of $d = 1$ m and a flow of $Q = 0.03$ m^3 s^{-1}. The friction coefficient after Strickler amounts to $k_{St} = 75$ m$^{1/3}$ s^{-1} and the gradient of the sewer is $J_E = 0.002$. You add at $x = 0$ and $t = 0$ some tracer material in the form of a Dirac pulse to the center of the channel. The tracer material has the same density as the water.

After what flow distance is the tracer material mixed over the entire cross section?
For this partially filled sewer the following hydraulic details result:

Fully filled: $Q_{full} = 1.05$ m^3 s^{-1} and $v_{full} = 1.34$ m s^{-1}.

Partially filled: $Q_{part} = 0.03$ m^3 s^{-1}, $v_{part} = 0.6$ m s^{-1}, $h_{part} = 0.12$ m, $h_{mean} = 0.077$ m
(mean depth), $W = 0.65$ m (width at surface), $u* = 0.04$ m s^{-1},

$D_{T,y} = 0.15 \cdot h_{mean} \cdot u* = 0.0005$ m^2 s^{-1} (Table 4.3).

The diffusion distance up to complete cross mixing amounts to W/2. From Eq. (4.17) it follows that $W/2 = (2 \cdot D \cdot t_Q)^{0.5}$ or the required time is $t_Q = W^2/(8 \cdot D)$ = 106 s.

In this time the water flows over $t_Q \cdot v = 64$ m, thus the water will be completely mixed between two man holes.

What approximate time course of the concentration do you expect after a straight flow distance of 2 km?
The dispersion coefficient amounts to approximately

$$D_D = 0.003 \cdot \frac{v^2 \cdot B^2}{h \cdot u*} = 0.15 \text{ m}^2\text{s}^1 \text{ (Table 4.4), which is much higher than the turbu-}$$

lent diffusion coefficient of $D_{T,x} = D_{T,y} = 0.0005$ m^2 s^{-1}.

The flow time over the length L is $t_L = L/v = 3330$ s. The concentration is in longitudinal direction normally distributed with $\mu_x = t_L \cdot v = 2000$ m and $\sigma_x = (2 \cdot D_D \cdot t_L)^{0.5} = 32$ m. Assuming the wave of tracer to be $6 \cdot \sigma_x$ wide (99%), we can observe an elevated tracer concentration during approximately 320 s.

Example 4.22: Sampling in laminar flow

If water runs laminarly over an inclined surface (e. g., in a trickling filter), then a parabolic velocity profile will develop with the maximum flow velocity towards the air and standing water against the fixed surface.

If the surface is covered with biomass which degrades pollutants, consumes O_2, and sheds CO_2, then a linear O_2 concentration profile is formed, with the saturation concentration towards the air and a reduced concentration at the surface of the biomass (for CO_2 the situation is just inverse).

In computing the transport of O_2 or CO_2 in the flowing water, we must consider the velocity as well as the concentration profile (Eqs. (3.4) and (4.1)). If the water is caught in a sampling container, then the concentration of O_2 or CO_2 in the water sample does not correspond to the locally averaged concentration in the flowing water but rather to the flow-weighted mean concentration, which can be obtained from:

$$C_{fw} = \frac{\int_h C(h') \cdot v(h') \cdot dh'}{\int_h v(h') \cdot dh'}.$$

For oxygen this flow-weighted mean would be higher than a local mean concentration, whereas for carbon dioxide it would be lower.

Example 4.23: Mass transfer is a directed vector variable

We indicate the mass transfer as a directed variable (vector).
With *advection* the direction results from the flow direction. The flow velocity v is valid for water as well as all materials.
Dispersion is parallel to the direction of flow. The dispersion coefficient is in turbulent flow equal for all materials.
With *molecular diffusion* the direction results from the concentration gradient. It can be different for different materials. Moreover, the diffusion coefficient is specific for each species.
Turbulent diffusion depends on the dimension of the current (depth and width) and the concentration gradient. Since turbulence is not isotropic, the turbulent diffusion coefficient differs for the lateral, longitudinal, and vertical diffusion but is equal for all species.
Sedimentation follows the external force field and depends on particle-specific parameters.

4.2.7 Numeric Dispersion

Numeric dispersion is not a physically occurring transport process, but an artifact which results from the numeric solution of the partial differential equations that describe transport.

There are various procedures for the numeric solution of the partial differential equations that describe transport. The numeric computation in the procedure with finite differences involves the computation of results for fixed grid points separated by small distances Δx in space and Δt in time. Figure 4.17a shows the one-dimensional shift of a change of concentration with pure advection. We can describe this situation with the following partial differential equation:

$$\frac{\partial C}{\partial t} = -v \cdot \frac{\partial C}{\partial x}.$$ (4.28)

The simplest, not really suitable, numeric solution of Eq. (4.28) has the form:

$$\frac{C(t,x) - C(t - \Delta t, x)}{\Delta t} = -v \cdot \frac{C(t - \Delta t, x) - C(t - \Delta t, x - \Delta x)}{\Delta x} \quad \text{and thus}$$

$$C(t,x) = C(t - \Delta t, x) - v \cdot \frac{\Delta t}{\Delta x} \cdot \big(C(t - \Delta t, x) - C(t - \Delta t, x - \Delta x) \big).$$ (4.29)

Figure 4.17b–d show the initial conditions at time $t = 0$ and the numeric results for two time steps Δt. The concentration front does not coincide exactly with a grid point after a time step Δt. The space element Δx is only partially filled in a time step Δt. This causes a flattening of the concentration gradient, which with increasing time will have a similar effect as diffusion or dispersion; in this case we speak of numeric dispersion.

Numeric dispersion disappears if $v \cdot \Delta t = \Delta x$ is always exactly maintained. Since the flow velocity v varies with space and time, this is typically not possible. If $v \cdot \Delta t > \Delta x$, then the value of C rises above the local concentration (which is thermodynamically impossible in reality) and will do so for all following time steps, which results in an increase of the concentration in a geometric series: numeric integration becomes unstable. The grid in time and space must be suffi-

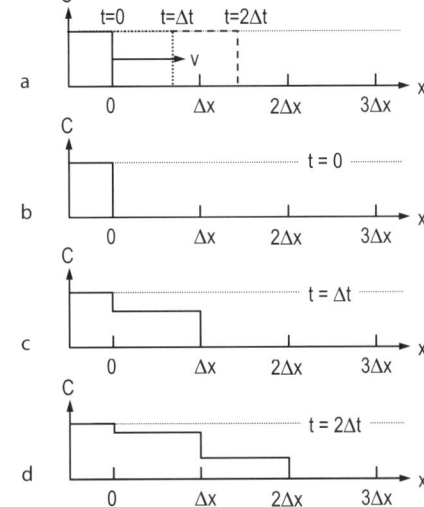

Fig. 4.17 Numeric dispersion: (a) a compound moves in reality with velocity v over two time steps Δt, (b) the initial condition for the numeric integration, and (c) the result after a time step Δt. The element Δx is only partially filled, but over the entire width. (d) Shows the result after two time steps: numeric dispersion is visible

ciently refined and, in addition, these steps must be chosen such that the *Courant criterion* is always obeyed:

$$\Delta t < \frac{\Delta x}{v} . \tag{4.30}$$

Numeric dispersion (an equivalent dispersion coefficient) for integration with Eq. (4.29) has the order of magnitude of:

$$D_{numDisp} < v \cdot \frac{\Delta x}{2} . \tag{4.31}$$

The decrease of Δx requires an appropriate decrease of Δt according to Eq. (4.30), thus the cost of computation rises with the square of the reduction of the numeric dispersion coefficient. Frequently numeric dispersion is of the same order of magnitude or even larger than real dispersion – this allows one to substitute numerical dispersion for real dispersion. Unfortunately there is no method which permits one to quantify numeric dispersion, except if the problem has an analytic solution.

Professional programs for the solution of the partial differential equations discussed here try to minimize the numeric dispersion by suitable numeric procedures. It can, however, only be eliminated for special cases.

Example 4.24: Extent of the numeric dispersion

You want to model the transport of a material in a water pipeline with a diameter of $D = 1$ m and a flow velocity of $v = 1.2$ m s^{-1} based on Eq. (4.29). You choose a fixed time step of $\Delta t = 1$ s.
How large is the numeric dispersion?
The Courant criterion Eq. (4.30) requires for this situation a local discretization of $\Delta x > 1.2$ m. You choose $\Delta x = 2$ m.
With Eq. (4.31) the numeric dispersion becomes $D_{num,Disp} < 1.2 \cdot 2 / 2 = 1.2$ m^2 s^{-1}.
According to Table 4.4 the effective dispersion coefficient is
$D_D = 5.93 \cdot d \cdot u^* \approx 5.93 \cdot 1 \cdot 0.05 \cdot 1.2 = 0.36$ m^2 s^{-1}.
Thus, numeric dispersion would be larger than the effective dispersion. Either you select an improved integration procedure or you choose a narrower grid. You do not have to introduce dispersion as an additional process.

4.2.8 Convection

Convection results from gradients of the density of the water. Different density can be caused by gradients of the temperature or the salt or particle content of the water.

Fig. 4.18 Density of water as a function of temperature (Linde, 1999)

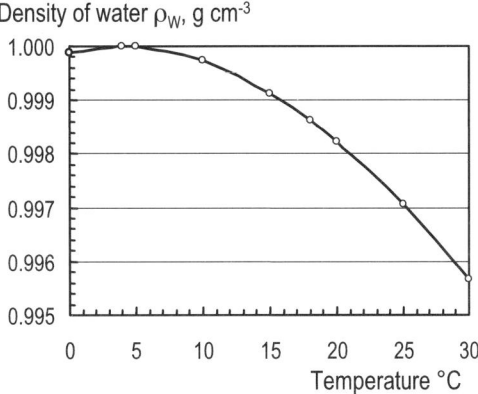

As water heats up, it expands and its density drops (above 4°C, see Fig. 4.18). Water regions with low density rise as a "cloud", whereas regions with higher density compensate this transport and fall. Whenever heat energy is supplied to the water or extracted from it, heat-driven transport can develop.

Example 4.25: Convection in the central heating systems

Warm water is less dense than cold water. In a central heating system where the heat source is in the cellar of the house, the water circulates automatically in the system. The hot, less dense water rises and returns as cold, denser water. If the heat source is under the roof, these density gradients must be overcome by using a circulation pump.

Example 4.26: Cooking with water

When heating water, hot water rises from the bottom to the top and thereby induces mixing in the pan. At the surface, steam is given up, the water cools, and sinks downward. One can observe the striae which are caused by the different refractive indexes of hot and cold water.
In the end steam bubbles develop, which mix the cooking water with mechanical energy, this process no longer corresponds to convection.

The yearly cycle in the temperature stratification of lakes (stagnation and mixing) is induced by convection: in autumn cold surface water drops through the lower, warmer layers.

Apart from gradients of the temperature, the salinity (salt content, seawater) and the content of fine suspended materials have an influence on the density of water. If water is introduced into a standing, stratified body of water, then the new water will flow into the layer that has approximately the same density as itself.

Example 4.27: Combined sewage discharge into a lake

During a summer thunderstorm combined sewer overflow is discharged into a lake which has a temperature of 23°C at the surface. Soon after the beginning of the thunderstorm, the combined sewage has a temperature of 17°C, in addition it carries approximately 200 g m^{-3} of fine, mineral suspended solids with a density of $\gamma_P = 2.5$ g cm^{-3}.

At which depth will the combined sewage stratify?

The density of the combined sewage amounts to approximately

$$\rho_{17°C} + 200 \text{ g m}^{-3} \cdot (\rho_P - \rho_W)/\rho_P = 998.800 + 0.120 = 998.920 \text{ kg m}^{-3},$$

which corresponds to the density at a temperature of 16.3°C.

Thus, the combined sewage will enter clearly below the surface and sink to the layer with a temperature of 15–17°C. This goes along with the observation that at the lake surface the hygienic conditions are barely impaired and hygienic quality requirements of swimming water can be maintained even after a thunderstorm in spite of the very high concentrations of bacteria in combined sewage.

Example 4.28: Common salt (NaCl) as a tracer solution

Common salt is frequently used as an economical tracer material which can be followed very easily by measuring the conductivity of the water. The salt is usually dosed in the form of a highly concentrated solution which may have a density 10% higher than that of water. In sewers there is a tendency for the dense salt solution to sink to the bottom of the channel and mix only very slowly with the wastewater. Mixing is much slower than by turbulent diffusion or dispersion since the density gradients require significant input of mixing energy to obtain a homogeneous concentration or density.

Examle 4.28: Convection in sedimentation basins

In sedimentation tanks we find water that still contains large quantities of suspended solids and has accordingly a higher density than the water already clarified. These density gradients may induce secondary density currents which affect the treatment performance. This effect is particularly pronounced in secondary clarifiers. Here the loading of the influent with activated sludge solids is particularly high.

4.2.9 Mass Transfer Coefficients

A mass transfer coefficient is a material- and system-specific parameter that simplifies the prediction of the transfer of materials across the laminar boundary layers at phase boundaries (water–air, water–solids, water–oil, etc.).

Many transformation processes are heterogeneous, i.e., they proceed with their characteristic rate only, if at least two phases (gas–liquid–solid) are present. Thus, materials must be exchanged over phase boundaries. Laminar boundary layers are formed at an interface because the water at the surface of a solid body does not move relative to the surface (or this movement is reduced at an interface towards a gas phase). Here molecular diffusion is the dominant transport process. Figure 4.19 shows schematically a concentration profile of a material that is mixed in the turbulent zone of the water whose concentration then decreases towards the surface of another phase, where it is consumed. Slow transport by molecular diffusion in the boundary layer stands in contrast to the rapid mixing by the turbulence in the liquid.

We do not usually model the hydrodynamic details at the phase boundaries; thus at the boundary layer in Fig. 4.19 we replace the laminar sublayer that progressively develops into a fully turbulent zone with a fictitious boundary layer in which only molecular diffusion occurs and a turbulent zone in which the mixing is very rapid. Experimentally the thickness $d_F = \Delta x$ of this fictitious boundary layer is not measurable. However, we can quantify the mass transfer which amounts to:

$$j_{F,A} = -D_A \cdot \frac{\Delta C_A}{\Delta x} = k_{F,A} \cdot \Delta C_A \text{ with } k_{F,A} = \frac{D_A}{\Delta x} = \frac{D_A}{d_F} \tag{4.32}$$

$j_{F,A}$ = flux of material A over the boundary layer $[M_A L^{-2} T^{-1}]$
D_A = molecular diffusion coefficient of material A $[L^2 T^{-1}]$
ΔC_A = difference of concentration of material A across the boundary layer $[M_A L^{-3}]$
$d_F = \Delta x$ = thickness of the fictitious boundary layer $[L]$
$k_{F,A}$ = mass transfer coefficient for material A $[L T^{-1}]$

Empirical relationships to estimate mass transfer coefficients are given in the literature based on dimensionless numbers (Reynolds number, Schmidt number, Froude number, etc.) and many experiments in different systems. Based on work by Froessling (1938), Levenspiel (1999) provides a relationship that allows the

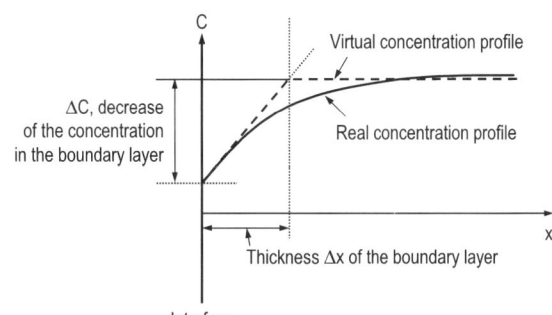

Fig. 4.19 Effective and linearized concentration profile of a material that is consumed at the phase boundary surface

estimation of the Sherwood number (Sh) for individual particles exposed to a liquid with flow velocity v (or settling velocity):

$$Sh = 2 + 0.6 \cdot Re_p^{1/2} \cdot Sc^{1/3} \quad or$$

$$\frac{k_{F,A} \cdot d_P}{D_A} = \frac{d_P}{d_F} = 2 + 0.6 \cdot \left(\frac{d_P \cdot v}{v}\right)^{1/2} \cdot \left(\frac{v}{D_A}\right)^{1/3}. \tag{4.33}$$

Sh = Sherwood number [–]
Sc = Schmidt number
Re_p = Reynolds number for individual particles
v   = kinematic viscosity of the water, $1.0034 \, mm^2 \, s^{-1}$ at 20°C

For water and small molecules (up to 250 Dalton), the Schmidt number Sc varies in the range 500–2000, and a typical value is 1000 (dimensionless).

For packed columns (sand filter, activated carbon, etc.) with $Re_p > 80$ the following relation applies (Levenspiel, 1999, after Ranz, 1952):

$$Sh = 2 + 1.8 \cdot Re_p^{1/2} \cdot Sc^{1/3} \quad or$$

$$\frac{k_{F,A} \cdot d_P}{D_A} = \frac{d_P}{d_F} = 2 + 1.8 \cdot \left(\frac{d_P \cdot u}{v}\right)^{1/2} \cdot \left(\frac{v}{D_A}\right)^{1/3}. \tag{4.34}$$

Example 4.28: Thickness of a boundary layer for mass transfer

A grain of activated carbon, with a diameter of 2 mm and a density of $\rho_P = 1700 \, kg \, m^{-3}$ sediments in standing water with 20°C. *What is the thickness of the boundary layer around this particle for the exchange of an organic compound A with a diffusion coefficient of $D_A = 1 \cdot 10^{-9} \, m^2 \, s^{-1}$?*
After Fig. 4.5 the sedimentation velocity at 20°C amounts to $v_S = 0.2 \, m \, s^{-1}$.

$v_{20°C} = 1 \, mm^2 \, s^{-1} = 1 \cdot 10^{-6} \, m^2 \, s^{-1}$.

After Eq. (4.33) $Sh = d_P/d_F = 2 + 0.6 \cdot 400^{0.5} \cdot 1000^{0.33} = 122$ or

$d_F = 2 \, mm/122 = 16 \, \mu m$.

The mass transfer coefficient is $k_{F,A} = D_A/d_F = 0.061 \, mm \, s^{-1} = 5.3 \, m \, d^{-1}$.

Example 4.29: Boundary layers in a filter bed

The activated carbon grain from Example 4.28 is inserted into the filter bed of an activated carbon filter which is operated with a hydraulic load of 15 m h⁻¹. *How thick is the boundary layer in this case?*
The effective water velocity with approximately 25% porosity in the filter material is $v_F = v_H/0.25 = 60 \, m \, h^{-1} = 0.017 \, m \, s^{-1}$.

After Eq. (4.34): $Sh = d_P/d_F = 2 + 1.8 \cdot 34^{0.5} \cdot 1000^{0.33} = 107$ and $d_F = 20 \, \mu m$.

The result is $k_{F,A}=4.3$ m d^{-1}. With a Reynolds number of only $Re_P=34$ this result lies, however, outside the range of the validity of Eq. (4.34) and is thus very inaccurate.

Example 4.30: Adsorption on an activated carbon grain

In the activated carbon filter of Example 4.29 drinking water that contains dissolved organic materials with a concentration $C_{A,0}$ is treated. The mean adsorption rate onto the activated carbon for this grain size is estimated to be:

$$j_A = k_A \cdot C_A = 3 \text{ md}^{-1} \cdot C_A .$$

Figure 4.20 show the geometrical situation of a single grain.
What is the observed rate of adsorption?
The flux of material into the boundary layer must correspond to the flux of material into the grain:

$$j_F = k_F \cdot (C_{A,0} - C_A) = j_A = k_A \cdot C_A \text{ thus } j_A = \frac{k_A \cdot k_F}{k_A + k_F} \cdot C_{A,0} .$$

With $k_A = 3$ m d^{-1} and $k_F = 4.3$ m d^{-1} the result is $j_A = 1.8$ m $d^{-1} \cdot C_{A,0}$. Thus, the boundary layer has a large influence on the adsorption rate, which is reduced by $(3-1.8)/3 = 40\%$. An increase of the filtration rate v and thus of Re_P would decrease the thickness of the boundary layer d_F and improve the adsorption process, but cause additional head loss.

Example 4.31: Typical mass transfer coefficients

Typical effective boundary layers have a thickness of $d_F = 5-100\ \mu m$, which with the diffusion coefficient of oxygen at 10°C results in

$$k_F = 1.2 \cdot 10^{-4} \text{ m}^2 \text{ d}^{-1}/d_F = 1-25 \text{ m d}^{-1}.$$

The higher value is typical for rising gas bubbles; the smaller value is in the range for laminar flow in thin water films.

Example 4.32: Oxygen limitation in a trickling filter

In a trickling filter the mass transfer coefficient is determined for the transport of oxygen from air over the laminar flowing water film to the surface of the biomass.

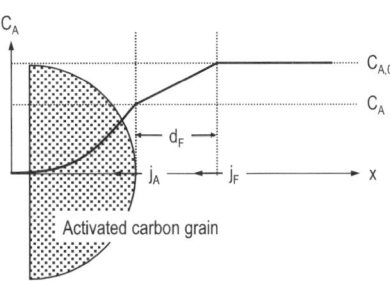

Fig. 4.20 Concentration profile in the laminar boundary layer and inside the activated carbon grain

The experimental result at 10°C is $k_F = 2$ m d^{-1}. The consumption of oxygen in the biofilm is proportional to $S_{O2}^{0.5}$ (where S_{O2} is the oxygen concentration at the surface of the biofilm), according to:

$$j_{O2,R} = -k_R \cdot S_{O2}^{0.5}, \text{ with } k_R = 4 g_{O2}^{0.5} \text{m}^{-0.5} \text{d}^{-1}.$$

The oxygen concentration at the air–wastewater boundary is $S_L = 10$ g$_{O2}$ m^{-3}.
How large is S_{O2}? How much oxygen does the biofilm consume?
Since the water film is very thin in the trickling filter, we can assume that the same amount of oxygen is transported into the water film as is consumed by the biofilm:

$$j_{O2,R} = -k_R \cdot S_{O2}^{0.5} = k_f \cdot (S_{O2} - S_L) \text{ or } k_F \cdot S_{O2} + k_R \cdot S_{O2}^{0.5} - k_F \cdot S_L = 0 \text{ and}$$

$$S_{O2}^{0.5} = \frac{-k_R \pm \sqrt{k_R^2 - 4 \cdot k_F^2 \cdot S_L}}{2 \cdot k_F} = 2.3 \text{ g}^{0.5} \text{m}^{-1.5} \text{ or } S_{O2} = 5.4 \text{ g m}^{-3}.$$

Thus, $j_{O2,R} = j_{O2,F} = 9.2$ g$_{O2}$ m^{-2} d^{-1}.

Chapter 5
Transformation Processes

For the representation of many simultaneous transformation processes the format of the stoichiometric matrix has established itself in environmental engineering sciences. It goes back to the introduction of the activated sludge model no.1 (Henze et al., 1987) and has found a mature and consistent format with the introduction of the activated sludge model no. 3 (Gujer et al., 2000).

The stoichiometric matrix is suitable for the introduction of sophisticated and extensive reaction systems in a clear and compact format that is especially geared towards the coding of simulation programs.

5.1 Case Study

The following case study is the basis for the introduction of the concepts in this chapter.

Heterotrophic microorganisms with the elementary composition $C_5H_7NO_2$ degrade a carbohydrate ($C_6H_{12}O_6$) in the presence of dissolved oxygen (O_2). In this growth process heterotrophic biomass is formed, and ammonium NH_4^+ is used as a nutrient and is integrated into the biomass. Bicarbonate (HCO_3^-) serves as the pH buffer.

In addition to growth a decay process of the biomass also takes place, according to which biomass is respired with oxygen by endogenous respiration and the nitrogen is released as ammonium.

In Table 5.1 these two processes are written in the typical way a chemist would define their stoichiometry.

In this chapter transformation processes, as shown in Table 5.1, will be defined in a format that initially may appear to be unnecessary complicated. However, once accepted, it will allow integrating extensive and complicated process combinations in a manageable form which is also easy to code into simulation programs.

Table 5.1 Case study: growth and decay of biomass. Stoichiometry defined in a format as typically used by chemists

Growth:
$C_6H_{12}O_6 + 2.45\,O_2 + 0.71\,NH_4^+ + 0.71\,HCO_3^- \rightarrow 0.71\,C_5H_7NO_2 + 3.16\,CO_2 + 5.29\,H_2O$
Decay:
$C_5H_7NO_2 + 5\,O_2 \rightarrow NH_4^+ + HCO_3^- + 4\,CO_2 + H_2O$

5.2 Transformation Written in Conventional Form

In transformation processes materials are converted from educts (raw materials) into products by either chemical reactions or processes catalyzed by living organisms, in particular microorganisms. We characterize such processes by defining state variables (material concentrations), stoichiometry (relationship between educts and products), and kinetics (the rate of the process).

Chemical reactions are sometimes presented in equations of the following form (see Table 5.1):

$$aA + bB \rightarrow cC + dD \quad (e.\,g.,\ 2\,H_2 + O_2 \rightarrow 2\,H_2O), \qquad (5.1)$$

where on the left we find the educts and on the right the products. Another, less common way of defining such a reaction reminds rather of a mathematical equation. It has the following form:

$$\nu_A \cdot A + \nu_B \cdot B + \nu_C \cdot C + \nu_D \cdot D = 0 \quad (e.\,g.,\ -2\,H_2 - O_2 + 2\,H_2O = 0). \qquad (5.2)$$

This way of writing does not correspond to a mathematical equation, but the + sign indicates a chemical operator with the meaning "reacts with" and "= 0" suggests that all materials participating in the reaction are included in the equation. The ν_i are called *stoichiometric coefficients*; they are positive for products and negative for educts. The reaction equation $-2\,H_2 - O_2 + 2\,H_2O = 0$ could be read as: we extract 2 mol of H_2 and 1 mol of O_2 from the environment and return 2 mol of H_2O.

Example 5.1: Stoichiometry in units of mass

In the decay process in Table 5.1 the stoichiometry is indicated in molecular units. *Which stoichiometry results in units of mass?*
The formula weights amount to:

$$C_5H_7NO_2 : 113g,\ O_2 : 32g,\ NH_4^+ : 18g,\ HCO_3^- : 61g,\ CO_2 : 44g,\ H_2O : 18g.$$

Writing the decay process in the format of Eq. (5.2) in units of mass results in:

$$-113\,g\,C_5H_7NO_2 - 160\,g\,O_2 + 18\,g\,NH_4^+ + 61\,g\,HCO_3^- + 176\,g\,CO_2 + 18\,g\,H_2O = 0.$$

In the sum the amounts of material add up to 0, which corresponds to the law of mass conservation.

In chemistry molar units are typically used for the definition of stoichiometric coefficients. For many reaction partners in environmental and technical processes, however, molar units are not suitable (what is a mole of biomass or a mole of pollutants?) and there is no reason to limit stoichiometry to molecular conversions. With the transition to stoichiometric relations which are not molecularly defined, the term *stoichiometry* obtains, however, a more empirical character.

The larger the absolute value of the stoichiometric coefficient for a material, the larger is its conversion in the course of the reaction. The turnover rates (reaction rates) for the different reaction partners are subject to the following relationship:

$$\frac{r_A}{v_A} = \frac{r_B}{v_B} = \frac{r_C}{v_C} = \frac{r_D}{v_D} = \rho \qquad (5.3)$$

r_i = reaction rate (transformation rate) of the material i. Its value is positive if the material is produced and negative if it is consumed $[M_i\, L^{-3}\, T^{-1}]$

v_i = stoichiometric coefficient for the material i $[M_i\, M_{i*}^{-1}]$. With the dimension given v_i indicates how much of material i is produced, per material i* that is converted. v_i is always positive for products and negative for educts

ρ = process rate $[M_{i*}\, L^{-3}\, T^{-1}]$

For one material (here marked with the index i*) we can freely select the dimensionless stoichiometric coefficient v_{i*} (see below), which fixes the dimension of the process rate ρ to $[M_{i*}\, L^{-3}\, T^{-1}]$ and also of all other stoichiometric coefficients to $[M_i\, M_{i*}]$. i* corresponds to the index of the material which arises in the denominator of the stoichiometric coefficients. It may be chosen differently for each process; in any one process it applies, however, to all materials.

The process rate ρ is *always positive*. For all materials i we have:

$$r_i = \rho \cdot v_i . \qquad (5.4)$$

The transformation rates r_i are positive for all products and negative for all educts. They are intensive variables which we introduce into the material balance Eqs. (3.11) or (3.15) for material i.

The process rate ρ is an intensive variable, which is characteristic for a process. It describes how the local environmental conditions (intensive state variables such as temperature, concentrations, and pH) affect the reaction rate. If the environmental conditions are fixed, then the four reaction rates r_i in Eq. (5.3) are determined. Equation (5.3) is then mathematically equivalent to four equations similar to Eq. (5.4), and altogether there are five unknowns (r_i and ρ). Thus, there is one degree of freedom.

In molecular processes we know the absolute value of the stoichiometric coefficients. They indicate how many molecules of a material participate in the reaction. In aggregated, global processes, such as are frequently used in microbial or technical models, these coefficients can only be specified in relative terms. Whether we express the oxygen consumption of growing biomass relative to the production of biomass (kg oxygen used per kg of biomass produced) or relative to the degradation of substrates (kg oxygen used per kg substrate consumed) depends on our experience and priorities. We can use the degree of freedom described above to select one stoichiometric coefficient per process freely. In the context of this text, we designate the appropriate material with the index i* and select v_{i*} as dimensionless with the value $+1$ if material i* is a product and -1 if i* refers to an educt. Thus, the process rate ρ and the stoichiometric coefficients v_i refer to the material i*, and the absolute value of r_{i*} becomes identical to ρ.

Example 5.2: Characterization of a transformation process

Microorganisms mineralize glucose with oxygen to carbon dioxide and water, according to:

Written in analogy to Eq. (5.1): $C_6H_{12}O_6 + 6\,O_2 \rightarrow 6\,CO_2 + 6\,H_2O$.

Written after Eq. (5.2): $-1 \cdot (C_6H_{12}O_6) - 6 \cdot (O_2) + 6 \cdot (CO_2) + 6 \cdot (H_2O) = 0$.

Equation (5.3) becomes: $\dfrac{r_{C_6H_{12}O_6}}{-1} = \dfrac{r_{O_2}}{-6} = \dfrac{r_{CO_2}}{6} = \dfrac{r_{H_2O}}{6} = \rho'$.

Frequently the consumption of oxygen is very simple to measure. We assume that it is $r_{O2} = -10\,molO_2\,m^{-3}\,d^{-1}$. We can freely select one stoichiometric coefficient. We select $i* = O_2$ and $v_{O2} = -1$ (dimensionless). Equation (5.3) now becomes:

$$\frac{r_{C_6H_{12}O_6}}{-1/6\,mol_{C_6H_{12}O_6}\,mol_{O_2}^{-1}} = \frac{r_{O_2}}{-1} = \frac{r_{CO_2}}{1\,mol_{CO_2}\,mol_{O_2}^{-1}}$$

$$= \frac{r_{H_2O}}{1\,mol_{H_2O}\,mol_{O_2}^{-1}} = \rho = 10\,mol_{O_2}\,m^{-3}\,d^{-1}$$

By Eq. (5.4) the absolute value and the units of all four reaction rates are given as:

$$r_{C_6H_{12}O_6} = -1.67\,mol_{C_6H_{12}O_6}\,m^{-3}d^{-1},$$

$$r_{O_2} = -10\,mol_{O_2}\,m^{-3}d^{-1},$$

$$r_{CO_2} = 10\,mol_{CO_2}\,m^{-3}d^{-1},$$

$$r_{H_2O} = 10\,mol_{H_2O}\,m^{-3}d^{-1}$$

It is important that we carefully compile the units of the stoichiometric coefficients. We proceed from Eq. (5.4) with r_{O2} [$molO_2\,m^{-3}\,d^{-1}$] and v_{O2} [–] to obtain the units of ρ as [$molO_2\,m^{-3}\,d^{-1}$]. With the units of the transformation rates r_i [$mol_i\,m^{-3}\,d^{-1}$] and with Eq. (5.4) the units of v_i can be obtained as [$M_i\,M_{i*}^{-1}$], or specifically [$mol_i\,molO_2^{-1}$].

5.3 Stoichiometric Matrix

The format of the so-called stoichiometric matrix combines several processes that simultaneously affect an array of materials. The format mediates in very compact form an overview of the interaction of the processes and can at small expenditure be coded into simulation programs.

If several transformation processes (here named by their index j) affect a material i, then the observed, or total transformation rate r_i as a result of all processes j is equal to the sum of the transformation rates in the individual processes:

$$r_i = \sum_j r_{j,i} = \sum_j \nu_{j,i} \cdot \rho_j . \tag{5.5}$$

Thus, the stoichiometric coefficients become the stoichiometric matrix, with the rows for each process j and the columns for each material i. The process rates form a vector ρ_j over all processes j and the observed transformation rates r_i are a vector over all materials i. In matrix notation this reads:

$$[r_i] = [\rho_j]^T \cdot [\upsilon_{j,i}] .$$

In this form stoichiometry and kinetics can easily be transferred into simulation programs.

Table 5.2 shows how a stoichiometric matrix for many materials and processes can be presented. The index i* stands for the material to which the stoichiometry and kinetics are related. i* can take a different value for each process j.

Example 5.3: Michaelis–Menten enzyme kinetics

Michaelis and Menten (1913) suggested the following mechanism for enzymatic reactions: substrate S connects itself with an enzyme E and forms an unstable enzyme substrate complex ES, which either disintegrates into a product P and the enzyme E or back to the substrate S and the enzyme E. In the traditional way of writing the following equations result:

$S + E \rightarrow ES$,

$ES \rightarrow S + E$,

$ES \rightarrow P + E$.

Table 5.2 Schematic representation of the stoichiometric matrix for a large reaction system

j	Verbal description of process	Materials i			Process rate ρ_j $[M_{i*}\, L^{-3}\, T^{-1}]$
		C_1 $[M_1]$	C_{i*} $[M_{i*}]$	C_i $[M_i]$	
1	Process name 1	$\nu_{1,1}$	$\nu_{j,i}\left[M_i M_{i*}^{-1}\right]$		ρ_1 = kinetic expression 1
2	Process name 2	$\nu_{2,1}$			ρ_2 = kinetic expression 2
j	Process name j	$\nu_{j,1}$	$\nu_{j,i*} = 1$	$\nu_{j,i}$	ρ_j = kinetic expression j
Observed transformation rate r_i		r_1	r_{i*}	r_i	$r_i = \sum_j \nu_{j,i} \cdot \rho_j$ $[M_i\, L^{-3}\, T^{-1}]$

The same mechanism, written as a stoichiometric matrix, has the form:

j	Substrate S	Enzyme E	Complex ES	Product P	Process rate ρ
1	-1	-1	$+1$		$k_1 \cdot S \cdot E$
2	$+1$	$+1$	-1		$k_2 \cdot ES$
3		$+1$	-1	$+1$	$k_3 \cdot ES$

The dimensions of the reaction constants are $[L^3 M^{-1} T^{-1}]$ for k_1 and $[T^{-1}]$ for k_2 and k_3.

The enzyme substrate complex decays very rapidly (k_2 and k_3 are very large), thus its concentration is small. From this we derive the assumption that the observed reaction rate is $r_{ES} \approx 0$ or:

$$r_{ES} = k_1 \cdot E \cdot S - k_2 \cdot ES - k_3 \cdot ES = k_i \cdot E \cdot S - (k_2 + k_3) \cdot ES.$$

With $r_S = -k_1 \cdot S \cdot E + k_2 \cdot ES$ and $E_0 = E + ES =$ total enzyme concentration, after elimination of E and ES one obtains:

$$r_S = -k_3 \cdot \frac{S}{\dfrac{k_2 + k_3}{k_1} + S} \cdot E_0 = -K \cdot \frac{S}{K_S + S} \cdot E_0 = -v_m \cdot \frac{S}{K_S + S}.$$

This equation corresponds to the equation for enzyme kinetics proposed by Michaelis and Menten.

Example 5.4: Acid–base equilibrium as a forward and a backward reaction

An acid–base equilibrium can be understood as a system of two reactions: the protonation of the base and the deprotonation of the acid. As an example the equilibrium of the carbonate, bicarbonate, and carbonic acid system is introduced:

(1) $CO_3^{2-} + H^+ \rightarrow HCO_3^-$	and	(2) $HCO_3^- \rightarrow CO_3^{2-} + H^+$
(3) $HCO_3^- + H^+ \rightarrow CO_2 + H_2O$	and	(4) $CO_2 + H_2O \rightarrow HCO_3^- + H^+$

Written in the format of a stoichiometric matrix without consideration of H_2O we obtain:

j		CO_3^{2-}	HCO_3^-	CO_2	H^+	Process rate ρ
1	Protonation	-1	$+1$		-1	$k_1 \cdot [CO_3^{2-}] \cdot [H^+]$
2	Deprotonation	$+1$	-1		$+1$	$k_2 \cdot [HCO_3^-]$
3	Protonation		-1	$+1$	-1	$k_3 \cdot [HCO_3^-] \cdot [H^+]$
4	Deprotonation		$+1$	-1	$+1$	$k_4 \cdot [CO_2]$

The dimensions of k_2 and k_4 are $[T^{-1}]$ and those of k_1 and k_3 are $[L^3 mol^{-1} T^{-1}]$. At equilibrium all observed transformation rates become $r_i = 0$ and thus:

$$r_{CO3} = k_2 \cdot [HCO_3^-] - k_1 \cdot [CO_3^{2-}] \cdot [H^+] = 0 \text{ or}$$

$$\frac{k_2}{k_1} = K_{equ} = \frac{[CO_3^{2-}] \cdot [H^+]}{[HCO_3^-]}.$$

This corresponds to the equilibrium constant for the dissociation of bicarbonate. Similarly the equilibrium constant for the dissociation of CO_2 can also be deduced from $r_{CO2} = 0$. If many different equilibrium reactions must be considered at the same time, the matrix representation provides an overview that can easily be worked on by algebra. From the literature frequently only equilibrium constants K_{equ} are known, not, however, the rate constants k_j. One of the two rate constants must then be selected freely. If these equilibrium reactions are to be inserted into material balance equations which will be solved numerically, then these rate constants must be chosen to be rather large such that the simulation rapidly reaches an approximate equilibrium (e. g., for $k_2 > 10^6 \, d^{-1}$ and $k_1 = k_2/K_{equ}$).

Case study

The case study in Table 5.1 with growth and decay of biomass is introduced in Table 5.3 in the format of a stoichiometric matrix. Substrate S_S and biomass X_H are measured in terms of COD. Dissolved oxygen S_{O2} is expressed as mass concentration. The symbols S and X stand for dissolved or particulate material concentrations, respectively. Ammonium, bicarbonate, carbon dioxide, and water are not included in the model, based on the assumption that they are of minor importance. The process rates follow the frequently used Monod kinetics for growth, and the decay is first order for processes such as endogenous respiration. The stoichiometric coefficients for biomass were selected dimensionless as +1 or −1 ($i* = 3 =$ biomass), which also defines the units of the process rate ρ_j. The yield Y (g_{COD} assimilated in biomass per g_{COD} substrate degraded) can be computed from the chemical stoichiometry in Table 5.1 to be $Y = 0.59 \, g_{COD,Biomass} \, g^{-1}_{COD,Substrate}$ (see also Example 5.20). The stoichiometric coefficient for oxygen $v_{O2} = -(1-Y)/Y$ is obtained with the help of the conservation law for theoretical oxygen demand (TOD, see Sect. 5.7.3).

Table 5.3 A simplified version of the processes in Table 5.1 in the format of a stoichiometric matrix

	Material i	1 Oxygen S_{O2}	2 Substrates S_S	3 Biomass X_H	Process rate ρ_j
j	Process	$g_{O2} \, m^{-3}$	$g_{COD} \, m^{-3}$	$g_{COD} \, m^{-3}$	$g_{COD,BM} \, m^{-3} \, d^{-1}$
1	Growth	$-\dfrac{1-Y}{Y}$	$-\dfrac{1}{Y}$	1	$\rho_1 = \mu_m \cdot \dfrac{S_S}{K_S + S_S} \cdot X_H$
2	Decay	−1		−1	$\rho_2 = b \cdot X_H$
Observed transformation rate r_i			$r_i = \sum_j v_{j,i} \cdot \rho_j$		

The observed, net transformation rates for the three considered materials yield with Eq. (5.5):

$$r_{O2} = -\frac{1-Y}{Y} \cdot \mu_m \cdot \frac{S_S}{K_S + S_S} \cdot X_H - 1 \cdot b \cdot X_H \,,$$

$$r_S = -\frac{1}{Y} \cdot \mu_m \cdot \frac{S_S}{K_S + S_S} \cdot X_H \,,$$

$$r_H = 1 \cdot \mu_m \cdot \frac{S_S}{K_S + S_S} \cdot X_H - 1 \cdot b \cdot X_H \,. \tag{5.6}$$

5.4 Kinetics

Kinetics indicates, how the environment, i. e., the state variables, influence the rate of a transformation process.

The process rate ρ_j is defined by a kinetic expression (see Table 5.3). The form of the kinetics is process specific and must be defined together with the definition of the process. Table 5.4 summarizes the frequently used forms of the kinetic expressions. Monod kinetics is an approach used to describe the growth (reproduction) of microorganisms with concentration X as a function of their substrate concentration S. The simple inhibition term describes the effect of a material that slows down the growth of the organisms with increasing concentration S. Figure 5.1 shows the effect of concentration on the various process rate terms.

The important characteristics of a transformation process are defined with its reaction kinetics and its stoichiometry. In Table 5.5 three kinds of processes are compared:

- A degradation process corresponding to a simple chemical reaction. First-order kinetics, which is used here, suggests that the substrate with the concentration C_S will, with probability $p = k_1 \cdot \Delta t$, be degraded in the next time interval Δt.
- In a catalytic process the quantity (concentration) of the catalyst C_C plays an important role: the more catalyst that is present, the faster the process proceeds. The quantity of the catalyst is not affected, however, by the process ($v_C = 0$).
- In an autocatalytic process the catalyst is a product of the process ($v_C > 0$). The longer the process proceeds in a closed system, the faster the reaction runs. A typical example of such a process is the growth of microorganisms. In the course of the degradation of the substrate (C_S) the microorganisms (C_C) increase, thus the process is accelerated: Exponential growth results as long as no other limitation occurs.

The course of the degradation of a material subject to the three different processes is shown in Fig. 5.2.

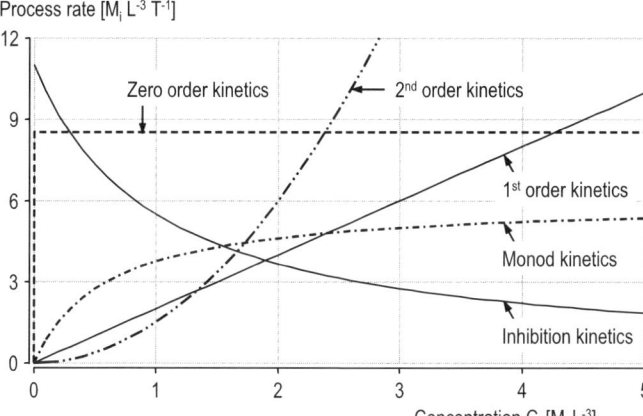

Process rate $[M_i\ L^{-3}\ T^{-1}]$

Fig. 5.1 Effect of the concentration C_i on the process rate ρ for the kinetic expressions in Table 5.4

Table 5.4 Examples of frequently used kinetic expressions

Type of reaction	Kinetics	Dimension of the parameters
0. order	$\rho = \begin{cases} k_0 & \text{if } C_i > 0 \\ 0 & \text{if } C_i = 0 \end{cases}$	$[k_0] = [M_i\ L^{-3}\ T^{-1}]$
1st order	$\rho = k_1 \cdot C_i$	$[k_1] = [T^{-1}]$
2nd order	$\rho = k_2 \cdot C_i^2$ or $\rho = k_2 \cdot C_1 \cdot C_2$	$[k_2] = [M_i^{-1}\ L^3\ T^{-1}]$
n^{th} order	$\rho = k_n \cdot C_i^n$	$[k_n] = [M_i^{1-n}\ L^{-3 \cdot (1-n)}\ T^{-1}]$ $[n] = [-]$
Monod kinetics	$\rho = \mu_m \cdot \dfrac{S}{K_S + S} \cdot X$	$[\mu_m] = [T^{-1}]$ $[K_S] = [M_S\ L^{-3}]$
Simple inhibition	$\rho = \mu_m \cdot \dfrac{K_I}{K_I + S} \cdot X$	$[\mu_m] = [T^{-1}]$ $[K_I] = [M_S\ L^{-3}]$
Competing substrates	$\rho = \mu_m \cdot \dfrac{S_1}{K_S + S_1} \cdot \dfrac{S_1}{S_1 + S_2} X$	$[\mu_m] = [T^{-1}]$ $[K_S] = [M_{S1}\ L^{-3}]$

Table 5.5 The interaction of stoichiometry and kinetics determines the type of the reaction

Type of reaction	C_S Educt	C_C Catalyst	Process rate ρ
Degradation	-1		$k_1 \cdot C_S$
Catalytic degradation	-1		$k_2 \cdot C_S \cdot C_C$
Autocatalytic process	-1	+1	$k_3 \cdot C_S \cdot C_C$

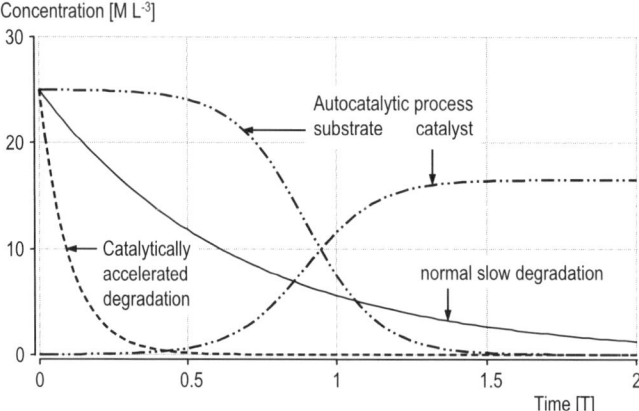

Fig. 5.2 Comparison of the degradation of a material in a closed system (batch reactor) in different reactions according to Table 5.5

5.4.1 Temperature Effects

Most chemical and biological processes are accelerated by increasing temperature. The Arrhenius equation is frequently used to adapt the rates from one temperature to the other one.

Our experience indicates that most transformation processes are accelerated with increasing temperature. The Arrhenius equation (5.7) is a frequently used model, in which reaction rates are adapted to different temperatures. The higher the temperature, the larger the kinetic energy of the individual molecules. In a collision of two molecules the probability that a reaction occurs therefore increases with increasing temperature. The Arrhenius equation has the form:

$$k = A \cdot \exp\left(-\frac{E_A}{R \cdot T}\right) \qquad (5.7)$$

k   = rate constant
A   = a constant with the units of k
E_A = activation energy of the reaction [kJ mol^{-1}]
R   = universal gas constant, $8.314 \cdot 10^{-3}$ kJ mol^{-1}K^{-1}
T   = temperature in K = °C + 273.15 K

For many processes which are of importance in urban water management, the concept of the activation energy, which is based on a molecular, mechanistic approach, is rather unsuitable. Thus, Eq. (5.7) is frequently applied in a simplified form:

$$k(T) = k(T_0) \cdot \exp\left(k_T \cdot (T - T_0)\right) \qquad (5.8)$$

Fig. 5.3 Experimentally derived
dependence of the growth rate of
nitrifying bacteria on temperature

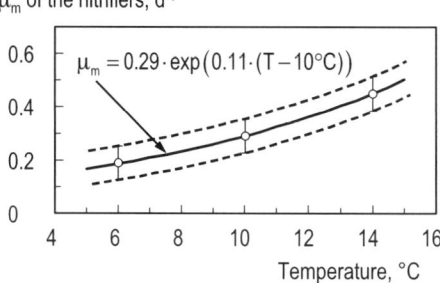

$k(T)$ = rate constant at temperature T
T_0 = reference temperature, for which the rate constant is known [°C]
k_T = temperature coefficient which indicates how strongly the reaction is accelerated per °C [°C^{-1}]

Example 5.5: Temperature step change of 10°C, $\theta_{10°C}$

A frequent assumption for the temperature dependence of a chemical reaction is that its rate doubles with any increase of the temperature of 10°C. From this assumption we derive:

$$\theta_{10°C} = \frac{k(T+10°C)}{k(T)} = 2 \quad \text{or} \quad r(T+10°C) = r(T) \cdot 2 = r(T) \cdot \exp(k_T \cdot 10°C),$$

which results in $k_T = 0.069°C^{-1}$.
For biological reactions, in particular growth rates of microorganisms, $\theta_{10°C}$ is often within the range 2–3 or $k_T = 0.069–0.11°C^{-1}$.
Figure 5.3 shows the increase of the maximum growth rate μ_m of the nitrifiers with temperature. For any 1°C change there is an increase of 11%, which results in $\theta_{10°C} = 3.0$.
This increase leads to the fact that we must increase the volume of the activated sludge tank by 11% for any 1°C-lower design temperature.

5.5 State Variables

State variables are the variables in the models that we use to characterize the behavior of natural and technical systems. In connection with transformation processes mainly the material concentrations, the pH value, and the temperature are of interest.

State variables are intensive variables that describe the development of a system as a function of the system variables space and time. Here primarily the material concentrations and the variables that affect the kinetics (temperature, pressure, and pH value) are of interest. In environmental engineering sciences, we pursue both chemically well-defined and chemically ill-defined species. Frequently we must rely on groups of materials which we can measure in terms of collective or group parameters (COD, TOC, BOD$_5$, TSS, TKN, P$_{tot}$, etc.) only; the properties of these materials are primarily defined by their method of analysis.

A state variable is defined by:

- a characterization of its meaning and importance (see Example 5.6)
- the indication of the units in which the concentration is provided. Examples are: g_{NO3}/m^3, mg NO_3^--N/l, mol NO_3^-/m^3, and mmol/l
- the designation of the method of the analysis procedure which is used to follow the variable. That is of particular importance with regard to collective parameters;
- information about the composition of the material relative to the conservative elements, for which the conservation laws are to be applied (composition matrix, see Sect. 5.6).

Example 5.6: Definition of state variables

The following text is an excerpt from the description of the activated sludge model no. 3, which defines some state variables of the model ASM3 (Gujer et al., 2000). ThOD = TOD = theoretical oxygen demand. COD = chemical oxygen demand, which can be determined with the help of the dichromate method.

1. S_{O2} [M(O$_2$) L^{-3}]: Dissolved oxygen, O$_2$. Dissolved oxygen can directly be measured and is subject to gas exchange. In stoichiometric computations S_{O2} is introduced as negative ThOD.
2. S_I [M(COD) L^{-3}]: Inert soluble organic material. The prime characteristic of S_I is that these organics cannot be further degraded in the treatment plants dealt with in this report. ... It can readily be estimated from the residual soluble COD in the effluent of a low loaded activated sludge plant.
3. S_S [M(COD) L^{-3}]: Readily biodegradable organic substrates (COD). This fraction of the soluble COD is directly available for consumption by heterotrophic organisms. ... S_S is preferentially determined with the aid of a bioassay (respiration test). Measuring the sum of $S_I + S_S$ in the form of the total soluble COD in wastewater as determined with 0.45 μm membrane filtration may lead to gross errors. This is due to the fact that some X_S (see later) in wastewater (e. g., starch) cannot adsorb to the small amount of biomass present in the influent and therefore contributes to the analytically determined soluble material.
4. S_{NH4} [M(N) L^{-3}]: Ammonium plus ammonia nitrogen $\left(NH_4^+ - N + NH_3 - N\right)$. For the balance of the ionic charges, S_{NH4} is assumed to be all NH_4^+. Because ASM3 assumes that organic compounds contain a fixed fraction of organic nitrogen ($t_{N,i}$, composition matrix), the influent $S_{NH4,0}$ cannot be observed directly (measured analytically) but should be computed from wastewater composition: Kjeldahl nitrogen – organic nitrogen. In the activated sludge reactors and in the effluent S_{NH4} is equivalent to observed concentrations. With the redox reference level chosen, S_{NH4} does not have a ThOD.

etc. Altogether the Model ASM3 covers 13 materials and 9 processes.

Example 5.7: Nitrogen in nitrification

Nitrification proceeds according to the following reaction equations:

$$NH_4^+ + 1.5\,O_2 \rightarrow NO_2^- + H_2O + 2\,H^+$$

$$NO_2^- + 0.5\,O_2 \rightarrow NO_3^-$$

If we express all materials in terms of mass, then from any 1 g NH_4^+ first 2.56 g NO_2^- are formed, which afterwards become 3.44 g NO_3^-. In the chemical unit of moles we obtain from 1 mol NH_4^+, 1 mol NO_2^- and afterwards 1 mol NO_3^-. Indicated as nitrogen, as is usual in the engineering sciences today, results: from 1 g NH_4^+–N we obtain 1 g NO_2^-–N, which then becomes 1 g NO_3^-–N. The concentrations of the different nitrogen forms are expressed as N; therefore the units of mg N/l or g N m^{-3} are typical.

Case study

The case study in Table 5.1 contains altogether seven state variables (different materials). The production of water (H_2O) is very small compared with the water flow; it cannot accurately be measured and is therefore not considered. Carbon dioxide (CO_2) has importance, e. g., for the computation of the pH value in activated sludge tanks. Such a computation requires, however, that we also follow the entire buffer system in detail, particularly the loss of CO_2 into the atmosphere as a consequence of aeration. Here we do without a statement about CO_2. There remain five state variables, which are characterized in Table 5.6. Based on their units, typical chemical parameters for these materials were introduced; this requires that we adapt the stoichiometric coefficients in Table 5.1 accordingly (see Example 5.20).

Table 5.6 Characterization of the state variables for the case study

i	State variable	Chemical composition	Measured variable	Concentration	Unit
1	Oxygen	O_2	O_2	S_{O2}	g_{O2} m^{-3}
2	Substrate	$C_6H_{12}O_6$	COD	S_S	g_{COD} m^{-3}
3	Ammonium	NH_4^+	N	S_{NH4}	g_N m^{-3}
4	Bicarbonate or alkalinity ALK	HCO_3^-	Mole H^+	S_{ALK}	mol_{ALK} m^{-3}
5	Biomass	$C_5H_7NO_2$	COD	X_H	g_{COD} m^{-3}

5.6 Composition of Materials

In view of the use of conservation laws, we must characterize the individual state variables such that we can convert their units into the units of conservative variables. The relevant information is contained in the so-called composition matrix.

To each state variable an analytical procedure is assigned, with which we characterize the state quantitatively. Frequently we must use group parameters that capture only a part of the characteristics of the materials, e. g., the COD concentration cannot provide us with any information on the nitrogen or the phosphorus content of the materials. With the knowledge about the composition of the individual materials we can convert the state variables into other units of measurement. We use the following equation:

$$C_{k,i} = C_i \cdot \iota_{k,i} \tag{5.9}$$

k = index of the new units of measurement [−]
$C_{k,i}$ = converted concentration in the units of the material k $[M_k \, L^{-3}]$
C_i = state variable i (concentration to be transformed) $[M_i \, L^{-3}]$
$\iota_{k,i}$ = conversion factor of the units of the material i into those of the material k $[M_k \, M_i^{-1}]$

Thus, for all state variables i we must obtain a conversion factor $\iota_{k,i}$, in order to convert them into the units of alternative measurements of interest. The result is the composition matrix $\iota_{k,i}$.

Example 5.8: Molar units of the composition

In the following table the materials of the case study in Table 5.1 are characterized with regard to their elementary composition. The table is to explain the principle; it does not add any new information.

	Material i	1 $C_5H_7NO_2$ Mole	2 O_2 Mole	3 NH_4^+ Mole	4 HCO_3^- Mole	5 CO_2 Mole	6 H_2O Mole
Element k							
1 C	Mole C	5	$\iota_{C,O2}=0$		1	1	
2 H	Mole H	7	$\iota_{H,O2}=0$	4	1		2
3 N	Mole N	1	$\iota_{N,O2}=0$	1			
4 O	Mole O	2	$\iota_{O,O2}=2$		3	2	1

Here the individual $\iota_{k,i}$ values have the unit $mol_k \; mol_i^{-1}$.

Case study

For the computation of the stoichiometric coefficients in the case study of Table 5.1 we will use three conservation laws: for theoretical oxygen demand

(TOD), for the element N, and for electrical charge. From the seven materials which are involved in the two processes, five are of interest (see Table 5.6).

In Table 5.7 the two processes *growth* and *decay* with the five remaining state variables are characterized. The two stoichiometric coefficients for the biomass X_H were selected as dimensionless and +1 resp. −1. For both processes we must obtain a further coefficient, based on our specialized knowledge. These are the coefficients for the substrate S_S, whereby we use for the decay process $v_{2,SS}=0$. This corresponds to our assumption (or knowledge) that in the decay process no substrate is converted. We can determine the remaining six unknown coefficients with the help of the conservation laws, for which we need to know the composition matrix $t_{k,i}$; in this case it contains only two parameters, which must be obtained experimentally ($i_{N,SS}$ and $i_{N,XH}$, the nitrogen content of the substrate and the biomass, respectively).

Graphically the composition matrix $t_{k,i}$ is arranged below the stoichiometric matrix in Table 5.7. To the left is the list of the regarded conservative variables k. For every one of these variables k the conversion factors $t_{k,i}$ are contained in the matrix for all state variables i:

- For $k=1$ (TOD) the resulting factors are −1 for oxygen and +1 for those materials which are characterized with the help of COD (see Example 5.19).
- From $k=2$ (N) we see that only substrate, ammonium, and biomass contain nitrogen. Oxygen, bicarbonate, water, and carbon dioxide do not contain nitrogen, therefore $t_{N,i}=0$ (see Example 5.9).
- For $k=3$ (charge) the values result from the fact that only ammonium and bicarbonate are electrically charged. The charge is expressed in moles of positive charge. One mole of NH_4^+–N contains 14 g N, therefore $i_{Charge,NH4}=1$ mole of charge/14 g N. One mole of bicarbonate carries 1 mol of negative charge, therefore $i_{Charge,HCO3}=-1$ mole of charge/mol of HCO_3^-.

The units/dimensions of the transformation factors result from the relationship of the unit of the conservative variables M_k to the measure of the state variables M_i as $M_k M_i^{-1}$.

Table 5.7 Symbolic stoichiometry and composition for the case study in Table 5.1

State variables	S_{O2} g O_2	S_S g COD	S_{NH4} g N	S_{HCO3} mol	X_H g COD	Process rate, Vector ρ_j with the units g COD_{XH} m^{-3} d^{-1}
j Processes			Stoichiometry			
1 Growth	$v_{1,O2}$	$-\dfrac{1}{Y_H}$	$v_{1,NH4}$	$v_{1,HCO3}$	+1	$\rho_1 = \mu_m \cdot \dfrac{S_S}{K_S + S_S} \cdot X_H$
2 Decay	$v_{2,O2}$		$v_{2,NH4}$	$v_{2,HCO3}$	−1	$\rho_2 = b \cdot X_H$
k Conservatives			Composition			
1 TOD in g	−1	1			1	
2 N in g		$i_{N,SS}$	1		$i_{N,XH}$	
3 Charge mol +			1/14	−1		

CAUTION: In Table 5.7 we have not included water and carbon dioxide (CO_2) as reaction partners. Since both these compounds have neither TOD nor a nitrogen content and they are also not charged, this does not interfere with the application of the conservation equations in the next section. If we drop reaction partners, we must, however, always make sure that this is the case. Here we cannot apply the conservation laws for carbon since CO_2 contains carbon.

Example 5.9: Composition of substrate and biomass in the case study

The composition matrix in Table 5.7 contains the two parameters $i_{N,SS}$ and $i_{N,XH}$, which are easily determined in the laboratory. This requires that in a sample of the substrate and the biomass both the COD and the organically bound nitrogen (TKN$-$ NH$_4^+$ –N) are determined. i_N values result from the relationship orgN/ COD. The two parameters have the units $g_N\,g^{-1}_{COD}$.
According to Table 5.1 the substrate has the composition of a carbohydrate ($C_6H_{12}O_6$) and thus contains no nitrogen, therefore $i_{N,SS}=0$.
The biomass has the composition $C_5H_7NO_2$. The formula weight of 113 g BM corresponds to 160 g COD (see Example 5.20) and contains 14 g N. $i_{N,XH}=14/160$ $=0.0875\,g_N\,g^{-1}_{COD,BM}$.
Thus, all elements of the composition matrix are available.

5.7 Conservation Laws

A chemical element remains unchanged in a chemical reaction (a transformation process). Thus, a conservation law applies to all elements: in chemical processes the mass of the individual elements remains constant.

If we capture with the state variables i all materials in which a certain element is contained, we can write the conservation law in mathematical terms as:

$$\sum_i v_{j,i} \cdot \iota_{k,i} = 0 \qquad (5.10)$$

$v_{j,i}$ = stoichiometric coefficient for the material i in process j [$M_i\,M_{i^*}^{-1}$]
$\iota_{k,i}$ = transformation factor (composition) for the material i into the units of the conservative variable k [$M_k\,M_i^{-1}$]

Example 5.10: Conservation law for nitrogen in the case study of Table 5.1

With the five state variables in Table 5.7 we capture all materials which contain nitrogen. Thus, we can apply Eq. (5.10) to both processes (growth and decay). Growth:

$$v_{1,O2}\cdot\iota_{N,O2}+ v_{1,SS}\cdot\iota_{N,SS}+ v_{1,NH4}\cdot\iota_{N,NH4}+ v_{1,HCO3}\cdot\iota_{N,HCO3}+ v_{1,XH}\cdot\iota_{N,XH}=0$$

or

$$-\frac{1}{Y_H} \cdot i_{N,SS} + \nu_{1,NH4} \cdot 1 + 1 \cdot i_{N,XH} = 0 \text{ and thus } \nu_{1,NH4} = \frac{1}{Y_H} \cdot i_{N,SS} - i_{N,XH} \cdot$$

Decay:

$$\nu_{2,O2} \cdot i_{N,O2} + \nu_{2,SS} \cdot i_{N,SS} + \nu_{2,NH4} \cdot i_{N,NH4} + \nu_{2,HCO3} \cdot i_{N,HCO3} + \nu_{2,XH} \cdot i_{N,XH} = 0$$

or

$$\nu_{2,NH4} \cdot 1 - 1 \cdot i_{N,XH} = 0 \text{ and thus } \nu_{2,NH4} = i_{N,XH}.$$

With the help of each applicable conservation law we can compute one stoichiometric coefficient for each process j. We save the expenditure of obtaining this coefficient experimentally from the ratio of two transformation rates. The determination of the composition matrix is frequently quite simple, see Example 5.9.

Since a conservation law applies to all elements, we can assign a freely chosen weight γ_k to the individual elements k and add the conservation laws for all elements k:

$$\sum_k \left[\gamma_k \cdot \sum_i \nu_{j,i} \cdot \iota_{k,i} \right] = 0$$

$$\sum_i \nu_{j,i} \cdot \left[\sum_k \gamma_{k,i} \cdot \iota_{k,i} \right] = 0 \tag{5.11}$$

γ_k = weight of the conservative element k

If we select the atomic weights of the elements k as the weights γ_k, then the conservation law for the mass results. However, since in many reactions we do not capture all materials involved with the state variables, Eq. (5.11) can only rarely be used in this form. In the case study of Table 5.7 we do not, e.g., capture the two materials water and carbon dioxide.

Example 5.11: Conservation law for mass in the decay process

In Table 5.1 the decay of biomass is characterized as follows:

$$C_5H_7NO_2 + 5\,O_2 \rightarrow NH_4^+ + HCO_3^- + 4\,CO_2 + H_2O.$$

According to Eq. (5.2) the stoichiometric coefficients ν_i are -1 and -5 for the educts and 1, 1, 4, and 1 for the products.

The composition matrix $\iota_{k,i}$ has the form:

Element k	γ_k		$C_5H_7NO_2$	O_2	NH_4^+	HCO_3^-	CO_2	H_2O
		ν_{Decay}	−1	−5	1	1	4	1
C	12		5			1	1	
H	1		7		4	1		2
N	14		1		1	3		
O	16		2	2			2	1
		$\Sigma\gamma_k \cdot \iota_{k,I}$	113	32	18	61	44	18

As the weights of the elements γ_k we select the atomic weights ($C = 12$, $H = 1$, $N = 14$, and $O = 16$). After Eq. (5.11) we compute the conversion factors ($\Sigma\gamma_k \cdot i_{k,i}$ = formula weight or molecular weight) and sum the product:

$$-1 \cdot 113 - 5 \cdot 32 + 1 \cdot 18 + 1 \cdot 61 + 4 \cdot 44 + 1 \cdot 18 = 0.$$

The mass is conservative in this process (any other sequence of weights γ_k could be used).

Conservation laws refer to individual transformation processes and their stoichiometry and not to the material or mass balance equation. The technical literature frequently does not clearly distinguish between the application of a conservation law and a mass balance equation. In a stoichiometric matrix the conservation laws apply to a row (index i); it connects different materials and points out how a single transformation process affects the different materials. The material or mass balance equation refers to a column (index j); it connects the different transformation and transport processes and points out how these affect a single material.

Example 5.12: Comparison of a mass balance with a conservation law

Based on Table 5.7 the following balance equation for the biomass X_H results for a completely mixed system with constant volume (stirred tank reactor, CSTR):

$$V \cdot \frac{dX_H}{dt} = Q \cdot \left(X_{H,in} - X_H \right) + \left(\mu_m \cdot \frac{S_S}{K_S + S_S} \cdot X_H - b \cdot X_H \right) \cdot V \text{ or}$$

$$\frac{dX_H}{dt} = \frac{Q}{V} \cdot \left(X_{H,in} - X_H \right) + r_{1,XH} + r_{2,XH} .$$

The conservation law for TOD for the first process (growth) has the form:

$$-1 \cdot \nu_{1,O2} - 1 \cdot \frac{1}{Y} + 1 = 0 \text{ or after multiplication with } \rho_1 : -r_{1,O2} + r_{1,SS} + r_{1,XH} = 0 .$$

In one case we sum up the transformation rates over the processes j, and in the other case over the materials i.

5.7.1 Conservation Law for Several Processes

Conservation laws are equally applicable for single as well as combined transformation processes.

Since Eq. (5.10) applies to each process, we can add this equation over several processes j:

$$\sum_j \sum_i \nu_{j,i} \cdot \iota_{k,i} = 0 . \tag{5.12}$$

Equation (5.12) becomes interesting after we compute the transformation rates $r_{j,i}$ with the aid of the process rates ρ_j:

$$\sum_j \left(\rho_j \cdot \sum_i \nu_{j,i} \cdot \iota_{k,i} \right) = \sum_j \sum_i r_{j,i} \cdot \iota_{k,i} = \sum_i \left(\iota_{k,i} \cdot \sum_j r_{j,i} \right) = \sum_i \iota_{k,i} \cdot r_i = 0 . \tag{5.13}$$

Equation (5.13) states that for an entire reaction system a conservation law applies to the observed reaction rates r_i. That is interesting because it is frequently difficult to separate the individual processes in an experiment.

Example 5.13: Conservation law for observed reaction rates

Equation (5.13) applied to the model in Table 5.7 and the conservation of TOD results in:

$$-r_{O2} + r_{SS} + r_{XH} = 0.$$

Thus, we can use the observed substrate degradation rate (r_{SS}) and the observed biomass production rate (r_{XH}) to compute the oxygen consumption rate (r_{O2}). Or in other words, from the three rates we can predict the one that is the most difficult to measure based on TOD conservation – a significant reduction of experimental effort.

5.7.2 Charge Balance

Apart from the conservation law for chemical elements a conservation law also applies to electrical charge: the positive and negative electrical charges that are converted in a transformation process compensate each other.

If we consider all electrically charged materials (ions) in the array of state variables, we can write the conservation law for electrical charge as:

$$\sum_i \nu_{j,i} \cdot \iota_{Charge,i} = 0 \tag{5.14}$$

$\nu_{j,i}$ = stoichiometric coefficient for material i in process j $[M_i\ M_{i*}^{-1}]$
$\iota_{Charge,i}$ = charge density of the material i $[mol_{Charge}\ M_i^{-1}]$. These values can be positive as well as negative.

Example 5.14: Use of the charge balance in the case study of Table 5.7

Table 5.7 contains all the electrically charged particles that are converted in the processes of the case study (Table 5.1). The application of the conservation law for the charge leads to the following equations:

Growth: $v_{1,NH4} \cdot 1/14 - v_{1,HCO3} = 0$.

Decay: $v_{2,NH4} \cdot 1/14 - v_{2,HCO3} = 0$.

Since the two coefficients for ammonium could be computed with the help of the conservation law for nitrogen (Example 5.10), the conservation of the electrical charge yields the stoichiometry for the alkalinity, which is represented here by its dominant species, bicarbonate HCO_3^-.

Example 5.15: Electrical neutrality as a conservation law

Aqueous solutions are electrically neutral. With the example of the ion balance of the data on the label of the Swiss mineral water Henniez (old, complete analysis, before 1986) this can be demonstrated:

Ion Index k	Concentration g m^{-3}	mol m^{-3}	Charge per Ion, $l_{Charge,k}$	Charge mol m^{-3}
Li^+	0.13	0.02	1	0.02
Na^+	7.49	0.33	1	0.33
K^+	1.09	0.03	1	0.03
Mg^{2+}	18.31	0.75	2	1.50
Ca^{2+}	98.30	2.45	2	4.90
Sr^{2+}	1.02	0.01	2	0.02
Sum of the positive charges (cations)				6.80
Cl^-	11.50	0.32	1	0.32
F^-	0.06	0.00	1	0.00
SO_4^{2-}	9.92	0.10	2	0.20
NO_3^-	23.20	0.37	1	0.37
HCO_3^-	360.13	5.90	1	5.90
Sum of the negative charges (anions)				6.79

The sum of the positive charges is equal to the sum of the negative charges. This type of charge balance can be used to check the accuracy of the analysis. The proton H^+ (pH value) is missing in this balance. With a concentration of approximately 10^{-7} mol/l or 10^{-4} mol m^{-3} it is of no importance; the same applies to OH^-.

Example 5.16: The charge balance in digested sludge is an indication for large transformation

The following example of a charge balance refers to the influent and effluent of a mesophilic digester at 35°C and 15 days average residence time on the laboratory scale. In this example the very significant transformation of the charges is characteristic, in particular the release of NH_4HCO_3 from the degradation of organic nitrogen compounds and the increase of the pH buffer capacity (HCO_3^-) in the digested sludge.

| Ion | mol Charge m^{-3} | | | |
| | Influent (raw) | | Effluent (digested) | |
	Cation	Anion	Cation	Anion
NH_4^+	4.4		28.9	
Ca^{2+}	3.8		5.9	
Mg^{2+}	0.7		1.0	
K^+	0.3		0.7	
HCO_3^-		4.0		37.5
$S^{2-}+HS^-$		0.7		0.3
Acetate$^-$		3.2		0.2
Propionate$^-$		0.8		0.0
other org. base$^-$		0.1		0.0
Total	9.2	8.8	36.5	38.0

The deviations in the sums of the anions and cations are partially due to analytic inaccuracy (sewage sludge), and partially attributed to a noncomprehensive list of the ions. The error amounts, however, in both samples to less than 5% (data: Thesis ETH No. 8958, M. Tschui, 1989).

5.7.3 Theoretical Oxygen Demand

With the theoretical oxygen demand (here TOD, sometimes ThOD, pronounced "thi oh di") we have a property available which is based on a theoretical concept and satisfies a conservation law. For organic compounds, it can be approximately determined analytically with the help of the group parameter chemical oxygen demand (COD).

We can freely choose the weights of the elements γ_k in Eq. (5.11). A particularly valuable way of selecting these weights orients itself towards the chemical analysis of the chemical oxygen demand (COD). In this analysis the organic materials are oxidized by $K_2Cr_2O_7$ to the final products H_2O, CO_2, SO_4^{2-}, Fe^{3+}, and NH_4^+ as far as possible. The final product of most forms of organically bound nitrogen is NH_4^+, which is remarkable since ammonium is a reduced form of nitrogen. From the consumption of dichromate for this oxidation an equivalent chemical oxygen demand is obtained. In order to compute the theoretical oxygen demand (TOD), the weights of the individual elements are specified such that the final products of the COD analysis do not have a TOD. Table 5.8 summarizes these weights.

Table 5.8 weights a mole of electrons that is donated in a redox reaction with +8 g TOD, or −8 g TOD if the electrons are accepted.

Table 5.8 Weights γ_k of the most frequently encountered elements for the computation of the theoretical oxygen demand of organic and some inorganic compounds in domestic wastewater treatment

Element		γ_{TOD}	Unit
Carbon	C	$+32$	g TOD mol^{-1} C
Nitrogen	N	-24	g TOD mol^{-1} N
Hydrogen	H	$+8$	g TOD mol^{-1} H
Oxygen	O	-16	g TOD mol^{-1} O
Sulfur	S	$+48$	g TOD mol^{-1} S
Phosphorus	P	$+40$	g TOD mol^{-1} P
Iron	Fe	$+24$	g TOD mol^{-1} Fe
Proton	H$^+$	0	g TOD mol^{-1} H$^+$
Negative Charge	–	$+8$	g TOD mol^{-1} Charge
Positive Charge	+	-8	g TOD mol^{-1} Charge

Example 5.17: TOD of a proton

If we want to compute the TOD of a proton, we obtain from Table 5.8:
TOD of H $+8$ gTOD mol^{-1}H
TOD of + charge -8 gTOD mol^{-1} positive charge
Total: 0 gTOD mol^{-1}H$^+$
The proton corresponds to the oxidized form of hydrogen; one electron has been donated.

Example 5.18: Oxidation capacity and oxygen demand

What is the TOD of nitrate NO_3^-?
The use of Table 5.8 results in:

$1 \cdot (-24$ gTOD mol^{-1}N$) + 3 \cdot (-16$ gTOD mol^{-1}O$) + 1 \cdot (8$ gTOD mol^{-1} negative charge$)$
$= -64$ gTOD mol^{-1} NO$_3^-$.

Nitrate can assume the role of oxygen in many microbiological processes (denitrification); accordingly nitrate is equivalent to an oxidation capacity that corresponds to a negative oxygen demand: -64 gTOD mol^{-1} NO$_3^-$ is equivalent to the well-known value of -4.57 gTOD g^{-1} NO$_3^-$ –N.

Example 5.19: Application of theoretical oxygen demand to the case study

In the case study of the aerobic degradation of organic materials (Table 5.1) the following stoichiometric equation is given for the growth process:

$C_6H_{12}O_6 + 2.45 O_2 + 0.71 NH_4^+ + 0.71 HCO_3^-$
$\rightarrow 0.71 C_5H_7NO_2 + 3.16 CO_2 + 5.29 H_2O.$

We obtain the following transformation coefficients $\iota_{TOD,i}$:

$C_6H_{12}O_6$	$6 \cdot 32 + 12 \cdot 8 - 6 \cdot 16$	$= 192$ g TOD mol^{-1}	Electron donor, pollutant
O_2	$-2 \cdot 16$	$= -32$ g TOD mol^{-1}	Electron acceptor
NH_4^+	$-24 + 4 \cdot 8 - 8$	$= 0$	Nutrient
HCO_3^-	$8 + 32 - 3 \cdot 16 + 8$	$= 0$	Counterion
$C_5H_7NO_2$	$5 \cdot 32 + 7 \cdot 8 - 24 - 2 \cdot 16$	$= 160$ g TOD mol^{-1}	Biomass
CO_2	$32 - 2 \cdot 16$	$= 0$	
H_2O	$2 \cdot 8 - 16$	$= 0$	

The conservation law for TOD states:

$$(-1) \cdot (192 \text{ g TOD}) + (-2.45) \cdot (-32 \text{ g TOD}) + (+0.71) \cdot (160 \text{ g TOD}) = 0.$$

We can freely choose one stoichiometric coefficient, and another one can be obtained from the TOD conservation; the third one we must determine experimentally. Since experiments are expensive, we select the cheapest experiment that results in a reliable value for one of the three stoichiometric coefficients. In the case study all values can of course be derived from the stoichiometric equations, which are based on experimental evidence.

If we select BOD_5 in place of COD for the characterization of the pollutants, then we must determine two stoichiometric coefficients experimentally, because BOD_5 is not subject to a conservation law.

Example 5.20: Computation of the Yield Y in COD units

In Table 5.1 the conversion of organic materials is indicated in chemical stoichiometry. In Table 5.3 materials involved are indicated as COD.
What is the value of the yield Y in Table 5.3?

Formula weight for substrate $C_6H_{12}O_6$:	180 g/mol
TOD per mole of substrate:	192 g TOD/mol

$\iota_{TOD,Substrate} = 192$ g TOD/180 g substrate $= 1.07$ g TOD g^{-1} substrate

Formula weight biomass $C_5H_7NO_2$:	113 g biomass/mol
TOD per mole of biomass:	160 g TOD/molm

$\iota_{TOD,Biomass} = 160$ g TOD/113 g biomass $= 1.42$ g TOD g^{-1} biomass

Transformation of units:
Thus, the yield coefficient becomes $Y = 0.71 \cdot 113 \cdot \iota_{TOD,Biomass}/(180 \cdot \iota_{TOD,Substrate})$ $= 0.59$ g TOD$_{Biomass}$/g TOD$_{Substrate}$ (0.71 originates from the stoichiometry in Table 5.1)

5.8 Summary

For small systems of transformation processes, the format of a stoichiometric matrix appears fastidious and complicated at first. The full value of this format becomes apparent only when large models are introduced as is the case for the simulation of biological wastewater treatment, sludge stabilization or even rivers and lakes.

In Table 5.9 the model of the case study from Table 5.1 is presented in the format of the stoichiometric matrix, for direct application. Since this representation covers kinetics, composition, and stoichiometry, it is far more informative than Table 5.1.

If we must identify the parameters of the model for a specific, practical situation on the basis of Table 5.7, only two parameters that cannot be derived from theoretical considerations are contained in the composition matrix: $i_{N,SS}$ and $i_{N,XH}$. These parameters are easily obtained in the laboratory. From the ten stoichiometric coefficients two can freely be selected (here $v_{1,XH}$ and $v_{2,XH}$). Two coefficients must be based on environmental engineering knowledge ($v_{1,SS}$ and $v_{2,SS}$), only one of them must be obtained experimentally (here $v_{1,SS}$ or Y). Six coefficients were computed with the help of the three conservation laws.

Table 5.7 provides an overview of the entire model that is easy to read once the format is accepted. The format reduces the expenditure for the identification of the stoichiometry to the absolutely necessary minimum. The representation is particularly suitable for models that are based on collective parameters, which are typically expressed in mass units, as is frequently used in the environmental engineering sciences.

Table 5.9 Computed stoichiometry and composition for the case study in Table 5.1 (see also Table 5.7)

State variables i	1 S_{O2} $g\ O_2$	2 S_S $g\ COD$	3 S_{NH4} $g\ N$	4 S_{HCO3} mol	5 X_H $g\ COD$	Process rate, Vector ρ_j with the unit g COD $m^{-3}\ d^{-1}$
j Processes		Stoichiometry				
1 Growth	−0.69	−1.69	−0.088	−0.0063	+1	$\rho_1 = \mu_m \cdot \dfrac{S_S}{K_S + S_S} \cdot X_H$
2 Decay	−1	0	0.088	0.0063	−1	$\rho_2 = b \cdot X_H$
k Conservatives		Composition				
1 TOD in g	−1	1			1	$\mu_m = 5\ d^{-1}$
2 N in g		0	1		0.088	$K_S = 5\ g\ m^{-3}$
3 Charge in mol			1/14	−1		$b = 0.2\ d^{-1}$

Chapter 6
Ideal Reactors

A reactor is an apparatus or a structure in which chemical, biological, and physical processes (reactions) proceed intentionally, purposefully, and in a controlled manner. In ideal reactors, the transport and mixing processes can be described mathematically exactly; this is in contrast to real, technical, built or natural reactors which must be modeled, but where the transport processes are only approximately known.

Ideal reactors are a theoretical proxy or concept which is analyzed instead of the real-world system to be simulated.

Contrary to natural systems, technical systems are planned and designed by engineers. The form, function, and characteristics of reactors (apparatuses, constructions) are frequently designed such that transport and mixing processes are easy to model; this allows one to accurately predict the expected performance. Ideal reactors are the models that are examined instead of real reactors. Whereas real reactors deviate in their behavior from the ideal reactors, they can frequently be described sufficiently accurately by ideal reactors.

Here the properties of ideal reactors, frequently used in water technology, are introduced and discussed. Chapter 8 will demonstrate how ideal reactors can be used to develop models of real reactors. Ideal reactors are not only used in the description of technical systems – they are equally applied to natural systems.

Ideal reactors are model systems for which the transport and mixing processes are exactly defined. They serve as abstract analogs of effective reactors. Their properties are chosen such that they can easily be described in mathematical terms.

6.1 Overview of Ideal Reactors

Ideal reactors differ regarding their influent and effluent, possible gradients of their state variables, and the geometry of their volume. Infinitely large internal mixing eliminates concentration gradients. For reactors with finite mixing (plug-flow reactors) the differential, intensive balance equation (3.15) must be applied, whereas well-mixed reactor compartments may be balanced with the aid of the extensive balance equation (3.11). Table 6.1 summarizes the properties of the ideal reactors that will be discussed in this chapter.

Table 6.1 Overview of the characteristics of the ideal reactors discussed here

Type of reactor	Influent and effluent	Volume	Directed flow	Internal mixing	Gradients of state variables
Batch reactor	no	constant	no	∞ large in x, y, z	no
Stirred tank reactor	yes, equal	constant	no	∞ large in x, y, z	no
Cascade of stirred tank reactors	yes, equal	constant	between sections	in sections ∞ large in x, y, z	between sections
Plug-flow reactor	yes, equal	constant	yes	∞ large in y, z Advection in x	in x direction
Turbulent plug-flow reactor	yes, equal	constant	yes	∞ large in y, z, advection and turbulence in x	in x direction
Stirred tank reactor with variable volume	yes, different amount	variable	no	∞ large in x, y, z	no

6.2 The Batch Reactor

The *batch reactor* has a constant volume which is so intensively mixed that inside the reactor gradients of state variables cannot occur in any direction. It has neither influent nor effluent, and exchange of material is very limited such that it does not affect the volume of the water (e. g., gas exchange over the free surface or the dosing of highly concentrated chemicals). Typical batch reactors are test tubes in the laboratory (frequently closed systems). Figure 6.1 schematically shows a batch reactor.

The material balance equation for water has the form:

$$V \cdot \frac{d\rho_W}{dt} + \rho_W \cdot \frac{dV}{dt} = r_W \cdot V = 0 \quad \text{or} \quad \frac{dV}{dt} = 0 \tag{6.1}$$

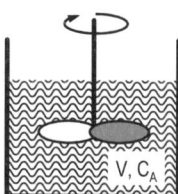

Characteristics of the batch reactor are:
- neither influent nor effluent
- constant volume
- homogeneous mixing, i.e. no gradients of intensive variables
- no steady state as long as a reaction occurs
- frequently a closed system

Fig. 6.1 Schematic representation and characteristics of a batch reactor

V = volume of the water in the reactor $[L^3]$
ρ_W = constant density of the water $[M_W L^{-3}]$
r_W = production of water $[M_W L^{-3} T^{-1}]$

With the assumptions that ρ_W is constant and that the production of water r_W can be neglected, a constant volume V results.

For any material A the mass balance equation (3.11) has the form:

$$V \cdot \frac{dC_A}{dt} + C_A \cdot \frac{dV}{dt} = r_A \cdot V \tag{6.2}$$

C_A = concentration of the material A $[M_A L^{-3}]$
r_A = production of the material A $[M_A l^{-3} T^{-1}]$

With constant volume this reduces to:

$$\frac{dC_A}{dt} = r_A . \tag{6.3}$$

Equation (6.3) corresponds to the general mass balance equation for a batch reactor. In the steady state C_A is constant and thus $r_A = 0$.

Example 6.1: Boiling eggs

A typical example of a batch procedure (reactor) is the boiling of three-minute eggs. Only at ambient temperature would a steady state be possible – eggs then, however, would not be boiled, i. e., the process of boiling would not be active.

Example 6.2: Reaction order in a batch reactor

How does the concentration of a material A in a batch reactor change when it is subject to a single reaction of zero $(r_A = -k_0)$, first $(r_A = -k_1 \cdot C_A)$ or second order $(r_A = -k_2 \cdot C_A^2)$?
General solution:

$$\frac{dC_A}{dt} = r_A = -k_n \cdot C^n \quad \text{with the initial condition } C_A(t=0) = C_{A,0}.$$

After separation of variables this results in:

$$\int_{C_{A,0}}^{C_A(t)} \frac{dC}{C^n} = -k_n \cdot \int_0^t dt' = -k_n \cdot t \quad \text{with the solutions (see Fig. 6.2):}$$

for $n = 1$: $C_A = C_{A,0} \cdot \exp(-k_1 \cdot t)$,

for $n \neq 1$: $\dfrac{1}{-n+1} \cdot \left(C_A^{-n+1} - C_{A,0}^{-n+1}\right) = -k_n \cdot t$,

for $n = 0$: $C_A = C_{A,0} - k_0 \cdot t$ with $C_A \geq 0$,

for $n = 2$: $\dfrac{1}{C_A} = \dfrac{1}{C_{A,0}} + k_2 \cdot t$.

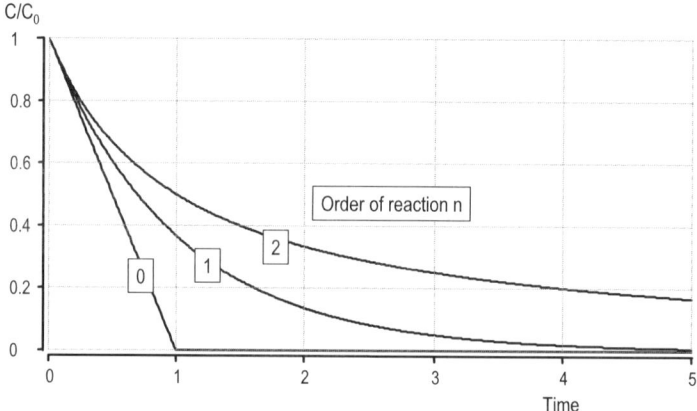

Fig. 6.2 Decrease of the concentration of a material in a batch reactor subject to different reaction kinetics (see Example 6.2)

Example 6.3: Biotechnology in the production of food

The production of a loaf of cheese or bread, a bottle of Champagne, etc. can be understood as a batch procedure. The cheese matures, different types of microorganisms are active in sequence, the condition of the cheese changes with time. The same applies to bread, or wine that ferments in the bottle, etc.

Example 6.4: Batch reactor with several materials and reactions

In a batch reactor the following equilibrium reactions are active:
$A + 2\,B \leftrightarrow C$, or expressed in matrix format:

	S_A	S_B	S_C	ρ
Forward reaction	-1	-2	1	$k_V \cdot S_A \cdot S_B^2$
Backward reaction	1	2	-1	$k_R \cdot S_C$

What is the form of the balance equations for the three materials A, B, and C?

$$\frac{dS_A}{dt} = \sum_j r_A = v_{1,A} \cdot \rho_1 + v_{2,A} \cdot \rho_2 = -1 \cdot k_V \cdot S_A \cdot S_B^2 + 1 \cdot k_R S_C \ ,$$

$$\frac{dS_B}{dt} = \sum_j r_B = v_{1,B} \cdot \rho_1 + v_{2,B} \cdot \rho_2 = -2 \cdot k_V \cdot S_A \cdot S_B^2 + 2 \cdot k_R S_C \ ,$$

$$\frac{dS_C}{dt} = \sum_j r_C = v_{1,C} \cdot \rho_1 + v_{2,C} \cdot \rho_2 = 1 \cdot k_V \cdot S_A \cdot S_B^2 - 1 \cdot k_R S_C \ :$$

three coupled, nonlinear, ordinary differential equations.

Fig. 6.3 Time course of the concentrations of three materials subject to coupled reactions in a batch reactor (see Example 6.4)

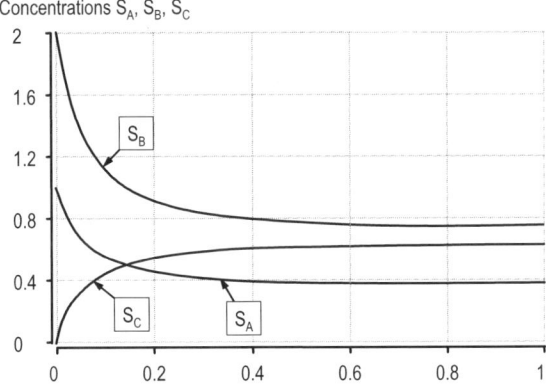

Figure 6.3 shows the development of the three concentrations. Clearly the three materials approach equilibrium or a steady state. Once steady state is reached, the observed reaction rates become zero, however, there is still transformation back and forth between the three materials.

6.3 The Continuous Flow Stirred Tank Reactor (CSTR)

Analogous to the batch reactor the *continuous flow stirred tank reactor* (*CSTR*) (Fig. 6.4) consists of an intensively mixed volume in which no gradients of state variables arise. Influent and effluent are equal, which results in a constant volume.

Since the content of a CSTR is ideally mixed, there are no gradients of intensive properties inside the reactor. This allows for a simple solution of the integrals for accumulation and reaction over the entire volume in the balance equation. In addition, the effluent concentrations must be equal to the concentrations in the reactor itself. The transport processes over the system boundaries result in $Q \cdot C_i$. A material balance equation, written for the material A, has the following form:

$$V \cdot \frac{dC_A}{dt} + C_A \cdot \frac{dV}{dt} = Q \cdot C_{A,in} - Q \cdot C_A + r_A \cdot V . \tag{6.4}$$

Applied to water with the constant density $C_{A,in} = C_A = \rho_W$ and without production ($r_W = 0$) this leads to $dV/dt = 0$ or constant volume. Thus, Eq. (6.4) is simplified for many materials to:

$$V \cdot \frac{dC_A}{dt} = Q \cdot C_{A,in} - Q \cdot C_A + r_A \cdot V . \tag{6.5}$$

This ordinary differential equation can further be simplified for the steady state to an algebraic equation:

$$0 = Q \cdot (C_{A,in} - C_A) + r_A \cdot V . \tag{6.6}$$

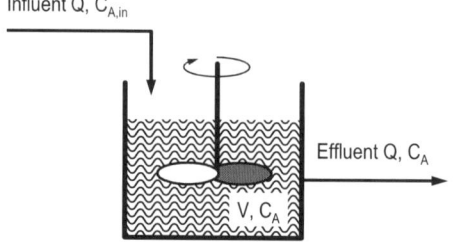

Influent Q, $C_{A,in}$

Effluent Q, C_A

V, C_A

Characteristics of the stirred tank reactor:

- amount of influent and effluent are identical
- constant volume
- complete mixing
- no concentration gradients
- concentrations in the effluent are identical to concentrations in the reactor
- steady state has interesting properties

Fig. 6.4 Schematic representation and characteristics of a continuous flow stirred tank reactor (CSTR)

First-order degradation processes (reactions) are frequent. For this situation we obtain for the steady state (with $r_A = -k \cdot C_A$):

$$\frac{C_A}{C_{A,in}} = \frac{1}{1 + k \cdot V / Q} = \frac{1}{1 + k \cdot \theta_h} \tag{6.7}$$

$\theta_h = V/Q$ = mean hydraulic residence time of the water in the reactor [T]

Example 6.5: The ice rink as a CSTR

An example of a stirred tank reactor is an ice rink. The users enter the ice rink by an entrance door, mix with the others and soon cannot be distinguished any more from all the other visitors. In each fixed time period a certain number of users leave the ice rink. The time individual users spend in the rink when they are leaving is not evident to the indifferent observer and is different for each individual user. For each visitor who leaves the field, the "reaction" proceeded to a different degree: the longer an individual stayed, the colder his or her fingers will be.

Example 6.6: Performance of a CSTR subject to different reaction order

What is the effluent concentration C_A from a CSTR at steady state, if material A is subject to a degradation process of zero, first, or second order?
$\theta_h = V/Q$ = mean hydraulic residence time
Zero order: From $0 = Q \cdot (C_{A,in} - C_A) - k_0 \cdot V$ follows: $C_A = C_{A,in} - k_0 \cdot \theta_h$, $C_A \geq 0$

First order: From $0 = Q \cdot (C_{A,in} - C_A) - k_1 \cdot C_A \cdot V$ follows: $C_A = C_{A,in} \cdot \dfrac{1}{1 + k_1 \cdot \theta_h}$

Second order: From $0 = Q \cdot (C_{A,in} - C_A) - k_2 C_A^2 \cdot V$ follows:

$$C_A = \frac{-1 + \sqrt{1 + 4 \cdot k_2 \cdot \theta_h \cdot C_{A,in}}}{2 \cdot k_2 \cdot \theta_h}$$

Example 6.7: Reactions in series

In a CSTR in a steady state two first-order reactions in series are active according to:

$$A \rightarrow B \rightarrow C.$$

Represented in the format of a stoichiometric matrix this leads to:

Process	S_A	S_B	S_C	ρ
1. reaction	-1	$+1$		$k_1 \cdot S_A$
2. reaction		-1	$+1$	$k_2 \cdot S_B$

What are the steady-state effluent concentrations of the three participating materials?

The three material balance equations, written for the steady state and with consideration of the stoichiometry and kinetics, result in:

$$0 = Q \cdot (S_{A,in} - S_A) - k_1 \cdot S_A \cdot V,$$

$$0 = Q \cdot (S_{B,in} - S_B) + k_1 \cdot S_A \cdot V - k_2 \cdot S_B \cdot V,$$

$$0 = Q \cdot (S_{C,in} - S_C) + k_2 \cdot S_B \cdot V.$$

These three algebraic equations can easily be solved for the unknown quantities S_A, S_B, and S_C.

Example 6.8: Recirculation in a CSTR

How does recirculation of the effluent back to the reactor affect the performance of a CSTR?

The balance for the material A in the reactor of Fig. 6.5 has the form:

$$V \cdot \frac{dC_A}{dt} = Q \cdot C_{A,in} + R \cdot C_A - Q \cdot C_A - R \cdot C_A + r_A \cdot V,$$

and after simplification this leads to:

$$V \cdot \frac{dC_A}{dt} = Q \cdot C_{A,in} - Q \cdot C_A + r_A \cdot V.$$

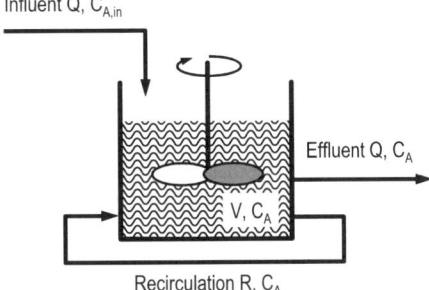

Influent Q, $C_{A,in}$

Effluent Q, C_A

V, C_A

Fig. 6.5 Schematic representation of a CSTR with external recirculation

Recirculation R, C_A

This simplified balance corresponds to Eq. (6.5). The recirculation does not have any influence on the performance of the ideal stirred tank reactor. Internal mixing (back mixing) is infinitely large, therefore external back mixing does not change the conditions in the reactor. Occasionally the CSTR is also called an ideal back mixed reactor.

6.4 A Cascade of Stirred Tank Reactors

The cascade of stirred tank reactors is not really an ideal reactor, but a series of frequently identical CSTRs and thus a combined reactor (Fig. 6.6). Since the model of the cascade of stirred tank reactors is frequently used in water technology, it is introduced separately here.

In the cascade no water is back-mixed from a later to an earlier reactor. A chain of infinitely many, infinitely small, stirred tank reactors corresponds to a conveyer belt or a plug-flow reactor (see Sect. 6.5).

Model development for a cascade proceeds in steps, as each individual reactor is understood as a subsystem. The resulting set of equations can be solved in the direction of flow from the front end to the rear end, given that no external recirculation exists.

A cascade of CSTRs is schematically represented in Fig. 6.6. Assuming equal volumes for all the subsystems, the balances for a material A have the following format:

$$V = V_{tot}/n$$

$$V \cdot \frac{dC_{A,1}}{dt} = Q \cdot (C_{A,in} - C_{A,1}) + r_{A,1} \cdot V \tag{6.8}$$

$$V \cdot \frac{dC_{A,i}}{dt} = Q \cdot (C_{A,i-1} - C_{A,i}) + r_{A,i} \cdot V \quad \text{for } i = 2...n$$

n = number of equal reactors in series [–]
V_{tot} = total volume of all reactors together [L^3]

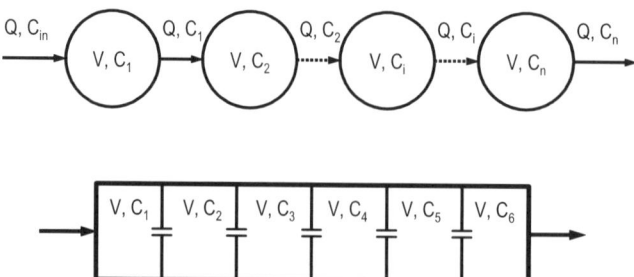

Fig. 6.6 Schematic representations of a cascade of stirred tank reactors

Example 6.9: Cascade at steady state

A cascade of CSTRs with n equal reactors in series is at steady state. A first-order reaction degrades material A ($r_A = -k \cdot C_A$).
How does the material concentration change along the cascade, and what influence does the number of stirred tank reactors have?
In the steady state the material balance for the ith reactor results in:

$$0 = Q \cdot (C_{A,i-1} - C_{A,i}) - k \cdot C_{A,i} \cdot V \quad \text{with} \quad V = V_{tot}/n$$

Or (see Eq. (6.7)): $\dfrac{C_{A,i}}{C_{A,i-1}} = \dfrac{1}{1 + k \cdot V/Q}$

V = V_{tot}/n = volume of individual reactors [L^3]
n = number of reactors in series [–]
V_{tot} = total volume of all reactors in the cascade [L^3]

$$\frac{C_2}{C_{in}} = \frac{C_2}{C_1} \cdot \frac{C_1}{C_{in}} = \frac{1}{1 + k \cdot V_1/Q} \cdot \frac{1}{1 + k \cdot V_2/Q} = \frac{1}{\left(1 + k \cdot \dfrac{V_{tot}}{n \cdot Q}\right)^2} = \left(1 + k \cdot \frac{V_{tot}}{n \cdot Q}\right)^{-2}$$

and similarly $\dfrac{C_i}{C_{in}} = \left(1 + k \cdot \dfrac{V_{tot}}{n \cdot Q}\right)^{-i}$ resp. $\dfrac{C_n}{C_{in}} = \left(1 + k \cdot \dfrac{V_{tot}}{n \cdot Q}\right)^{-n}$.

For $n \to \infty$ this leads to:

$$\frac{C_\infty}{C_{in}} = \lim_{n \to \infty} \left(1 + k \cdot \frac{V_{tot}}{n \cdot Q}\right)^{-n} = \exp\left(-k \cdot \frac{V_{tot}}{Q}\right).$$

This is the solution for a corresponding plug-flow reactor (see Sect. 6.5 and Example 6.14). On the basis of the derived equations C_n/C_{in} is found as a function of $k \cdot V_{tot}/Q$:

$k \cdot V_{tot}/Q$	n = 1	2	5	10	∞
			C_n/C_0		
1	0.500	0.444	0.402	0.386	0.368
2	0.333	0.250	0.186	0.162	0.135
5	0.167	0.082	0.031	0.017	0.007
10	0.091	0.028	0.004	0.001	$<10^{-4}$
100	0.010	$<10^{-3}$	$<10^{-6}$	$<10^{-10}$	$<10^{-43}$

With increasing reaction ($k \cdot V_{tot}/Q$) or decreasing concentration of the effluent (C_n/C_{in}) the difference in performance between stirred tank reactors and cascades of stirred tank reactors increases. Only partitioning into several subsystems makes it possible to reach a high efficiency.

Example 6.10: Disinfection of surface water for 300,000 inhabitants

In the course of the production of drinking water, $144{,}000\,\mathrm{m^3\,d^{-1}}$ ($100\,\mathrm{m^3\,min^{-1}}$) of water are to be disinfected. The realistic goal is to reduce the number of germs in the influent by a factor of 10^8 ($C_{out}/C_{in} = 10^{-8}$). The disinfection process can be described as a first-order reaction; accordingly $r_G = -k \cdot C_G$ ($C_G = [\#\mathrm{germs\ L^{-3}}]$). For the selected disinfection procedure it is estimated that $k = 5\,\mathrm{min^{-1}}$.

What is the volume of the corresponding disinfection reactor, designed as a cascade of n equal stirred tank reactors?

From Example 6.9: $\dfrac{C_n}{C_{in}} = \left(1 + k \cdot \dfrac{V_{tot}}{n \cdot Q}\right)^{-n}$.

The necessary volume V_{tot} to reach $C_{out}/C_{in} = 10^{-8}$ as a function of n is:

n	V_{tot} in m^{-3}	$\theta_h = V_{tot}/Q$
1	$2 \cdot 10^9$	38 yr
2	$4 \cdot 10^5$	2.8 d
3	28,000	0.2 d
5	3900	39 min
10	1100	11 min
∞	370	<4 min

The partitioning of a disinfection reactor into a cascade of CSTRs is thus extremely effective. With only one stirred tank reactor the inlet is mixed into the entire reactor and must accordingly immediately be diluted with the reactor content to reach the 10^8 times smaller effluent concentration. The entire reaction must proceed with the low effluent concentration. In a cascade approach, some disinfection close to the inlet can proceed rapidly with an increased concentration of germs, whereas the inlet to the last reactor contains very few germs and must therefore hardly be diluted.

Example 6.11: Cascade with recirculation

In a cascade of two stirred tank reactors with external recirculation (Fig. 6.7) at steady state, we observe a first-order degradation process ($r = -k \cdot C$).

How does the recirculation of the effluent of the second reactor to the first reactor affect the performance of the cascade?

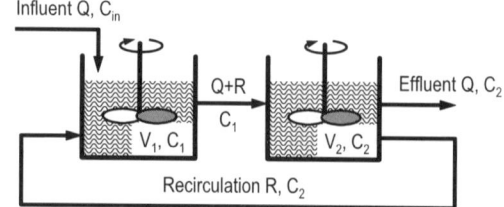

Fig. 6.7 A cascade of two stirred tank reactors with recirculation

The material balances in steady state are:

for the first reactor: $0 = Q \cdot C_{in} + R \cdot C_2 - (Q + R) \cdot C_1 - k \cdot C_1 \cdot V$,

for the second reactor: $0 = (Q + R) \cdot (C_1 - C_2) - k \cdot C_2 \cdot V$.

After substitution of C_1:

$$\frac{C_2}{C_{in}} = \frac{1 + R/Q}{\left(1 + R/Q + k \cdot V_{tot}/(2 \cdot Q)\right)^2 - (R/Q) \cdot (1 + R/Q)}.$$

For $R = 0$ this equation becomes the equation for two CSTRs in series as derived in Example 6.9.

For $R \rightarrow \infty$ the equation simplifies to

$$\frac{C_2}{C_{in}} = \frac{1}{1 + k \cdot V_{tot}/Q}.$$

This corresponds to the performance of a single CSTR with the same total volume. Thus, the recirculation of the effluent causes the performance of the cascade to lie between that of a single reactor and that of the cascade without recirculation: recirculation corresponds to partial back mixing.

Example 6.12: A fish ladder is a cascade of stirred tank reactors

In rivers with man-made obstacles, occasionally fish ladders (fish passes) are built to allow fish to climb against the current. Individual chambers of mixed, rather still water are fed by narrow, fast flowing connections. Fish can rise actively through the ladder against the current, while water and the materials contained in water can only flow downstream.

6.5 The Plug-Flow Reactor

The plug-flow reactor differs in principle from the ideal reactors discussed so far. As internal transport process in the direction of flow only advection is considered, whereas complete mixing is assumed transverse to the direction of flow (Fig. 6.8). Thus, gradients of state variables (intensive variables) can occur along the centerline of the reactor (Fig. 6.9).

Characteristics of the plug flow reactor are:
- in- and effluent are equal
- constant volume
- complete mixing only transverse to the direction of flow, i.e. gradients of intensive variables along the centerline are possible
- the steady state is interesting

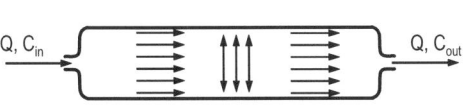

Fig. 6.8   Schematic representation and characteristics of a plug-flow reactor

Fig. 6.9 Representation of a differential element of a plug-flow reactor

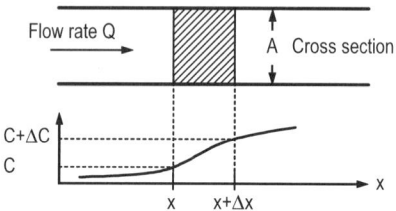

Since gradients of material concentrations are possible inside the reactor, gradients for the reaction rate also frequently result. For the derivation of a material balance equation, we must therefore select a small (differential) subsystem and afterwards integrate the balance equation over the entire reactor (Fig. 6.9).

A material balance equation for the small reactor element in Fig. 6.9 has the following form:

$$\Delta x \cdot A \cdot \frac{\partial C}{\partial t} = Q \cdot C(x) - Q \cdot C(x + \Delta x) + r \cdot \Delta x \cdot A$$

or with $\Delta C = C(x+\Delta x) - C(x)$:

$$\Delta x \cdot A \cdot \frac{\partial C}{\partial t} = -Q \cdot \Delta C + r \cdot \Delta x \cdot A \ .$$

With the transition of $\Delta x \to \partial x$ as well as $\Delta C \to \partial C$ and division by $A \cdot dx$ for the case of constant cross-sectional area A a partial differential equation results:

$$\frac{\partial C}{\partial t} = -\frac{Q}{A} \cdot \frac{\partial C}{\partial x} + r = -u \cdot \frac{\partial C}{\partial x} + r \tag{6.9}$$

$u(t)$ = velocity of flow $(A/Q(t))$ $[L\ T^{-1}]$

In the direction of flow x the flux of material A amounts to $j_A(x) = u \cdot C_A(x)$. Thus, Eq. (6.9) is can easily be derived from Eq. (3.15).

Frequently reactors that we model as plug-flow reactors do not have a pronounced, exactly defined longitudinal axis. If the reactor is *operated hydraulically at steady state* (Q = constant), it is reasonable to copy the longitudinal coordinate x onto a space–time coordinate τ:

$$\tau(x) = \int_0^x \frac{A}{Q} \cdot dx' = \int_0^x \frac{1}{u} \cdot dx' = \frac{x}{u} \quad \text{and} \quad d\tau = \frac{dx}{u} = \frac{A}{Q} \cdot dx$$

After substituting τ for x, Eq. (6.9) becomes:

$$\frac{\partial C}{\partial t} = -\frac{\partial C}{\partial \tau} + r = -u \cdot \frac{\partial C}{\partial x} + r \tag{6.10}$$

t = time, real time, absolute time [T]
τ = space–time, time since entering into the reactor [T]

The two kinds of time must strictly be differentiated: t refers to the time on a clock, e.g., 11:15 am; τ is a local coordinate and describes a time difference, the time elapsed since some material (water) entered the reactor, e.g., 12 min. *In a steady state only the derivatives with respect to the real time t become equal zero, not, however, those with respect to the space–time τ.*

If we designate the length of the reactor by L, then the maximum value of τ becomes:

$$\tau_{max} = \int_0^{\tau_{max}} d\tau = \int_0^L \frac{dx}{u(x)} = \int_0^L \frac{A(x)}{Q} \cdot dx \ .$$

In a steady state, Q is constant and $\int F \cdot dx = V$ corresponds to the total volume V of the reactor. Thus:

$$\tau_{max} = \frac{V}{Q} = \theta_h \ .$$

τ_{max} is used to compute the effluent concentration of a plug-flow reactor at steady state. In addition Eq. (6.10) can be simplified too:

$$\frac{dC_A}{d\tau} = r_A \quad \text{with} \quad 0 \le \tau \le \tau_{max} = V/Q. \tag{6.11}$$

The comparison of Eq. (6.11) with Eq. (6.3) shows an analogy: the plug-flow reactor responds in space–time τ as the batch reactor does in real time t. The plug-flow reactor may be seen as a conveyer belt on which a row of batch reactors are transported.

Example 6.13: The ski lift as a plug-flow reactor

An example of an ideal plug-flow reactor is the ski lift (Fig. 6.10). Each individual skier is exposed to the environment and his or her fingers become ever colder (reaction). At any time (real time) the ski-lift transports skiers who have spent a different period of time (space–time) on the handle (they have differently cold fingers). An observer standing outside of the lift always (real time t) sees skiers pass by who have spent the same amount of time on the lift (space–time τ), i.e., they all have equally cold fingers. If the speed of the ski lift is increased (no longer a steady state), then the picture perceived by the outside observer changes slowly as skiers pass by with ever warmer fingers. The skiers themselves note the increased speed only when leaving the ski-lift: they enjoy warmer fingers as the rate of cooling may be constant over space–time but $\tau_{max} = L/u$ is decreased.

Example 6.14: The plug-flow reactor in steady state with different reaction order

How large is the effluent concentration of a material A from a plug-flow reactor at steady state, if the material is degraded in a reaction of zero ($r_A = -k_0$), first ($r_A = -k_1 \cdot C_A$), or second ($r_A = -k_2 \cdot C_A^2$) order?

Fig. 6.10 The ski lift as an ideal plug-flow reactor

The general format of the stationary balance equation is:

$$\frac{dC_A}{d\tau} = r_A = -k_n \cdot C_A^n \quad \text{with} \quad C_A = C_{A,in} \quad \text{for} \quad \tau = 0.$$

After separation of the variables this leads to:

$$\int_{C_{A,in}}^{C_{A,out}} \frac{dC}{C^n} = -k_n \cdot \int_0^{\tau_{max}} d\tau = -k_n \cdot \tau_{max} = -k_n \cdot V/Q = -k_n \cdot \theta_h.$$

$\tau_{max} = V/Q = \theta_h$ = mean hydraulic residence time [T]

With the solutions:

for zero order (n = 0): $C_{A,out} = C_{A,in} - k_0 \cdot \theta_h$ and $C_{A,out} \geq 0$,

for first order (n = 1): $C_{A,out} = C_{A,in} \cdot \exp(-k_1 \cdot \theta_h)$,

for second order (n = 2): $C_{A,out}^{-1} = C_{A,in}^{-1} - k_2 \cdot \theta_h$.

These results are similar to the results for the batch reactor (Example 6.2, page 103), but they are based on space–time τ (the time since entrance into the reactor) rather than real time (t–t_0, the time since the beginning of the experiment).

6.6 Plug-Flow Reactor with Turbulence

In the direction of flow the plug-flow reactor with turbulence is subject to advection and a limited degree of mixing due to turbulence.

The ideal plug-flow reactor excludes any mixing of the water along the direction of flow of the reactor, e.g., as a consequence of turbulence. The behavior of real reactors is frequently affected by turbulence and dispersion, either as a consequence of turbulent flow conditions or based on system requirements, such as aeration, or flocculation processes which are not possible without turbulence. The

Characteristics of the turbulent plug flow reactor
- in- and effluent are equal
- constant volume
- ideal mixing transverse to direction of flow
- limited mixing along the direction of flow
- the steady state is interesting

Fig. 6.11 Schematic representation of a turbulent plug-flow reactor

Fig. 6.12 Transverse pro-files of the flow velocity in an ideal plug-flow reactor (*left*) and in a plug-flow reactor with turbulence overlaid (*right*)

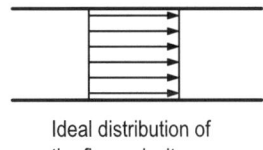

Ideal distribution of the flow velocity

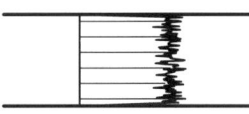

Fluctuations of the flow velocity due to turbulence

model of the plug-flow reactor with turbulence (Fig. 6.11) permits the influence of this additional mixing to be quantified.

Figure 6.12 illustrates the distributions of the advection velocities over the cross section in an ideal plug-flow reactor and in a plug-flow reactor with overlaid turbulence. Due to the fluctuations of the flow velocity, caused by turbulence, a longitudinal mixing of the flowing water is observed. The resulting longitudinal mixing may be described with the aid of Fick's first law, as long as the size of eddies is significantly smaller than the characteristic dimension of the reactor (spatial extension) (see Sect. 4.2.5):

$$j_{T,A} = -D_T \cdot \frac{dC_A}{dx} \tag{6.12}$$

$j_{T,A}$ = specific flux of material A as a consequence of turbulence $[M_A \, L^2 \, T^{-1}]$
D_T = turbulent diffusion coefficient $[L^2 \, T^{-1}]$ (sometimes dispersion-coefficient). D_T is independent of the material and characterizes the amount of turbulence or dispersion in the direction of flow

Figure 6.13 illustrates a small element of a plug-flow reactor with turbulence. A material balance for the section from x to x + Δx has the form:

$$\Delta x \cdot A \cdot \frac{\partial C}{\partial t} = Q \cdot C(x) - Q \cdot C(x + \Delta x) - A \cdot D_T \cdot \frac{\partial C}{\partial x}\Big|_x + A \cdot D_T \cdot \frac{\partial C}{\partial x}\Big|_{x+\Delta x} + r \cdot \Delta x \cdot A \, .$$

After the transition of $\Delta x \to \partial x$ and $C(x+\Delta x) - C(x) \to \partial C$ and division by A·dx (volume), we obtain for the case of constant A and D_T the following partial differential equation:

$$\frac{\partial C}{\partial t} = -u \cdot \frac{\partial C}{\partial x} + D_T \cdot \frac{\partial^2 C}{\partial x^2} + r \, . \tag{6.13}$$

u   = Q/A = local, average flow velocity, advection $[L \, T^{-1}]$.

Fig. 6.13 A small element of a plug-flow
reactor with turbulence as the basis for the
development of a differential mass balance
equation

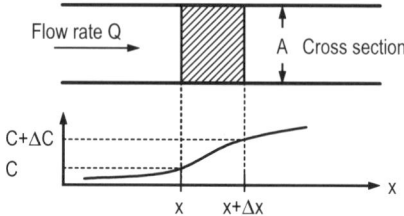

Equation (6.13) corresponds to Eq. (3.15) with $j_x = u \cdot C - D_T \cdot \partial C / \partial x$ (see Example 3.9).

For steady state an ordinary, second-order differential equation results:

$$\frac{d^2C}{dx^2} = \frac{u}{D_T} \cdot \frac{dC}{dx} - \frac{r}{D_T} . \tag{6.14}$$

The solution of Eq. (6.14) requires two boundary conditions, which are very case specific and must be formulated depending upon the system to be modeled. In Fig. 6.14 on the left a system is shown which is open for turbulence against the environment, i.e., individual eddies can reach back into the influent beyond the system border and thus it is possible that material is transported against the direction of flow into the influent. This system corresponds to a river section or a section of a turbulent pipe or a section of an aerated reactor. We define such a system as *open for turbulence*. Figure 6.14 on the right shows a system where turbulence may neither reach into the inlet nor the effluent, but is contained within the reactor. We define this system as *closed for turbulence*. Technical reactors with an influent and an effluent pipe are typically closed systems. An example is given in Fig. 6.11 (see also Example 4.19).

Figure 6.15 displays the length profile of material A which is degraded in a turbulent plug-flow reactor. The reactor is closed for turbulence; it is fed through a narrow, fast-flowing pipe. The two boundary conditions at the positions $x = 0$ and $x = L$ result from the following considerations.

In the supply line, advection prevails (controlled flow) and the mass flux of material just outside of the reactor amounts to

$$J_{A,in} = Q \cdot C_{A,in} .$$

Within the reactor transport by turbulence is added. The mass flux just inside the reactor amounts to:

$$J_{A,0} = Q \cdot C_{A,0} - D_T \cdot A \cdot \frac{dC_A}{dx}\bigg|_{x=0} .$$

A balance just over the influent leads to the condition $J_{A,in} = J_{A,0}$ or:

$$\frac{dC_A}{dx}\bigg|_{x=0} = -\frac{Q}{D_T \cdot A} \cdot \left(C_{A,in} - C_{A,0}\right) = -\frac{u}{D_T} \cdot \left(C_{A,in} - C_{A,0}\right) . \tag{6.15}$$

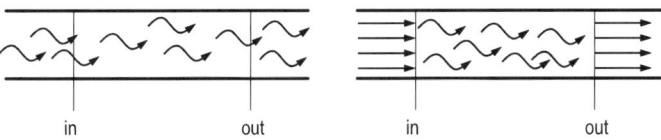

Fig. 6.14 Boundary conditions of systems with turbulence. On the *left* an open system, in which the turbulence extends beyond the system boundaries; on the *right* a closed system, in which advection dominates in the influent and effluent and turbulence is confined to the reactor volume

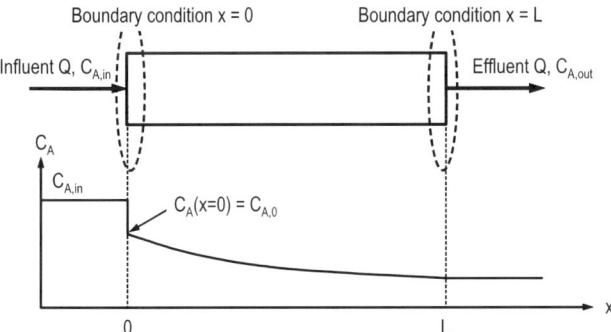

Fig. 6.15 Length profile of the concentration of material A in a turbulent plug-flow reactor which is closed for turbulence

Equation (6.15) corresponds to a combined boundary condition for dC_A/dx and $C_{A,0}$. It is interesting that, with increasing turbulence (D_T), an ever-larger step of the concentration from $C_{A,in}$ to $C_{A,0}$ results, because back mixing from the reactor to the head wall dilutes the influent. As $D_T \to \infty$ Eq. (6.15) requires $dC_A/dt \to 0$. The plug-flow reactor becomes a CSTR and the influent concentration $C_{A,in}$ is directly diluted to the discharge concentration $C_{A,out}$.

On the side of the effluent Eq. (6.16) can be deduced similarly:

$$\left. \frac{dC_A}{dx} \right|_{x=L} = \frac{Q}{D_T \cdot A} \cdot \left(C_{A,L} - C_{A,out} \right). \tag{6.16}$$

If the material A is degraded $dC_A/dx < 0$ must apply (C_A decreases along the flow direction). According to Eq. (6.16) $dC_A/dx < 0$ is only possible; if $C_{A,out} > C_{A,L}$, which is physically impossible, no soluble material can be concentrated without an input of energy. Therefore the only possible boundary condition, if the reactor is closed for turbulence, becomes:

$$\left. \frac{dC_A}{dx} \right|_{x=L} = 0 \quad \text{and} \quad C_{A,L} = C_{A,out}. \tag{6.17}$$

The balance equation (6.14) must be solved with the help of the upper and lower end boundary conditions (6.15) and (6.17). Since the two boundary condi-

Concentration C in g m⁻³

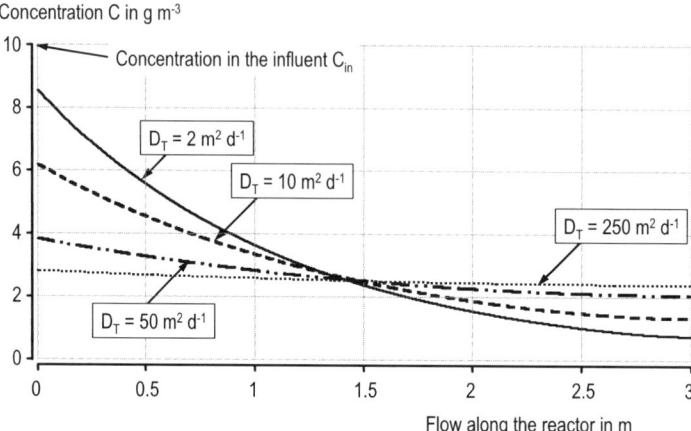

Fig. 6.16 Examples of a length profile of the concentration of a material that is degraded in a first order reaction in a turbulent plug-flow reactor. The parameter is the dispersion: the turbulent diffusion coefficient D_T

tions do not apply to the same location, this is a split boundary-value problem. The numeric solution of such a situation requires an iterative procedure, in order to satisfy both conditions (see Example 6.17). The length profile of the concentration of a material that is degraded in a first-order reaction is displayed for a plug-flow reactor with different turbulence D_T in Fig. 6.16. The increase of the step of the concentration on the influent side is clearly visible. As the turbulence D_T increases, the concentration profile increasingly approaches the totally mixed situation of a CSTR.

In a plug-flow reactor which is open for turbulence, the boundary conditions must be selected according to the specific situation. Frequently only the concentration in the influent is known, whereas information on the concentration gradient dC_A/dx is missing or is not accurate enough. For the numerical solution of the balance equation it is possible to choose a substantial virtual extension of the reactor and then use the boundary condition of a closed reactor, as shown in Fig. 6.17. The length profile in the virtual reactor can be computed iteratively, such that the concentration gradient in the influent is adjusted until the effluent concentration is met.

Example 6.15: A demonstration as a plug-flow reactor with turbulence

An example of a plug-flow reactor with turbulence is a demonstration. The participants all walk in the same direction (advection). With increasing time, a group of people who participate in the demonstration will ever more probably lose contact with each other and will be distributed along the course (turbulence). Contrary to the ski lift (Example 6.13) as an example of the ideal plug-flow reactor, the velocity of the participants is not fixed with the same accuracy.

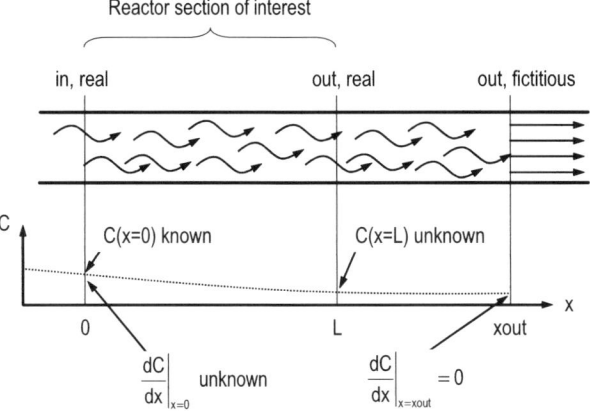

Fig. 6.17 Fictitious extension and choice of the boundary conditions for an open plug-flow reactor with turbulence, with inaccurately defined boundary conditions

Fig. 6.16 Sampling points in a plug-flow reactor with turbulence. In the influent (1) and in the effluent (4) advection is dominant; inside the reactor (2) and (3) turbulence is to be considered

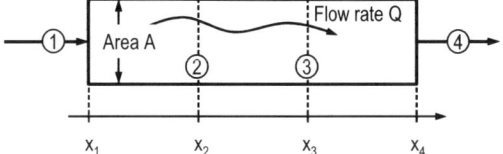

Example 6.16: Balance for subsystems in a reactor with turbulence

In a plug flow-reactor at steady state, aeration produces a large amount of turbulence (Fig. 6.18). In an experiment samples are taken from the influent, along the reactor, and from the effluent.

How can the degradation of material A in the subsections of the reactor be determined?

The stationary material balance for the reactor section from x_1 to x_2 has the following form:

$$0 = Q \cdot C_{A,1} - Q \cdot C_{A,2} + D_T \cdot A \cdot \left.\frac{dC_A}{dx}\right|_{X2} + r_{A,\text{Average},1>2} \cdot A \cdot (x_2 - x_1) .$$

For the reactor compartment from x_2 to x_3 we have:

$$0 = Q \cdot C_{A,2} - Q \cdot C_{A,3} + D_T \cdot A \cdot \left(-\left.\frac{dC_A}{dx}\right|_{X2} + \left.\frac{dC_A}{dx}\right|_{X3} \right) + r_{A,\text{Average},2>3} \cdot A \cdot (x_3 - x_2) ,$$

and similar for the third reactor compartment from x_3 to x_4:

$$0 = Q \cdot C_{A,3} - Q \cdot C_{A,4} - D_T \cdot A \cdot \left.\frac{dC_A}{dx}\right|_{X3} + r_{A,\text{Average},3>4} \cdot A \cdot (x_4 - x_3) .$$

Thus, without knowing D_T, we cannot determine the reaction rate from the measured concentrations $C_{A,2}$ and $C_{A,3}$ in the individual sections. The determination of D_T is discussed in Sect. 7.4.4 (see Example 8.4). In addition to the characterization of the turbulence the concentration gradients dC_A/dx must be available at the sampling points 2 and 3.

Since in the influent and effluent advection greatly outweighs the turbulence, and because with high velocities of flow in the pipe concentration gradients are small, we can write a balance over the entire reactor and determine the average reaction rate without knowing D_T:

$$0 = Q \cdot (C_{A,1} - C_{A,4}) + r_{A,\text{Average},1>4} \cdot V .$$

In reactors, in which laminar flow conditions prevail (e.g., trickling filters), the procedure shown here is inappropriate, because large concentration differences over the cross section are observed, and a local sample cannot characterize the concentration over the whole cross section.

This example points out that the transport processes must be characterized at sampling points, or else experimental results cannot be interpreted in the framework of material balances. *Here errors are frequently made.* The intuitively written, *wrong material balance*, in the form of

$$0 = Q \cdot (C_{A,1} - C_{A,2}) + r_{A,\text{Average},1>2} \cdot A \cdot (x_2 - x_1) \quad \text{can lead to large errors.}$$

Example 6.17: Implementation of a turbulent plug-flow reactor in Berkeley Madonna (BM)

The balance equation (6.14) with the boundary conditions (6.15) and (6.17) for a stationary plug-flow reactor which is closed for turbulence, is to be implemented in BM. The material is degraded in a first–order reaction.

Since for a closed reactor the boundary conditions refer to two different locations (beginning and end), a numeric forward integration can find a solution only iteratively. In BM the so called „shooting method" is implemented. On the basis of selected boundary conditions at the beginning, the condition at the end is computed (shot). Subsequently, the estimation of the initial condition is improved, until all boundary conditions are met within the tolerance.

With the following code, Fig. 6.16 was computed:

```
{Plug flow reactor with dispersion in steady state}
METHOD RK4            ; Integration with fourth-order Runge–Kutta
STARTTIME = 0         ; Influent side of the reactor
STOPTIME=3            ; End (length) of the reactor in m
DT = 0.0002           ; Time step becomes the length step in m
RENAME Time = x       ; Integration over the space in m
init C = C0           ; Boundary condition for C, within the reactor, Fig. 6.15
init C' = u/D*(C0-Cin) ; Boundary condition for C', Eq. (6.15)
C" = C'*u/D-r/D       ; Balance equation in steady state, Eq. (6.14)
C0 = 8                ; Estimated value for boundary condition of C in g m⁻³
Cin=10                ; Inlet concentration in g m⁻³, in the influent pipe, Fig. 6.15
```

Fig. 6.17 Example of the design of a plug-flow reactor with turbulence. The flow is directed by the inserted guidance walls, so that a longitudinal current develops. This configuration of the reactor can be retrofitted to an existing fully mixed basin

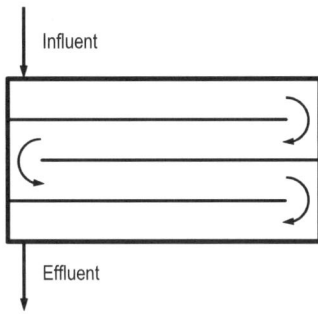

u = 10	; Flow velocity inside the reactor in m d^{-1}
D = 2	; Turbulence coefficient in $m^2 d^{-1}$ (must be ≥ 2 in order to obtain a result)
k = 10	; Rate constant for degradation d^{-1}
r = –k*C	; Reaction rate g $m^{-3} d^{-1}$

The solution requires the application of the module *Boundary Value ODE* under the menu Option Model. Here C' at the end of the reactor (x = 3) can be set to 0 (Eq. (6.17)). The parameter which must be adapted is C0. The numeric routines are only stable, if $D/(u·L) > 0.05$. If this value becomes smaller, the equations for the plug-flow reactor would approximately be applicable.

Example 6.18: Technical, plug-flow reactor with turbulence

A completely mixed basin is later to be converted into a plug-flow reactor with turbulence. For this the flow in the basin is directed with guidance walls as shown in Fig. 6.19, which decreases back mixing and therefore allows concentration gradients to develop.

Nonstationary plug-flow reactor with turbulence

The balance equation for a turbulent plug-flow reactor has the form of a partial differential equation (Eq. (6.13)) which cannot be simplified for the non-steady state. A possible approach to solving this equation numerically is to dissolve the reactor into a large number of subsystems of completely mixed compartments. The turbulence or back mixing is modeled with the help of a recirculation from a reactor compartment to the preceding one (Fig. 6.20). The larger the recirculation, the larger the back mixing or the modeled turbulence coefficient D_T becomes. In Example 6.19 the numeric implementation of this model in Berkeley Madonna is demonstrated.

The parameters of this model are the number of compartments or nodes n and the recirculation R. They are related to the turbulent diffusion coefficient D_T and the flow through the reactor Q. The details are discussed in Sect. 7.4.5, where

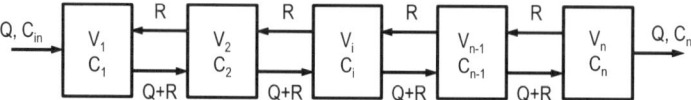

Fig. 6.18 Transformation of a plug-flow reactor with turbulence into a series of completely mixed compartments

methods are introduced based on the hydraulic residence time distribution to characterize the turbulence.

The boundary conditions for the solution of Eq. (6.13) are Eqs. (6.15) and (6.17). These are also apparent in Fig. 6.20. In the influent a jump of the concentration from the influent pipe $C_{A,in}$ to the effluent of the first compartment $C_{A,1}$ results, which is also required from Eq. (6.15). In the effluent the concentration $C_{A,out}$ is equivalent to the concentration in the last compartment $C_{A,n}$ which again satisfies Eq. (6.17).

Example 6.19: Numeric modeling of a closed turbulent plug-flow reactor

In Berkeley Madonna the reactor shown in Fig. 6.20 can be modeled as follows:

```
{Turbulent plug-flow reactor, not at steady state, tested}
METHOD RK4            ; Integration with fourth-order Runge–Kutta
STARTTIME = 0         ; Beginning of the forward integration
STOPTIME=3            ; End of the integration d
DT = 0.0002           ; Time step d
DTout = 0.01          ; Output interval d
n = 25                ; Number of partial reactors –
Vtot = 500            ; Total volume of the reactor m³
V = Vtot / n          ; Volume of a partial reactor, all of equal size m³
Q = 1000              ; Influent m³ d⁻¹
R = 5000              ; Recirculation between partial reactors m³ d⁻¹
Cin = Cm+AC*sin(2*pi*time/f)
                      ; Example of a time-dependent inlet concentration g m⁻³
Cm = 100              ; Average value of the inlet concentration g m⁻³
AC = 50               ; Amplitude of the inlet concentration gm⁻³
f = 1                 ; Frequency of the variation of the inlet concentration d⁻¹
rC[1..n] = -k*C[i]    ; Degradation 1. order in all partial reactors g m⁻³ d⁻¹
k = 10                ; Reaction constant d⁻¹
INIT C[1..n] = 1      ; Initial values for the concentrations g m⁻³
d/dt(C[1]) = (Q*Cin+R*C[2]–(Q+R)*C[1])/V+rC[1]
                      ; Balance for first reactor (closed for turbulence)
d/dt(C[2..n–1]) = ((Q+R)*(C[i–1]–C[i])+R*(C[i+1]–C[i]))/V+rC[i]
                      ; Balance for reactors 2…n-1 (open)
d/dt(C[n]) = ((Q+R)*(C[n–1]–C[n]))/V+rC[n]
                      ; Balance for the last reactor (closed)
Cout = C[n]           ; Effluent concentration g m⁻³
```

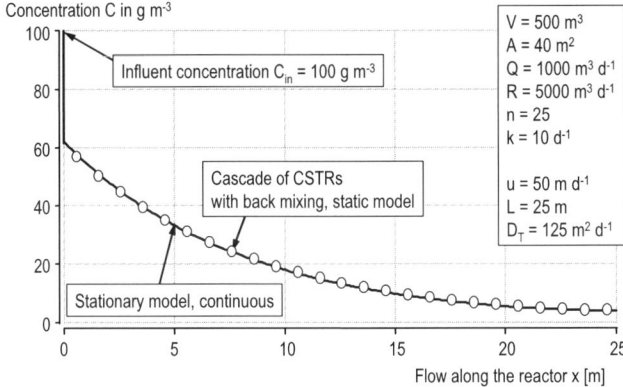

Concentration C in g m⁻³

Influent concentration C_{in} = 100 g m⁻³

Cascade of CSTRs
with back mixing, static model

Stationary model, continuous

Flow along the reactor x [m]

$V = 500$ m³
$A = 40$ m²
$Q = 1000$ m³ d⁻¹
$R = 5000$ m³ d⁻¹
$n = 25$
$k = 10$ d⁻¹

$u = 50$ m d⁻¹
$L = 25$ m
$D_T = 125$ m² d⁻¹

Fig. 6.19 Comparison of the prediction of the stationary model (Eq. (6.14), Example 6.17) with the discretized version of the dynamic model (Example 6.19, Eq. (7.39)

Figure 6.21 compares the results for the stationary model obtained from the direct integration of the steady-state balance equation (Eq. (6.14)) with the steady-state prediction of the discretized reactor model according to Fig. 6.20. Discretization is based on $n = 25$ nodes (subsystems) and the recirculation rate is computed from the turbulent diffusion coefficient with Eq. (7.39). With $n = 25$ discretization steps there is hardly any difference between the two solutions; thus the discrete model (code in Example 6.19) is a very good approximation of the turbulent plug-flow reactor.

6.7 Sequencing Batch Reactor

The stirred tank reactor with variable volume or more frequently the *sequencing batch reactor* (SBR), has different influent and effluent water flows Q_{in} and Q_{out} and thus a variable volume. Frequently the influent and effluent are subject to periodic, time-controlled variations. During a first period of a cycle the influent may fill the reactor and the water is stored. Afterwards without influent the reaction may proceed as far as required, and in the end the reactor is (partially) emptied. The reactor volume is intensively mixed, so that no gradients of state variables arise (Fig. 6.22).

The material balance for water permits the volume of the reactor to be calculated:

$$\frac{dV \cdot \rho_W}{dt} = Q_{in} \cdot \rho_W - Q_{out} \cdot \rho_W + r_W \cdot V. \tag{6.18}$$

The density ρ_W of the water is constant, and the production of water in the course of the reactions can be neglected ($r_W = 0$). Thus:

$$\frac{dV}{dt} = Q_{in} - Q_{out}. \tag{6.19}$$

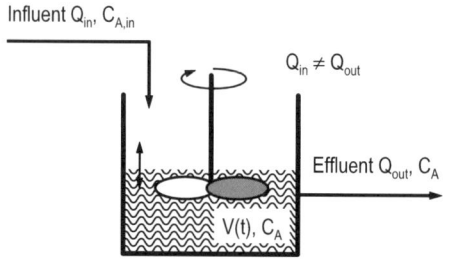

Influent Q_{in}, $C_{A,in}$

$Q_{in} \neq Q_{out}$

Effluent Q_{out}, C_A

$V(t)$, C_A

Characteristics of an SBR:
- influent and effluent may differ
- variable volume
- ideal mixing
- effluent concentration = concentration
  in the reactor
- the steady state corresponds to a CSTR
- open system

Fig. 6.20 Schematic illustration and characteristics of a sequencing batch reactor (SBR)

The material balance equation for material A results in:

$$\frac{dV \cdot C_A}{dt} = V \cdot \frac{dC_A}{dt} + C_A \cdot \frac{dV}{dt} = Q_{in} \cdot C_{A,in} - Q_{out} \cdot C_A + r_A \cdot V ,$$

which combined with Eq. (6.19) yields:

$$\frac{dC_A}{dt} = \frac{Q_{in}}{V(t)} \cdot \left(C_{A,in} - C_A \right) + r_A . \tag{6.20}$$

Although the form of Eq. (6.20) is identical to the balance for a stirred tank reactor (Eq. (6.5)), it can only be solved in combination with Eq. (6.19). In the steady state, Eq. (6.19) vanishes and Eq. (6.20) becomes identical to the balance for the stirred tank reactor.

If the balance equation is to be solved analytically, it is frequently advantageous to follow the change of mass $M_A = V(t) \cdot C_A(t)$ rather than the change of C_A (see Example 6.21):

$$\frac{dM_A}{dt} = Q_{in} \cdot C_{A,in} - Q_{out} \cdot C_A + r_A \cdot V(t) . \tag{6.21}$$

Figure 6.23 shows the course of the concentration of material A, which is degraded in a first-order reaction in an SBR. During the phase with influent, the concentration increases rapidly but does not reach the effluent. Already during filling, degradation and dilution become visible. After the rapid increase of the volume and the concentration at the beginning of a cycle, a reaction time follows with equally rapid degradation of the material A. Once the effluent starts, the reaction has proceeded as far as possible. The efficiency of the reactor is excellent and approximates that of a plug-flow reactor.

Example 6.20: The washing machine as an SBR

An example of an SBR is the washing machine. In sequence, different program steps are processed, which proceed with different filling levels of the machine. The sequence is optimally coordinated with the requirements of the washing process.

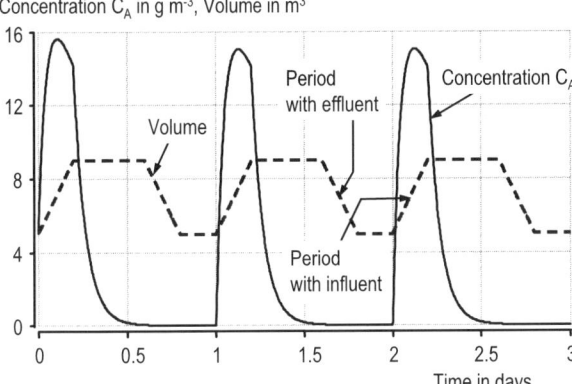

Fig. 6.21 Change of the concentration and the volume in a sequencing batch reactor (SBR) with a first-order degradation process over three cycles (see also Example 6.23)

Example 6.21: Application of the material balance of an SBR

An SBR is observed during a filling procedure with constant influent and without effluent. At the beginning of the cycle the reactor is empty. A first-order degradation reaction proceeds ($r_A = -k \cdot C_A$).
How does the concentration of the material A in the reactor change, if the inlet concentration remains constant?

From Eq. (6.19) follows, with $Q_{out} = 0$ and $V(0) = 0$: $V = Q_{in} \cdot t$

Equation (6.21) becomes $\dfrac{dM_A}{dt} = Q_{in} \cdot C_{A,in} - k \cdot M_A$

with the solution: $-\dfrac{1}{k} \cdot \ln \dfrac{Q_{in} \cdot C_{A,in} - k \cdot M_A(t)}{Q_{in} \cdot C_{A,in} - k \cdot M_A(0)} = t$.

With the conditions $V(0) = 0$ and $V(t) = Q_{in} \cdot t$ this results in:

$$C_A = C_{A,in} \cdot \frac{1 - \exp(-k \cdot t)}{k \cdot t} .$$

The result is shown in Fig. 6.24. The mass M_A increases, until the reaction (degradation) just compensates the influent.

Example 6.22: The activated sludge process in an SBR

The SBR technology is increasingly frequently being used for biological wastewater treatment based on the activated sludge process. Operation is typically in timed cycles. At the beginning of a cycle, the reactor is partly filled with concentrated activated sludge, which is then diluted increasingly with influent, whereby the reactor does not yet have any effluent. The reactor may be equipped with a swimming device for the addition of dissolved oxygen. After the filling period a reac-

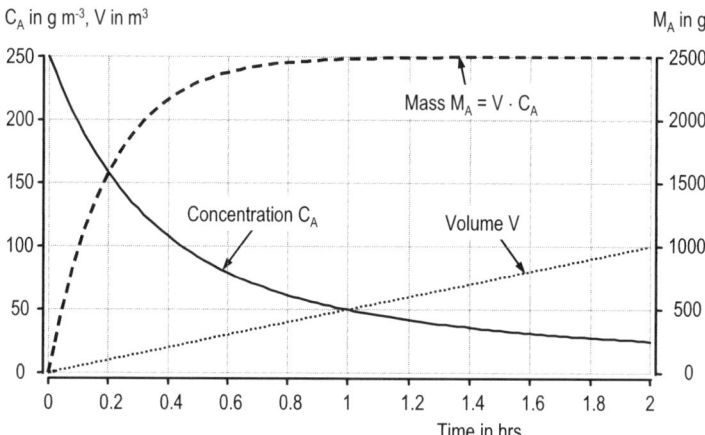

Fig. 6.22 Progress of the volume as well as the concentration and the mass of material A in an SBR with constant influent and a first order degradation reaction (see also Example 6.21 for an analytical solution)

tion period with neither influent nor effluent may follow. Once the degradation of the pollutants has proceeded far enough, the aeration equipment is stopped and, due to the lack of turbulence, the sludge begins to settle. Finally the supernatant (the treated wastewater) is decanted and the concentrated sediment is available for the next cycle.

This type of reactor is particularly applicable if the wastewater is generated inter-mittently, for instance in small municipalities or to deal with waste from indus-tries. Since different environmental conditions may be created in the reactor dur-ing the reaction phase (aerobic, anoxic, anaerobic), the system can be optimized for highly diverse tasks. In the laboratory the reactor is particularly suitable for the investigation of diverse processes and phenomena.

Today SBR technology which is structurally simple to realize, also is increasingly being used for large plants. In this case several such reactors are operated in paral-lel such that it is always possible to accept influent.

Example 6.23: Implementation of an SBR in Berkeley Madonna

The model of an SBR is to be implemented in Berkeley Madonna. The influent and the effluent are not to overlap temporally, and the material in the influent is degraded in a first-order reaction.

Figure 6.23 was computed with the following code:

{Implementation of an SBR, tested}

```
METHOD RK4              ; Integration with fourth-order Runge–Kutta
STARTTIME = 0           ; Begin of the forward integration
STOPTIME=3              ; End of the integration in d
DT = 0.002              ; Time step in d
Vmin = 5                ; minimal volume m³
```

Qin = if mod(time,1) < 0.2 THEN 20 ELSE 0
Qout = if (mod(time,1) > 0.6) AND V > Vmin THEN 20 ELSE 0
                              ; Cyclic formulation of influent and effluent in $m^3\ d^{-1}$
                              ; mod(time,1) = remainder to the division time/1, 1 is
                              the period in d
r = –k*C   ; Degradation rate in $g\ m^{-3}\ d^{-1}$
k = 15   ; Rate constant in d^{-1}
Cin = 100   ; Material concentration in the influent in $g\ m^{-3}$
INIT V = 5   ; Initial volume in m^3
d/dt(V) = Qin-Qout   ; Balance for the water
INIT C = 5   ; Initial value of the concentration in $g\ m^{-3}$
d/dt(C) = Qin*(Cin–C)/V+r ; Balance for the material

6.8 Completely Mixed or Plug-Flow Reactor?

For many degradation reactions plug-flow type reactors result in better performance than the completely mixed (back mixed) stirred reactors. For many processes,
however, mixing is a central element of the process, e.g., with aeration (gas exchange), generation of turbulence for the acceleration of flocculation and precipitation. However, mixing contradicts the characteristics of the plug-flow reactor so
that intermediate solutions with defined mixed ranges (e.g., a cascade of stirred
tank reactors or plug-flow reactors with turbulence) are used. These solutions combine the advantages of mixing with the extra performance offered by plug-flow.

Also recirculation (i.e., return of sludge and internal recirculation in the activated sludge process) leads to back mixing and brings the behavior and performance of a plug-flow reactor closer to those of a mixed system (see Example 6.11).

Back mixed systems are a necessity for autocatalytic processes. Back mixing
brings the catalyst from where it is produced back to the influent of the reactor,
where it is most required. Many microbial processes are autocatalytic, thus completely mixed or back mixed reactors or systems with recirculation flow are typical for biological treatment systems.

6.9 Summary

Table 6.1 summarizes the various characteristics of the ideal reactors discussed
here. The most important differences arise from the degree of mixing: in ideally
mixed reactors and reactor sections no gradients of intensive variables can develop,
the balance equations are ordinary differential equations, which for the steady state
degenerate into algebraic equations. In reactors that are completely mixed only
transverse to the direction of flow, gradients in the direction of flow result: this
leads to balance equations in the form of one-dimensional partial differential equations, which for the steady state simplify to ordinary differential equations.

Chapter 7
Hydraulic Residence Time Distribution

By using the distribution of the hydraulic residence time in a reactor (the residence time distribution, RTD) we characterize the mixing and the internal transport processes in a reactor. The comparison between the theoretically computed and experimentally determined RTD helps us to develop mathematical models of real reactors.

For ideal reactors the flow and mixing conditions are exactly defined, which permits us to deduce theoretically accurate model equations (material balances). However, in real, built reactors these characteristics always deviate to some degree from ideal conditions, e.g., as a consequence of short-circuiting, lack of turbulence, macroscopic internal currents, and stagnant zones.

The amount of mixing in a reactor has a great influence on its performance. In the derivation of a mathematical model for the description of a real reactor the internal transport processes (mixing and advection) must therefore be known as exactly as possible. This allows us to choose models that approximate the behavior of real reactors as closely as possible. Here simple methods are discussed which permit to at least partly characterize these processes. These methods are, however, insufficient for a comprehensive characterization of mixing and internal transport processes. When we map a reactor onto its hydraulic residence time distribution, information that relates to the space time or location where mixing takes place is lost. Combined with the inspection of an existing reactor or based on the draft of the reactor to be built, we succeed, however, in choosing an adequate model for internal mixing processes.

The residence time distribution has an analogy in hydrology: the *unit hydrograph* (more exactly the unit impulse response, Maniak, 1997) is a partial illustration of the discharge behavior of a complex watershed and corresponds, for sufficiently small time steps, to the RTD of the precipitation that is discharged. A unit hydrograph can be assigned to a catchment area, but from a unit hydrograph the exact form of the watershed cannot be reconstructed, as some information is lost. Exactly the same applies to reactors: a residence time distribution can be assigned to them, but the exact geometrical and hydraulic conditions of the reactor cannot be reconstructed from this information.

Example 7.1: The unit hydrograph versus the RTD of a reactor

The unit hydrograph represents a probability distribution of the delay time of a raindrop between falling on the catchment area and leaving this area. The discharge coefficient expresses the fact that only a fraction of the precipitation appears in the effluent; the integral underneath the unit hydrograph is therefore not unity but rather corresponds to the discharge coefficient.

The RTD corresponds to the distribution of the probability of the delay with which a drop in the influent leaves a reactor again. Since each drop in the influent must at some time leave the reactor, the integral under the RTD is unity.

Example 7.2: What information is lost when determining a RTD?

From a unit hydrograph we cannot derive how far above the measuring point two side arms of the rivers flow together. The information about the time of mixing of the water is lost in the experiment. We must reconstruct this information, e. g., from a geographical map.

In the same way from an RTD we cannot derive when the water that flows along different paths through the reactor is mixed. We must derive this information, e. g., from an inspection of the real reactor or a drawing of the planned reactor.

Example 7.3: The residence time distribution at the ski-lift and on a conveyer belt

On a ski-lift, all users spend exactly the same amount of time to overcome a certain elevation difference; the residence time distribution contains exactly one time. The same applies to a conveyer belt that transports gravel. In this case too individual stones spend exactly the same amount of time on the belt. Despite the same distribution of the residence time, we cannot derive the exact form of the transport system: the ski-lift transports discrete packages (e. g., always two persons per handle), while on the conveyer we observe continuous transport.

7.1 RTD: A Spectrum of Retention Times

The hydraulic RTD or spectrum of retention times is a probability distribution that indicates for each possible residence time in the system the probability that this residence time will apply to a specific water particle in the influent.

To determine the exact distribution of all the flow velocities in a reactor system with high spatial and temporal resolution places excessive demands on our resources. In addition we could barely handle the large amount of data collected. However, we can learn a lot if we know for how long an individual small water package ("a molecule") remains in the reactor. Since water is only produced in insignificant quantities in the reactors, only the transport processes affect this time, whereas transformation processes may be excluded. Moreover, we characterize only the steady state, in which the transport processes (mixing and advection) remain unchanged.

Different water packages travel through the reactor along different pathways; in the effluent they therefore have a different age (delay). If we follow a large number of water packages, we can derive a probability for a certain retention time. For all water packages together we can therefore develop a distribution of this probability, as known from statistics: the hydraulic residence time distribution or short RTD is a probability density function.

An example of a residence time distribution $f(\tau)$ is given in Fig. 7.1. The probability p that a water package leaves the reactor after a retention time within the narrow range from τ to $\tau + \Delta\tau$, is:

$$p(\tau, \tau + \Delta\tau) = f(\tau) \cdot \Delta\tau \; [-] \tag{7.1}$$

Density of the residence time f in h^{-1}

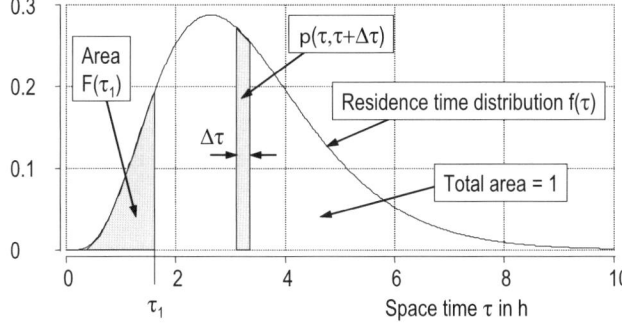

Fig. 7.1   Example of a residence time distribution (RTD)

Cumulative frequency of the residence time F [-]

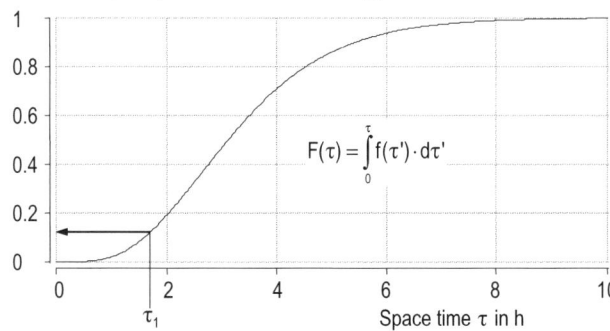

$$F(\tau) = \int_0^\tau f(\tau') \cdot d\tau'$$

Fig. 7.2   Cumulative frequency plot for the residence time distribution in Figure 7.1

$f(\tau)$ = Probability density for the residence time τ $[T^{-1}]$
τ = $t_{out}-t_{in}$, Residence time, delay from the influent to the effluent [T]

The probability that a water package leaves the reactor in the time interval from 0 to τ, i. e., with a retention time of $\leq \tau$, is (Fig. 7.2):

$$F(\tau) = \int_0^\tau f(\tau') \cdot d\tau' \tag{7.2}$$

$F(\tau)$= cumulative frequency of the retention time [–]

The probability that a water package will leave the reactor again (at steady state) is:

$$F(\infty) = \int_0^\infty f(\tau) \cdot d\tau = 1. \tag{7.3}$$

If the internal transport processes (mixing) in the reactor are in steady state, $F(\infty)=1$ must always be reached, frequently, however, only after a long period of time, i. e., asymptotically.

According to Eq. (7.2) we can obtain $f(\tau)$ from $F(\tau)$ as:

$$f(\tau) = \frac{dF}{d\tau}. \tag{7.4}$$

Example 7.4: The ski marathon as an experiment for the determination of the retention time

In the Engadiner Ski Marathon over 10,000 sportsmen and sportswomen participate. They all start at the same time. At the destination the retention time of each individual participant on the course is determined and communicated in the form of a rank. If we determine the number of participants for each minute interval, we receive the spectrum or the probability of the runtimes (we must standardize with the number of participants). If we plot rank divided by the number of participants against the runtime, we obtain the cumulative frequency (see Fig. 7.3).

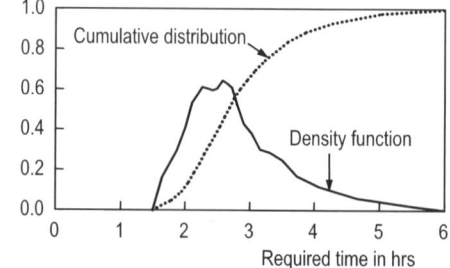

Fig. 7.3 Distribution of the run time in the Engadiner Ski Marathon 2003 (St. Moritz, Switzerland)

7.2 Characterization of Residence Time Distributions

Residence time distributions are frequency distributions. We characterize them with statistical indicators.

In statistics we characterize empirically determined frequency distributions with their average (expected) value, their variance, and if necessary higher moments.

The *zeroth moment* corresponds to the probability that a particle leaves the reactor again; it corresponds to Eq. (7.3).

The *average value* or the *expected value* is the first moment of $f(\tau)$, which is defined as:

$$\tau_m = \int_0^{\infty} \tau \cdot f(\tau) \cdot d\tau . \tag{7.5}$$

The variance (σ^2, second central moment) and the *standard deviation* (σ) of $f(\tau)$ are defined as:

$$\sigma^2 = \int_0^{\infty} (\tau - \tau_m)^2 \cdot f(\tau) \cdot d\tau = \int_0^{\infty} \tau^2 \cdot f(\tau) \cdot d\tau - \tau_m^2 . \tag{7.6}$$

With the aid of the two parameters τ_m and σ^2 a residence time distribution is only partially characterized. Different distributions can result in identical values of these parameters. Using only two parameters leads to a loss of information which can be reduced by using higher moments (skewness, kurtosis, etc.). Here we will only use τ_m and σ.

For the steady state the following (theoretically exact) relation always applies:

$$\tau_m = \frac{V}{Q} \equiv \theta_h \tag{7.7}$$

θ_h = mean hydraulic residence time [T]

This fact supports us in:

- testing observed, measured RTDs $f(\tau)$ for their reliability, or
- determining the water volume V or the flow rate Q for a reactor (see Example 7.5).

τ_m in Eq. (7.7) is based on the integration of $f(\tau)$ for $\tau = 0$ to ∞, see also Eq. (7.5). If $f(\tau)$ is based on experimental observation, then the experiment or the measurements are frequently stopped with $\tau \ll \infty$. The computed volume V will then only partly include the reactor sections with very small flow through (so-called dead volume). This is not a problem in principle, but rather a methodical shortcoming which must, however, be considered.

Example 7.5: Water volume in a trickling filter

A trickling filter with an internal biofilm surface of $A = 100{,}000\ m^2$ is fed with a constant flow rate of $Q = 2000\ m^3\ d^{-1}$. Experimentally $\tau_m = \theta_h = 10$ min is determined.

How much water is contained inside the trickling filter?

How thick is the water film that runs down the trickling filter?

The results will only be approximate, because the water content of the biomass in the biofilm increases the residence time. This water is, however, exchanged only very slowly with the water running down the filter, namely by molecular diffusion, i. e., it behaves nearly as dead volume. The experiment will hardly be long enough, and measurements will barely be sensitive enough to observe this effect.

$$\tau_m = \theta_h = V/Q = 10\ \text{min and therefore } V = \theta_h \cdot Q = 13.9\ m^3.$$

The average thickness of the water film is $V/A = 13.9\ m^3/100{,}000\ m^2 = 0.14$ mm.

In a technical trickling filter this is the simplest method to determine the water volume; which is important in many ways; e. g., the weight of the wet trickling filter in operation depends on this variable.

7.3 Experimental Determination of an RTD

We derive the hydraulic residence time distribution of a real reactor from simple experiments with nonreactive tracer substances.

There are many possible experiments that permit the hydraulic residence time distribution of a real reactor to be derived. Here only the simplest ones are introduced.

Since we cannot follow individual small water packages through a reactor, all the presented procedures are based on the marking of water packages by materials that are easy to measure with sufficient temporal resolution. Since a residence time distribution characterizes only the transport and mixing processes in the reactor, we must select a marking substance (a tracer), which does not change due to reaction and which is subject to the same transport processes as the water itself.

7.3.1 Tracer Substances

Common salt (NaCl), lithium and bromide salts, as well as fluorescent dyes are frequently used as tracers.

With a tracer we want to mark water molecules and follow them through the reactor in order to derive the residence time of the molecule. The requirements that a tracer substance has to fulfill are:

- It must have the same transport characteristics as water (molecular diffusion not too different, no sedimentation or flocculation).
- It must not affect the mixing processes and the flow conditions (temperature, density currents, viscosity of the water).
- It must not react or adsorb onto solids or reactor walls.
- It should be cheap. In technical reactors we frequently use large amounts of tracer material.
- It should have a high solubility in water and must be easy and cheap to analyze accurately in the water.
- It should not be toxic, because it will normally be discharged into the environment (in large amounts).
- The background concentration (the natural concentration in the water) should be small and constant in order not to distort the results of the experiment.

Among others materials the following *tracer substances* are frequently applied:

- NaCl, which can be measured online through the conductivity of the water with high temporal resolution.
- Li^+ and Br^-, which both have small background concentrations and for which reliable analytic methods are available. Br^- has the disadvantage that in the course of ozonation (disinfection of drinking water) the carcinogenic bromate is produced.
- Fluorescent dyes, which can be measured in very small concentrations if they do not adsorb to the solids and variable pH, and temperature levels do not falsify the signal.
- Radioactive or stable isotopes, which can be measured very sensitively. Radioactive tracers can be used with special permission only. Occasionally natural, stable isotopes can be used or radioactive isotopes are set free in an accident.

The choice of the best suited tracer substance depends on the question and on the matrix in which it has to be analyzed (clean or polluted water, online or in the laboratory, in operation, in the presence of high solids concentrations with the danger of adsorption, etc.). The available analytical procedures must ensure a sufficient temporal resolution.

7.3.2 Experimental Procedure

Herein experiments are described based on the addition of a pulse of a tracer substance, on a step change of an influent concentration, or on a random pollutograph which may be naturally observed.

Method 1: Addition of a Pulse of the Tracer Substance (Dirac Pulse)

Method 1 is the most frequently applied method for the characterization of technical systems.

At time t_0 a well-known quantity E_A of the tracer substance A is dosed into the influent of a reactor, which is approximately operated in the steady state. The background concentration $S_{A,0}$ of the tracer is small and known. The added amount of the tracer depends on the sensitivity of the analysis method; it should be large enough such that, e. g., 1% of E_A/V (reactor volume) can still be differentiated from the background concentration $S_{A,0}$. The concentration $S_A(t-t_0)-S_{A,0}$ in the effluent of the reactor is a measure for the fraction of the marked water package, which leaves the reactor with a space time (delay, age) of $\tau = t - t_0$.

The addition of the tracer substance and the measured effluent concentration are plotted in Fig. 7.4. As a check of the measurements it should be examined whether the integral of $Q \cdot (S_A(t-t_0)-S_{A,0})$ over the experimentation time corresponds with the necessary accuracy to the dosed amount of material E_A. Since the integral of $f(\tau)$ over τ from 0 to ∞ is unity, we can obtain $f(\tau)$ from the following standardization (Fig. 7.4):

$$f(\tau) = \frac{Q \cdot \left(S_A(t-t_0) - S_{A,0} \right)}{E_A} \quad \text{with } \tau = t - t_0. \qquad (7.8)$$

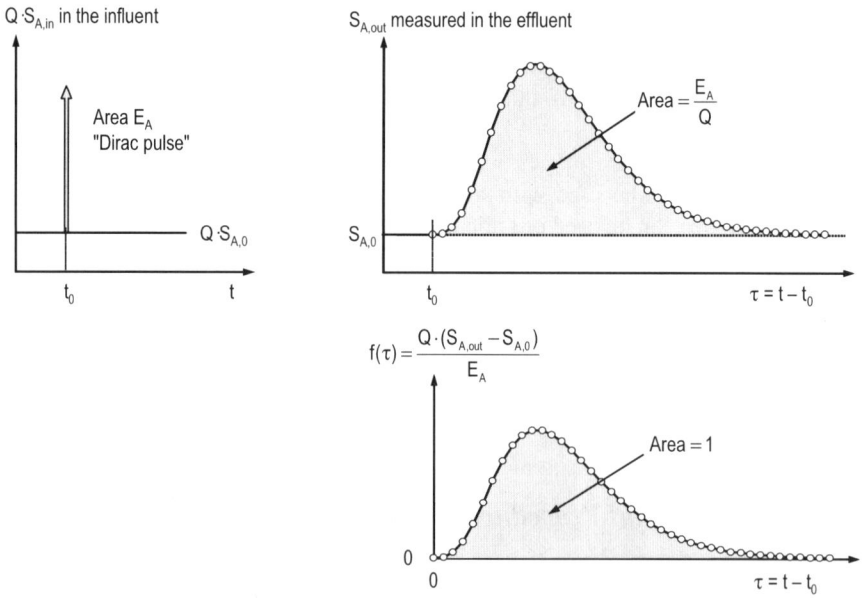

Fig. 7.4 Determination of the residence time distribution from an experiment with the addition of a marking substance to the influent as a pulse at time t_0

From each sampling point we obtain a result for $f(\tau)$ and Eqs. (7.5) and (7.6) can now be used to obtain τ_m and σ^2. Because the area underneath $f(\tau)$ is normalized to unity (Eq. (7.8)) the following equations apply in an experiment with pulse addition of the tracer:

$$E_A = \int_0^\infty Q \cdot (S_A - S_{A,0}) \cdot dt \, , \tag{7.9}$$

$$F(\tau) = \frac{\int_0^\tau Q \cdot (S_A - S_{A,0}) \cdot dt}{E_A} \, , \tag{7.10}$$

$$\tau_m = \frac{\int_0^\infty (t - t_0) \cdot Q \cdot (S_A - S_{A,0}) \cdot dt}{E_A} \, , \tag{7.11}$$

$$\sigma^2 = \frac{\int_0^\infty (t - t_0 - \tau_m)^2 \cdot Q \cdot (S_A - S_{A,0}) \cdot dt}{E_A} = \frac{\int_0^\infty (t - t_0)^2 \cdot Q \cdot (S_A - S_{A,0}) \cdot dt}{E_A} - \tau_m^2 \, . \tag{7.12}$$

Example 8.3 shows a simple code to estimate the three first moments E_A, τ_m, and σ^2 from an observed time series with the aid of Berkeley Madonna.

In biological wastewater treatment processes recirculation is frequently used. This requires special attention in the planning and execution of the experiment. Recirculation can lead to multimodal distributions of the residence time which cause difficulties in the interpretation of the results (Fig. 7.5).

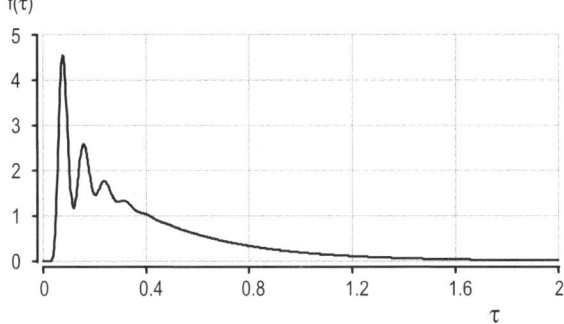

Fig. 7.5 Example of a residence time distribution in a reactor with recirculation. The periodic peak concentrations are a consequence of the feedback loop

Method 2: Step Change of the Inlet Concentration

In the influent to a reactor the concentration of the tracer is subject to a step change at time t_0; this can be achieved, e. g., by dosing and mixing a concentrated, constant mass flow of the tracer substance. In the effluent the concentration of the tracer is then an indication of how much inlet water with elevated tracer concentration is contained in the effluent. Assuming that the inlet concentration increases at time t_0 from $C_1(t < t_0)$ to $C_2(t > t_0)$ according to Fig. 7.6, the cumulative frequency distribution $F(\tau)$ can now be obtained from:

$$F(\tau) = \frac{C_{out}(t - t_0) - C_{out}(t < t_0)}{C_{in}(t > t_0) - C_{in}(t < t_0)} = \frac{C_{out}(\tau) - C_1}{C_2 - C_1}. \tag{7.13}$$

The function $f(\tau)$ arises as a result of numeric differentiation of $F(\tau)$ (see Eq. (7.4)). Unfortunately numeric differentiation is frequently very inaccurate.

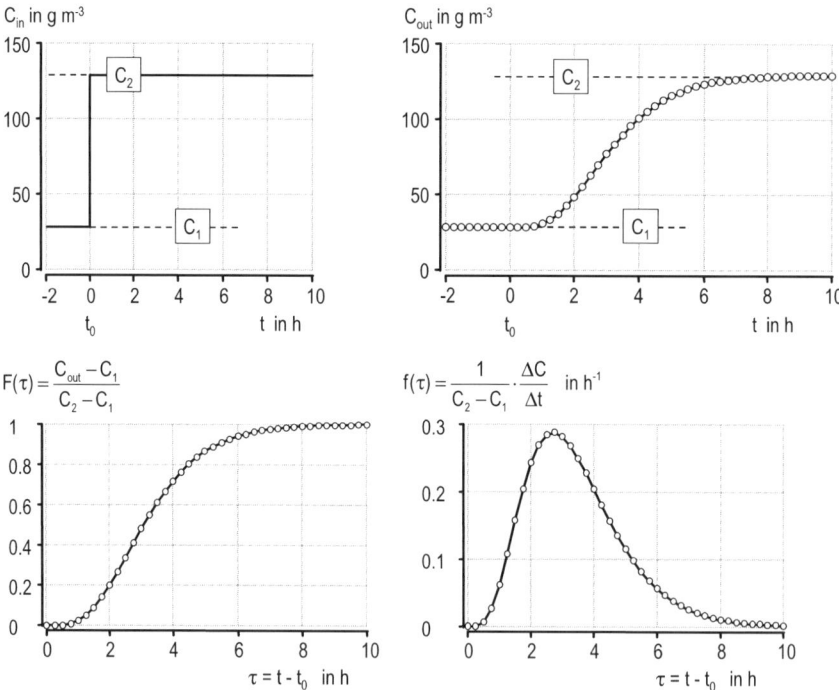

Fig. 7.6 Illustration of the determination of the residence time distribution with method 2: raising the concentration of a tracer substance in the influent from C_1 to C_2 over a longer period of time and then observing the breakthrough of the increased concentration in the effluent followed by standardizing and differentiating the results

Method 3: Use of Naturally Occurring Tracer Substances

This method is theoretically only applicable to systems that are closed for turbulence (see also Sect. 6.6 and Fig. 6.14), i. e., at the sampling points advection must far outweigh dispersion. Therefore, this method is only suitable for the characterization of reactors with narrow influent and effluent pipes or reactors subject to plug flow with small turbulence, in particular sewers and rivers with low dispersion.

In systems that are closed for turbulence, the course of the concentration of a tracer in the effluent $C_{out}(t)$ for any course of the tracer concentration in the influent $C_{in}(t)$ is the result of a convolution which we can compute with the aid of a convolution integral:

$$C_{out}(t) = C_{in} * f = \int_0^t f(t-t') \cdot C_{in}(t') \cdot dt' = \int_0^t C_{in}(t-t') \cdot f(t') \cdot dt'. \qquad (7.14)$$

Convolution shifts and weights with progressing time t' the inlet concentration according to the hydraulic residence time distribution $f(\tau)$, whereby τ is replaced by the time since entrance into the reactor $\tau = t - t'$.

For a convolution integral some interesting mathematical properties apply (the proof may be conducted after Laplace transformation):

$C*f = f*C,$ commutative law

$(C*f_1)*f_2 = C*(f_1*f_2),$ associative law

$\tau_m(C*f) = \tau_m(C) + \tau_m(f),$ (7.15)

$\sigma^2(C*f) = \sigma^2(C) + \sigma^2(f).$ (7.16)

Equations (7.15) and (7.16) show that, for systems that are closed for turbulence, the mean hydraulic residence time τ_m and the variance σ^2 of the signal of a tracer substance in the effluent $C_{out}(t)$ result from the addition of the appropriate values in the influent and the contribution of the hydraulic residence time distribution. Schematically this result is shown in Fig. 7.7.

Equations (7.15) and (7.16) have many applications (see Figs. 7.7 and 7.8):

- They permit the statistical characteristics of the effluent signal ($t_{m,Cout}$ und σ^2_{Cout}) to be calculated from the statistical characteristics of an inlet signal ($t_{m,Cin}$ and σ^2_{Cin}) and the residence time distribution ($\tau_{m,f(\tau)}$ and $\sigma^2_{f(\tau)}$).
- They can be used to compute the statistical characteristics of the RTD of a series of equal reactors from the characteristics of a single reactor.
- They permit the residence time distribution to be characterized based on the characteristics of an influent and an effluent signal.

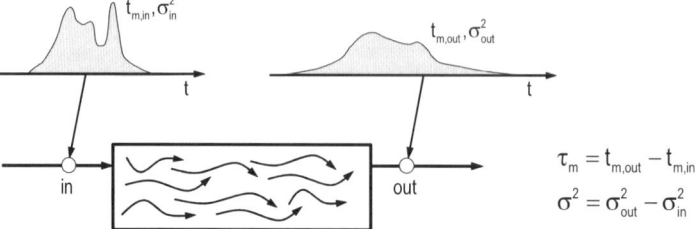

Fig. 7.7 Use of a random (naturally occurring) signal in the input for the estimation of the statistical characteristics of the residence time distribution of a reactor that is closed for turbulence or a section in a river with low dispersion

Fig. 7.8 Schematic representation of the addition of the statistical characteristics of the residence time distribution

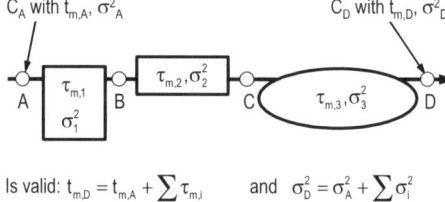

Example 7.6: Convolution with Berkeley Madonna

The convolution integral for the following situation is to be computed:

$$f(\tau) = \frac{1}{\theta_h} \cdot \exp(-\tau/\theta),$$ which is equivalent to $f(\tau)$ for a CSTR.

$$\begin{array}{l} C_{in} = 0 \quad \text{for} \quad t < 0 \\ C_{in} = 1 \quad \text{for} \quad t \geq 0 \end{array}$$, which represents a step change in the influent at time $t = 0$

We expect the cumulative residence time distribution $F(\tau)$ for a CSTR (Fig. 7.9). In Berkeley Madonna we can solve this task with the following code:

```
{Solving a convolution integral, tested}
METHOD RK4                     ; Integration with fourth-order Runge–Kutta
STARTTIME = 0                  ; Convolution starts at t = 0
STOPTIME = 4                   ; Convolution ends with STOPTIME, this is a vari-
                                 able, see below
DT = 0.02                      ; Time step
tPrime = STOPTIME-Time         ; The variable t'
Cin = Step(1,0)                ; Influent concentration, jump at time = 0 from 0 to 1
f = 1/theta*exp(−tPrime/theta) ; The second function with the time t'
theta = 1                      ; Mean hydraulic residence time of the reactor
Init Cout = 0                  ; Initialization of the convolution integral, lower
                                 limit of the integration
d/dt(Cout) = Cin*f             ; Convolution
```

In the computation STOPTIME assumes the role of t in Eq. (7.14) (upper limit of the integration). In Berkeley Madonna we can vary STOPTIME in a parameter plot (menu parameter, e. g., 201 steps between 0 and 4). In a figure of C_{out} against STOPTIME we obtain the result $C_{out}(t-t_0)$.
To obtain the parameter plot, the integral must be computed 201 times, which is not very efficient.

Example 7.7: Addition of the statistical characteristics of an RTD

For a single CSTR is valid: $\tau_m = \theta_h = V/Q$ and $\sigma^2 = \theta_h^2$.
What are the characteristics of a cascade of stirred tank reactors with the same volume, but three reactors in series?
For each individual reactor: $\tau_{m,i} = (V/3)/Q = \tau_m/3$ and $\sigma_i^2 = ((V/3)/Q)^2 = \sigma^2/9$.

For the entire cascade: $\tau_{m,tot} = 3 \cdot \tau_{m,i} = \tau_m$ and $\sigma_{tot}^2 = 3 \cdot \sigma_i^2 = \sigma^2/3$.

Thus, we expect that the residence time distribution for the cascade is distributed more closely around the mean residence time than in a single reactor.

Example 7.8: Accidental discharge into a river

In an accident, a nondegradable material is discharged into a river which has a regular cross section over many kilometers. The exact site of the accident is unknown.
At two sampling stations which lie approximately 10 km apart the pollutograph is determined, and the mean t_m and variance σ^2 of the passing time of the pollutant cloud are determined:

Upstream station $t_{m,u} = 07{:}15$ hrs $\sigma_u^2 = 1 \, h^2$

downstream station $t_{m,d} = 11{:}45$ hrs $\sigma_d^2 = 3 \, h^2$

Where is the most probable site of the accident, if we assume that the pollutant was discharged to the river during only a few minutes?
The distribution of the residence time in the river section between the two sampling points has the following characteristics:

$$\tau_m = t_{m,u} - t_{mo} = 11^{45} - 7^{15} = 4.5 \, h,$$
$$\sigma^2 = \sigma_d^2 - \sigma_u^2 = 2 \, h^2.$$

The variance σ^2 increases proportional to the mean flow time τ_m (addition). Thus:

$$\frac{t_{m,u} - t_{m,accident}}{\sigma_u^2 - \sigma_{accident}^2} = \frac{t_{m,d} - t_{m,u}}{\sigma_d^2 - \sigma_u^2}.$$

With $\sigma_{accident} = 0$ the resulting time is $t_{m,accident} = 05{:}00$ hrs. The accident scene is approximately 2.25 h or 5 km above the upstream sampling point. If the discharge lasted some time ($\sigma_{accident} > 0$), then the introduction site of the accident must lie further down the river.

Method 4: Numerical Identification of the Residence Time Distribution

This method is suitable for the evaluation of an experiment in which we cannot affect the addition of the marking substance; it is mentioned here as a reference to another environmental engineering discipline.

From a measured variable concentration of a tracer substance in the influent and in the effluent of a system (in analogy to method 3), the linear transfer function between effective runoff (influent) and effluent can be determined similarly as one would determine the unit hydrograph in a hydrologic watershed. The unit hydrograph curve or better unit pulse answer (Maniak, 1997) corresponds to the distribution of the residence time, if the correct dimensions of the signals are used.

The numeric, matrix-based methods for the computation of a unit hydrograph curve are documented in many hydrology textbooks. The principle corresponds to the reversal of the convolution (deconvolution).

Method 5: Parameterization and Parameter Identification

Methods 1–4 require that the experiment is run over a rather long period; typically about 99% of the tracer material has to reach the effluent during the experiment. An alternative that allows the experiment to be stopped earlier is to choose a hydraulic model for the reactor. We then obtain either a parameterized analytical equation for the RTD or we implement the model in order to simulate the experiment numerically. We can then adjust the parameters of the model such that the simulation best fits the observed time series (see Sect. 8.3.2). If required, the RTD is then obtained from simulation with the identified model.

7.4 Residence Time Distributions of Ideal Reactors

Our goal is to develop mathematical models of real reactors based on ideal reactors. Apart from the experimental determination of the RTD of real reactors, this requires that we know and characterize the theoretically exactly defined residence time distributions of the ideal reactors.

In ideal reactors the mixing characteristics and flow conditions are mathematically defined; they can therefore not experimentally be determined but must be mathematically derived. The procedure is similar to the experimental process: We add as an initial condition at time $t = 0$ the quantity of $E_A = 1$ (dimensionless) of the tracer substance A and then compute the progress of the concentration of this material in the effluent ($C_{A,out}$). This concentration corresponds to the theoretical residence time distribution $f(\tau)$.

The RTD is only uniquely defined for reactors at a hydraulic steady state (influent = effluent = constant). Thus, only the stirred tank reactor, the cascade of stirred tank reactors and the plug-flow reactor with and without turbulence are discussed here. It is also possible to derive a hydraulic residence time distribution for an SBR, however, with the time of the addition of the tracer substance an additional parameter results.

7.4.1 RTD of a Stirred Tank Reactor (CSTR)

Model assumption: At time t_0 we add to the influent of the stirred tank reactor the quantity of $E_A = 1$ of the tracer material A, which does not have a background concentration and which is immediately distributed over the entire reactor. We derive the initial condition for this experiment with: $S_{A,0} = S_A(t_0) = E_A/V = 1/V$. The probability that a water package already left the reactor after the time t is:

$$F(\tau) = \frac{E_A - V \cdot S_A(t_0 + \tau)}{E_A} = 1 - V \cdot S_A(\tau).$$ (7.17)

(initial amount – remaining amount) / initial amount
From Eq. (7.8) it follows that:

$$f(\tau) = f(t - t_0) = \frac{Q \cdot S_A(t - t_0)}{E_A} = Q \cdot S_A(\tau).$$ (7.18)

Tracer substances are not subject to transformation reactions. In the influent we find the tracer only at time t_0, which is already considered with the initial condition. The balance equation for the tracer substance A has the form:

$$V \cdot \frac{dS_A}{dt} = -Q \cdot S_A.$$ (7.19)

The solution of Eq. (7.19) considering the initial condition $S_{A,0} = 1/V$ results in:

$$S_A = S_{A,0} \cdot \exp\left(-\frac{Q \cdot (t - t_0)}{V}\right) = S_{A,0} \cdot \exp\left(-\frac{t - t_0}{\theta_h}\right) = \frac{1}{V} \cdot \exp\left(-\frac{\tau}{\theta_h}\right),$$

and with Eq. (7.18):

$$f(\tau) = \frac{1}{\theta_h} \cdot \exp\left(-\frac{\tau}{\theta_h}\right),$$ (7.20)

respectively Eq. (7.17):

$$F(\tau) = 1 - \exp(-\tau/\theta_h).$$ (7.21)

The two functions are shown in Fig. 7.9. The statistical characteristics of the function $f(\tau)$ for the stirred tank reactor are:

Mean hydraulic residence time $\tau_m = \theta_h = V/Q$ (7.22)

Standard deviation $\sigma = \theta_h$. (7.23)

Frequency of residence time f(τ) [T⁻¹]

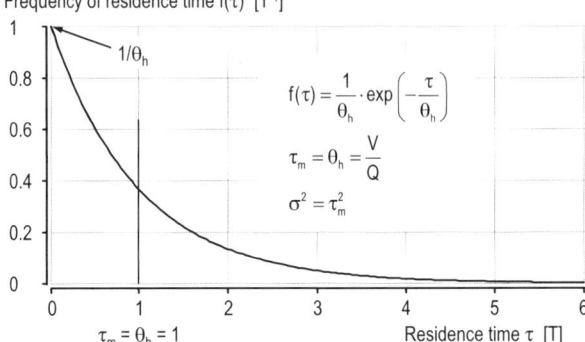

Cumulative frequency of residence time F(τ) [-]

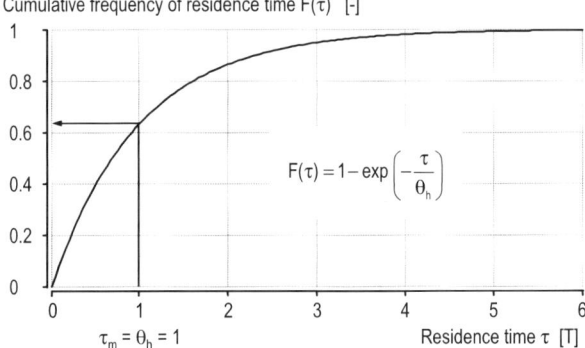

Fig. 7.9 *Above*: residence time distribution f(τ) for a CSTR. *Below*: cumulative residence time distribution F(τ) for the ideal stirred tank reactor with $\theta_h = V/Q = 1$

Example 7.9: Interpretation of an RTD

What is the most probable residence time of a water molecule in a stirred tank reactor?

The function f(τ) has its maximum value at $\tau = 0$, therefore the most probable residence time is $\tau = 0$. Complete mixing leads to an apparent short-circuiting.

Example 7.10: Interpretation of the cumulative residence time distribution

A stirred tank reactor in the steady state has an influent of $100\,m^3\,h^{-1}$.
What size CSTR is required such that 90% of the water remain in the reactor for more than 1 hour?

$F(1\,h) = 1 - 0.9 = 0.1$ (10% of the water has left the reactor after 1 h).

From Eq. (7.21) it follows that $-\tau/\theta_h = \ln(1 - F(\tau))$, and therefore for $\tau = 1\,h$, $\theta_h = 9.5\,h$ or $V = Q \cdot \theta_h = 950\,m^3$.

7.4.2 *Cascade of Stirred Tank Reactors*

The cascade of stirred tank reactors is a frequently used model for reactors whose residence time distribution indicates substantial dispersion or turbulence.

If we apply a similar procedure as for a stirred tank reactor to a series of equal CSTRs with a total volume of $V_{tot} = n \cdot V_i$ the following initial conditions result after adding the tracer substance to the influent:

$$C_{A,1}(0) = \frac{E_A}{V_1} = \frac{E_A \cdot n}{V_{tot}} = \frac{n}{V_{tot}} \quad \text{and} \quad C_{A,2..n}(0) = 0 .$$

The balance equations have the form:

$$\frac{dC_{A,1}}{dt} = -\frac{n}{\theta_h} \cdot C_{A,1} \quad \text{with} \quad \theta_h = \frac{V_{tot}}{Q}$$

$$\frac{dC_{A,i}}{dt} = \frac{n}{\theta_h} \cdot (C_{A,i-1} - C_{A,i}) \quad \text{for} \quad i = 2 \dots n .$$

The solution of this system of n coupled, linear differential equations with the initial conditions stated above exists and may be obtained with the aid of Laplace transformation. It has the form of the Erlang distribution, which consists of the sum of exponentially distributed random numbers. For $E_A = 1$ (dimensionless) this results in $f(\tau)$:

$$f(\tau) = \frac{n}{\theta_h} \cdot \frac{1}{(n-1)!} \cdot \left(\frac{n \cdot \tau}{\theta_h}\right)^{n-1} \cdot \exp\left(-\frac{n \cdot \tau}{\theta_h}\right) . \tag{7.24}$$

The average or expected value τ_m is $\tau_m = \theta_h = \dfrac{V_{tot}}{Q}$. $\tag{7.25}$

The variance of $f(\tau)$ is $\sigma^2 = \dfrac{\theta_h^2}{n}$. $\tag{7.26}$

The maximum value of $f(\tau)$ lies at $\tau(f_{max}) = \theta_h \cdot \dfrac{n-1}{n}$. $\tag{7.27}$

The maximum value of $f(\tau)$ is $f_{max} = \dfrac{n}{\theta_h} \cdot \dfrac{(n-1)^{n-1}}{(n-1)!} \cdot \exp(1-n)$. $\tag{7.28}$

Figure 7.10 shows $f(\tau)$ derived from Eq. (7.24), together with the cumulative frequency $F(\tau)$ for different cascades. For the cumulative frequency a rather complicated, analytical solution is available; Fig. 7.10 is, however, based on the numeric integration of the residence time distribution $f_n(\tau)$ (see Example 7.11). Figure 7.10 reveals that, with an increasing number of reactors, the RTD $f(\tau)$ becomes increasingly tightly distributed about θ_h until finally in the limit $n \to \infty$ no

Frequency of residence time f(τ) [T⁻¹]

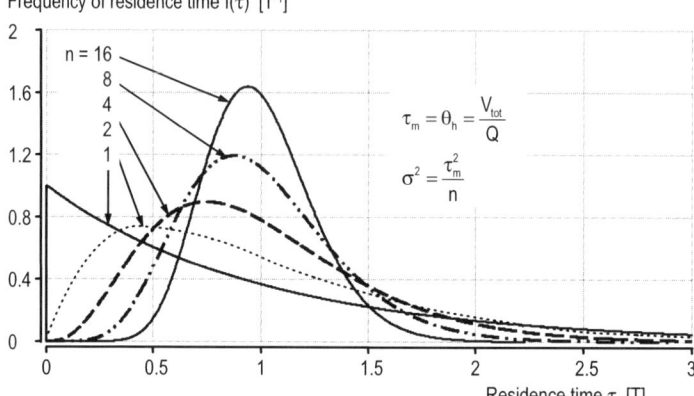

$$\tau_m = \theta_h = \frac{V_{tot}}{Q}$$

$$\sigma^2 = \frac{\tau_m^2}{n}$$

Cumulative frequency of residence time F(τ) [-]

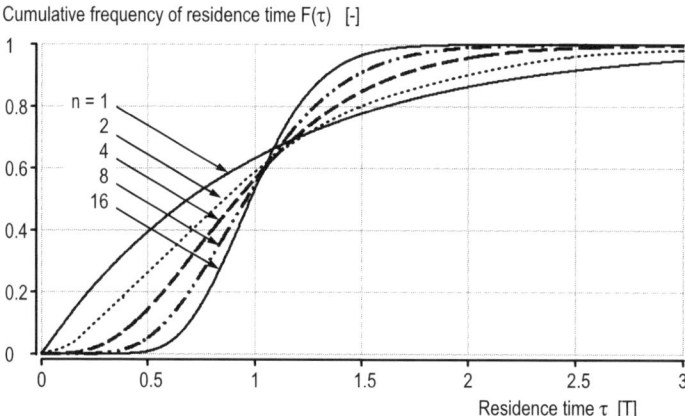

Fig. 7.10 Distribution of the residence times $f(\tau)$ (above) and the cumulative residence time $F(\tau)$ (below) for a cascade of equal stirred tank reactors with constant total volume and a total hydraulic residence time of $\theta_h = 1$. The number of stirred tank reactors is n

deviation (variance) from θ_h remains. In addition, the shape of the residence time distribution approaches a Gaussian normal distribution with increasing n.

The RTD in the effluent of each individual reactor of a cascade is shown in Fig. 7.11. With increasing reactor index the total volume of the system increases. It becomes evident that the maximum value f_{max} of the function $f(\tau)$ is positioned at $\tau_m \cdot (n-1)/n$.

Example 7.11: Simulation of the residence time distribution of a cascade of CSTRs with BM

The following code was used to compute Fig. 7.10:

```
{RTD of a cascade of CSTRs, tested}
METHOD RK4                          ; Integration with fourth-order Runge–Kutta
STARTTIME = 0                       ; Begin of the simulation
```

Frequency of residence time f(τ) [T⁻¹]

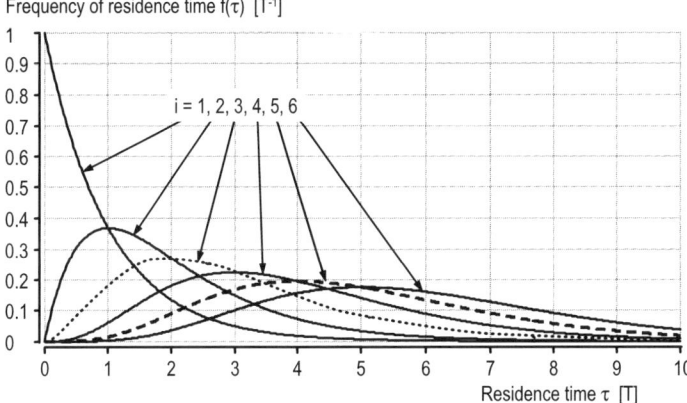

Cumulative frequency of residence time F(τ) [-]

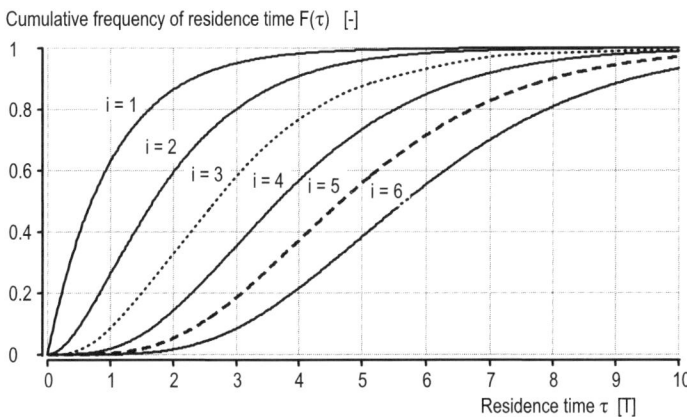

Fig. 7.11 Diagram similar to Fig. 7.10, but for identical volumes of the individual reactors, thus with increasing total volume

```
STOPTIME = 3                    ; End of the simulation
DT = 0.02                       ; Time step d
n = 4                           ; Number of reactors in series
Vtot = 1                        ; Total volume of the cascade m³
Q = 1                           ; Flow rate m³ d⁻¹
th = Vtot/Q                     ; θₕ = τₘ = Residence time in the cascade d
init f[1..n]  = if i = 1 then 1*n/th else 0  ; Initial conditions d⁻¹
d/dt(f[1..n]) = if i = 1 then −f[1]*n/th else (f[i−1]−f[i])*n/th    ; Balance equations
fn = f[n]                       ; Effluent concentration, copy for illustration
init Fkum = 0                   ; Cumulative RTD
d/dt(Fkum) = fn
```

The balance equations for the first and the later reactors are combined into one statement to allow the choice of n = 1.

Example 7.12: Computation of the variance of the residence time distribution of a cascade

For an individual stirred tank reactor Eqs. (7.22) and (7.23) yield $\tau_m = \theta_h$ and $\sigma^2 = \theta_h^2$.

Both the average residence times of a combination of reactors which are closed for turbulence, and the variance of the residence time distributions are additive (Eqs. (7.15) and (7.16)). If we assume that all stirred tank reactors of a cascade have the same volume, then:

$$\tau_{m,1} = \theta_{h,1} = \frac{V_1}{Q} \text{ and } \tau_{m,tot} = \sum_{i=1}^{n} \tau_{m,i} = n \cdot \tau_{m,1} ,$$

$$\sigma_1 = \theta_{h,1} \text{ and } \sigma_{tot}^2 = \sum_{i=1}^{n} \sigma_i^2 = n \cdot \sigma_1^2 = n \cdot \theta_{h,1}^2 = n \cdot \frac{\theta_{h,tot}^2}{n^2} = \frac{\theta_{h,tot}^2}{n} .$$

This result corresponds to Eq. (7.25).

Example 7.13: The secondary clarifier as cascade of stirred tank reactors

Secondary clarifiers in the activated sludge process are divided into different zones, in particular an upper clear water zone and a lower thickening zone, both, however, with undefined volume. With the help of a tracer experiment the residence time distribution of the clear water zone is determined, yielding $\tau_m = 1.5$ h and $\sigma^2 = 1.1$ h^2. During the experiment the clarified effluent was flowing at 500 m^3 h^{-1}. *How large is the volume of the clear water zone? Which cascade of stirred tank reactors could approximately copy the mixing conditions of this zone?*
The equation $\tau_m = \theta_h = V_{Clearwater}/Q_{out}$ results in $V_{Clearwater} = 750$ m^3.
From $\sigma^2 = \theta_h^2 /n$ it results that $n = 2.05 \approx 2$.
Thus, the clear water zone could be modeled, e. g., by using two stirred tank reactors in series with a volume of 375 m^3 each. Of course it is good practice to compare the measured with the theoretical residence time distribution. This may provide ideas on how to improve the model.

7.4.3 Plug-Flow Reactor

For the plug-flow reactor the cumulative frequency of the residence time distribution $F(\tau)$ can be derived intuitively on the basis of the following experiment (method 2, Fig. 7.6): at time t_0 the concentration of the tracer substance in the influent is suddenly increased, and the development of the concentration in the effluent is measured. The results of such an experiment are provided in Fig. 7.12. With Eq. (7.13) $F(\tau)$ may be obtained from this information. Furthermore, $f(\tau)$ can be deduced by taking the differential of $F(\tau)$ according to Eq. (7.4).

In the plug-flow reactor all water molecules have exactly the same residence time $\tau_m = \theta_h = V/Q$. The variance of $f(\tau)$ is therefore $\sigma^2 = 0$.

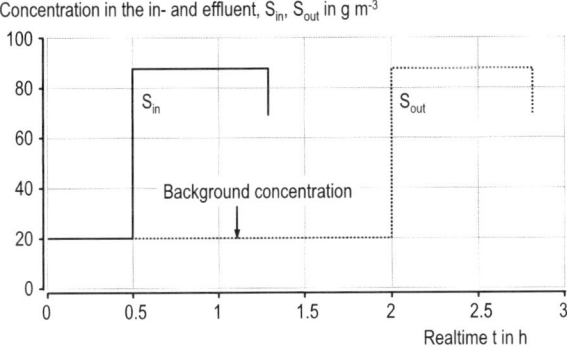

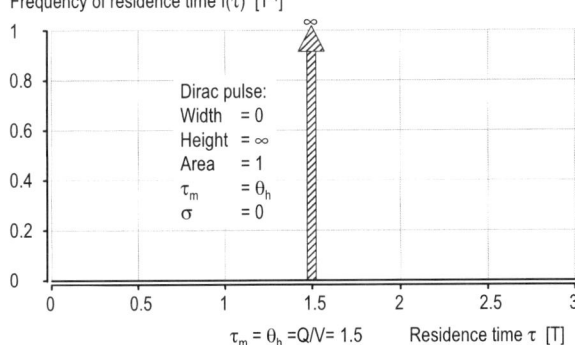

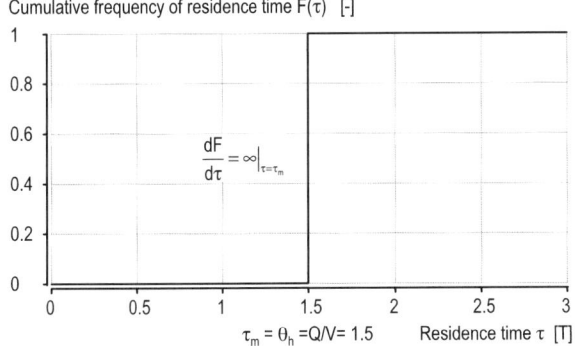

Fig. 7.12 *Upper*: progression of the concentration after a step change in the influent to an ideal plug-flow reactor with $\theta_h = 1.5\,h$. *Centre*: distribution of the residence time of the plug-flow reactor $f(\tau)$. *Lower*: cumulative distribution of the residence time $F(\tau)$

The residence time distribution of a plug-flow reactor does not have any temporal (horizontal) expansion. However, the integral underneath $f(\tau)$ is: $\int f(\tau) \cdot d\tau = 1$. We speak of a Dirac pulse $\delta(\tau_m)$, which manifests itself in the discontinuity or the jump of $F(\tau)$ at $\tau = \theta_h$.

7.4.4 Plug-Flow Reactor with Turbulence

The computation of the residence time distribution of a plug-flow reactor with turbulence requires the solution of a partial differential equation that heavily depends on its boundary conditions. The solution differs quite strongly for reactors which are either open or closed for turbulence.

The balance equation for the tracer which is added at time t_0 to the influent of the reactor, has the following form (Eq. (6.13), without reaction):

$$\frac{\partial C_A}{\partial t} = -u \cdot \frac{\partial C_A}{\partial x} + D_T \cdot \frac{\partial^2 C_A}{\partial x^2} . \tag{7.29}$$

In the plug-flow reactor with turbulence, the distribution of the residence time depends on the type of the sampling points (i. e., the boundary conditions for the integration of the balance equation for the tracer, respectively the transport processes, at the sampling points (see Sect. 6.6 and Fig. 6.14). Levenspiel (1999) provides an in-depth discussion and additional literature.

Multiplying Eq. (7.29) by $\theta_h \equiv \dfrac{V}{Q} = \dfrac{L}{u}$ yields:

$$\theta_h \cdot \frac{\partial C_A}{\partial t} = -L \cdot \frac{\partial C_A}{\partial x} + \frac{L \cdot D_T}{u} \cdot \frac{\partial^2 C_A}{\partial x^2} \tag{7.30}$$

L   = length of the reactor [L]
u   = average flow velocity in a prismatic reactor [L T^{-1}]
D_T = turbulent diffusion coefficient [L^2 T^{-1}]

If $t^* = t/\theta_h$ and $x^* = x/L$ are defined as dimensionless coordinates for time and space, Eq. (7.30) may be written as:

$$\frac{\partial C_A}{\partial t^*} = -\frac{\partial C_A}{\partial x^*} + \frac{D_T}{u \cdot L} \cdot \frac{\partial^2 C_A}{\partial x^{*2}} . \tag{7.31}$$

The remaining parameters are combined in a dimensionless relationship:

$$\frac{D_T}{u \cdot L} = N_T = \text{Turbulence or dispersion number} . \tag{7.32}$$

The dimensionless number N_T characterizes the dispersion or turbulence (D_T/L) relative to the advection (u). It does not have its own name, but its reciprocal value is sometimes called the Peclet number in the literature. Since this name is assigned to another clearly defined dimensionless number, the name *turbulence or dispersion number* is used here for N_T.

In the ideal plug-flow reactor by definition $D_T = 0$ and thus $N_T = 0$. In the stirred tank reactor $D_T = \infty$ and thus $N_T = \infty$. The plug-flow reactor with turbulence lies between these two extremes.

The boundary conditions for the plug-flow reactor with turbulence depend for the influent ($x = 0$) and the effluent ($x = L$) on the local turbulence. For reactors that are closed for turbulence (no back mixing into the influent is possible, see Fig. 6.14), applies for $x < 0\ D_T = 0$ (advection prevails), while for open reactors there is no change of D_T at $x = 0$ (dispersion or back mixing into the influent pipe is possible).

Open Reactors

If we add to an open prismatic reactor at time $t_0 = 0$ at the location $x = 0$ a pulse of a tracer with the mass 1 (a Dirac pulse, $\delta(0)$), then the solution of Eq. (7.29) for a fixed time $\tau_0 = t - t_0$ along the x coordinate is equivalent to a Gaussian normal distribution of the concentration, as presented in Fig. 7.13:

$$f(x) = \frac{1}{\sqrt{4 \cdot \pi \cdot D_T \cdot \tau_0}} \cdot \exp\left(\frac{-(x - u \cdot \tau_0)^2}{4 \cdot D_T \cdot \tau_0}\right) \qquad [L^{-1}] \qquad (7.33)$$

Expected (mean) value of x: $x_m = \tau_0 \cdot u$ $\qquad$ [L]

Variance of f(x): $\qquad \sigma_x^2 = 2 \cdot D_T \cdot \tau_0 = 2 \cdot N_T \cdot L^2$ $[L^2]$

The length profile of the concentration at a certain time, as shown in Fig. 7.13, is difficult to observe: we would have to take many samples along the reactor, all at the same time. The presentation in Fig. 7.13 corresponds to the concept of the *frozen cloud*. Since the width of this cloud is considerable even with a small turbulence number N_T, this normal distribution deforms during the period it runs through a certain location x_0 of the reactor. This deformation has the consequence that the time-dependent concentration curve that passes by a specific location becomes skewed. For open reactors the result is given by Eq. (7.34); examples are shown in Fig. 7.14:

$$f(\tau) = \frac{u}{\sqrt{4 \cdot \pi \cdot D_T \cdot \tau}} \cdot \exp\left(-\frac{(x_0 - u \cdot \tau)^2}{4 \cdot D_T \cdot \tau}\right) \ [T^{-1}] \qquad (7.34)$$

Expected (mean) value of τ: $\tau_m = (1 + N_T) \cdot \dfrac{x_0}{u}$ but $\theta_h = \dfrac{x_0}{u}$ $\qquad (7.35)$

Variance of f(τ): $\qquad \sigma_\tau^2 = \left(2 \cdot N_T + 8 \cdot N_T^2\right) \cdot \left(\dfrac{x_0}{u}\right)^2$ $\qquad (7.36)$

Equations (7.35) and (7.36) were first derived by Levenspiel and Smith (1957).

The larger the flow distance x_0, the smaller becomes N_T, and the RTD f(τ) approaches a normal distribution. For $N_T < 0.01$ we can typically assume a normal distribution.

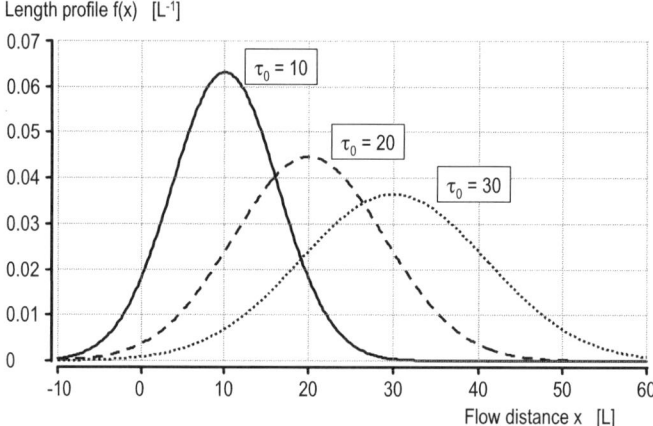

Fig. 7.13 Length profile of the concentration of a tracer which was added at time $t=0$ at $x=0$ as a Dirac pulse $\delta(0)$. The concentrations are given at different times in an open reactor (e. g., a river) as a so-called frozen cloud. Interestingly the tracer mixes above $x=0$

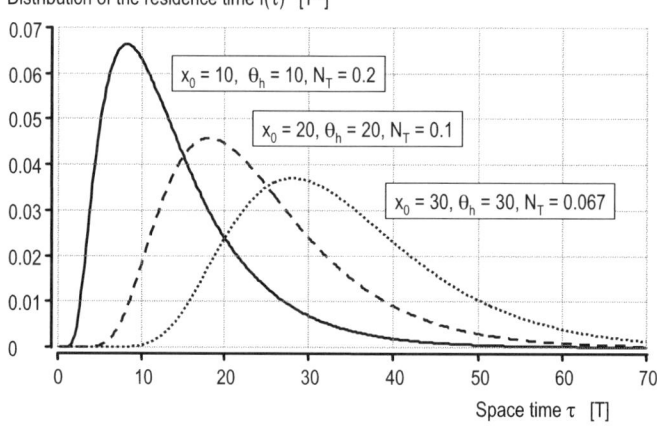

Fig. 7.14 Residence time distribution in an open plug-flow reactor with turbulence. The tracer is added at time $\tau=0$ at $x=0$ and observed at specific locations x_0 over time

   It is interesting that, according to Eq. (7.35), the expected value of the mean delay of the tracer in the reactor τ_m becomes larger than θ_h. The reason is the fact that with the addition of the tracer to an open reactor, turbulence transports a part of the tracer upstream, against the direction of flow. This amount of tracer which leaves the reactor upstream, will reach the lower boundary with an additional delay (see Fig. 7.13 for the amount mixed upstream). The real average residence time between input and output locations remains $\theta_h=x_0/u$. In Fig. 7.15, some individual particles are traced over time and space. Some particles leave and return to the reactor several times before they are finally transported to downstream of the observed system.

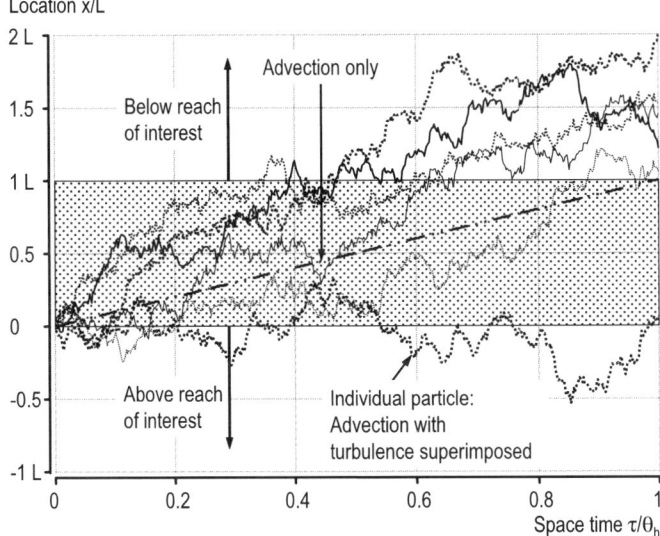

Fig. 7.15 Trace in space and time of six water packages (particles) in an open plug-flow reactor. The particles are partially above the observed upper control point; some leave the system beyond the lower control point only to be brought back by turbulence (simulation based on a random walk)

Closed Reactors

For reactors that are closed for turbulence no analytic solution for the course of the concentration over space or time is available. However we can obtain accurate numeric solutions (Fig. 7.16). In addition van der Laan (1958) showed that for this case the variance can be obtained analytically:

$$\frac{\sigma^2}{\tau_m^2} = 2 \cdot N_T - 2 \cdot N_T^2 \cdot \left(1 - \exp(-1/N_T)\right), \tag{7.37}$$

$$\tau_m = \theta_h = \frac{V}{Q} = \frac{L}{u}. \tag{7.38}$$

Since in a closed reactor no materials can diffuse back into the influent, Eq. (7.38) applies.

The numeric simulation of closed plug-flow reactors with dispersion is introduced in Sect. 7.4.5.

Example 7.14: Dispersion in an activated sludge tank

In activated sludge plants, Murphy and Boyko (1970) experimentally derived for aeration tanks with deep strip aeration the following relationship (units adapted):

$f(\tau) \cdot \theta_h$ [-]

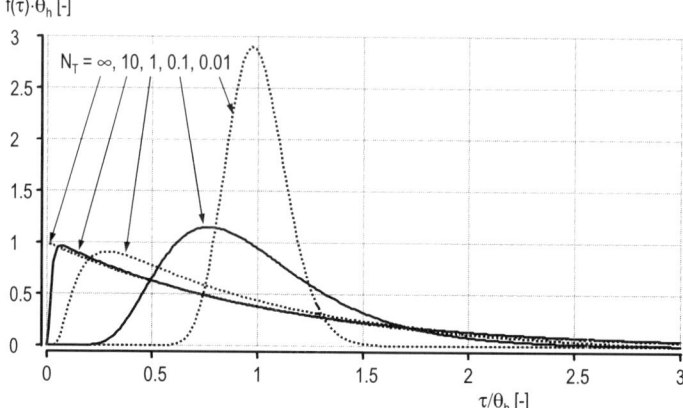

Fig. 7.16 Numerically computed distribution of the residence time in the closed plug-flow reactor with turbulence as a function of N_T, cf. Figure 7.17. The expected value $\tau_m = \theta_h$

$f(\tau) \cdot \theta_h$ [-]

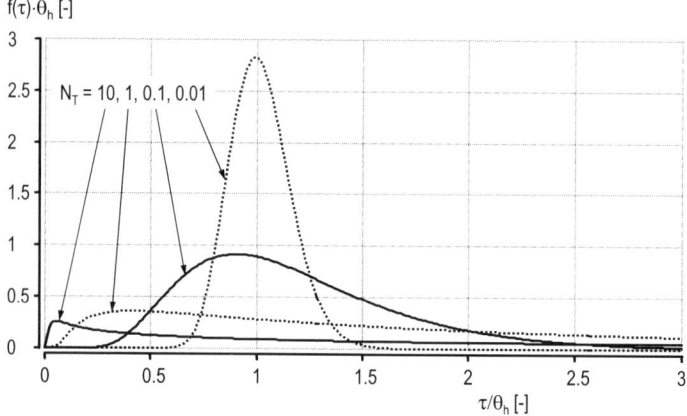

Fig. 7.17 Delay time (not residence time distribution) of a pulse in an open plug-flow reactor with turbulence (analytic solution). The expected value of τ is larger than θ_h (see Eq. (7.35))

$D_T = 8.21 \cdot B^2 \cdot A^{0.346}$ in $m^2 h^{-1}$

B = width of the tank in m

A = aeration rate in m^3 air m^{-3} tank h^{-1}

An aeration tank is 4 m wide, 4 m deep, and $2 \cdot 25$ m long (800 m^3 volume). The aeration rate is $A = 2 \ m^3 \ m^{-3} \ h^{-1}$. The total water flow rate (including return sludge) is 800 $m^3 \ h^{-1}$.

How can the tank be modeled as a cascade of stirred tank reactors?

$D_T = 167 \ m^2 h^{-1}$ $u = Q/(\text{width} \cdot \text{depth}) = 50 \ m \ h^{-1}$

$N_T = D_T/u \cdot L = 0.067$ $\theta_h = V/Q = 1 \ h$

From Eq. (7.37) we obtain $\sigma^2/\theta_h^2 = 0.130$.

For the cascade of stirred tank reactors Eq. (7.26) applies: $\sigma^2/\theta_h^2 = 1/n$, and therefore

$\quad$ n = 1/0.130 = 7.7.

Thus, seven or eight stirred tank reactors in series would have a similar residence time distribution and thus comparable transport processes. Per half reactor, four CSTRs in series could be used as a model. D_T can be used in order to characterize the transport processes, e. g., if samples are to be drawn within the reactor (see Example 6.16).

7.4.5 Numeric Simulation of Turbulence in a Plug-Flow Reactor

The model of the turbulent plug-flow reactor is frequently applied to rivers, sewers, and plug-flow type reactors (tanks). The application of this model typically requires numeric simulation.

The mass balance for the turbulent plug-flow reactor is a partial differential equation which must be discretized for numeric solution. Figure 7.18 shows a possibility for this discretization, in which we integrate the behavior of the discrete subsystems over time. To model the turbulence in subsections by back mixing R, it is necessary to link the geometric variables of the discrete model with the model of turbulent transport.

$\quad$ Equation (4.18) models mass transfer as a consequence of dispersion or turbulence as follows:

$$j_T \cdot A = -D_T \cdot \frac{\partial C}{\partial x} \cdot A \approx -D_T \cdot \frac{\Delta C}{\Delta x} \cdot A \; .$$

With the geometric definitions from Figure 7.18 the numeric model leads to:

$$R \cdot (C_{i-1} - C_i) = -R \cdot \Delta C = -D_T \cdot \frac{\Delta C}{\Delta x} \cdot A = -D_T \cdot \frac{n \cdot \Delta C}{L} \cdot A \text{ and } D_T = \frac{R \cdot L}{A \cdot n} \; .$$

$$(7.39)$$

Δx $\;=$ L/n = *length* of a single discrete element [L]
n $\quad$ = number of elements or partial reactors [–] (discretization steps)

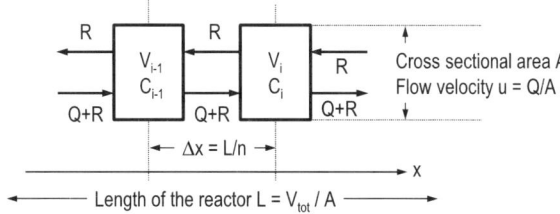

Fig. 7.18 Geometric definitions in the discrete model of a prismatic turbulent plug-flow reactor

Thus, the turbulence number N_T becomes with $Q = A \cdot u$:

$$N_T = \frac{D_T}{u \cdot L} = \frac{R}{n \cdot Q}.$$

(7.40)

Equation (7.40) yields the required internal back mixing R that is necessary to model the transport of material as a consequence of turbulence or dispersion. However the cascade of stirred tank reactors which is overlaid adds additional dispersion and causes an increase of the variance of the residence time distribution. Dispersion D_T is additive; we can therefore first subtract the contribution to dispersion from the cascade ($N_{T,C}$) and afterwards predict the necessary recirculation rate R from the residual $N_{T,R} = N_T - N_{T,C}$. Figure 7.19 indicates how increasing discretization and therefore increasing cost of computation reduces the contribution of $N_{T,C}$ to the total N_T. Figure 7.19 and Table 7.1 are based on Eqs. (7.26) and (7.37).

The following sequence of computations results (see Example 7.17):

- In an experiment or from the literature we determine the residence time distribution $f(\tau)$ or the turbulent diffusion coefficient D_T in the reactor.
- With Eq. (7.37) or (7.32) we compute the turbulence number $N_{T,total}$ for the reactor.

Table 7.1 Equivalent turbulence number $N_{T,C}$ of cascades of CSTRs with n reactors

n	$N_{T,C}$	n	$N_{T,C}$	n	$N_{T,C}$
1	∞	6	0.092	15	0.035
2	0.391	7	0.077	20	0.026
3	0.211	8	0.067	25	0.020
4	0.146	9	0.059	50	0.010
5	0.113	10	0.053	100	0.005

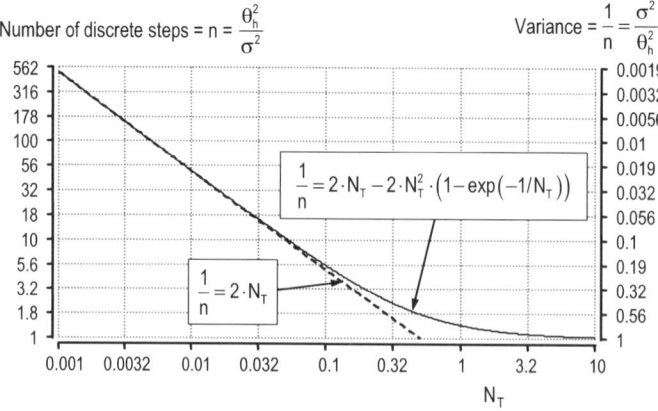

Fig. 7.19 Relationship between the turbulence number N_T and the number of stirred tank reactors n in a cascade closed for turbulence (Eqs. (7.26) and (7.37))

- We select a number of discretization steps n which suits our requirements. We then determine the turbulence number of $N_{T,C}$ for the associated closed cascade (Table 7.1).
- With Eq. (7.40) we determine the required internal recirculation R from the difference $N_{T,R} = N_{T,total} - N_{T,C}$. If $N_{T,R}$ is negative we must increase the number of discretization steps or else we will always simulate based on too high a dispersion rate.

Thus, we determined all necessary elements for the discretization of the turbulent plug-flow reactor. We must adapt the recirculation to the different operating conditions (aeration intensity, flow rate, external recirculation) and compile the appropriate model equations. In Example 7.15 the code for the simulation of the residence time distribution of a turbulent plug-flow reactor is given. It can easily be extended to include nonstationary flow or overlaid transformation processes.

Example 7.15: Code for the simulation of a closed turbulent plug-flow reactor

The following code simulates the residence time distribution of a turbulent plug-flow reactor. The three different balance equations are required in order to correctly include the boundary conditions.

{ Simulation of closed turbulent plug-flow reactor, tested }

```
METHOD RK4          ; Integration with fourth-order Runge–Kutta
STARTTIME = 0       ; Beginning of simulation
STOPTIME = 2        ; End of simulation h
DT = 0.002          ; Time step h
n = 25              ; Number of discretization steps
Q = 2500            ; Influent and effluent flow m³ h⁻¹
Vtot = 2500         ; Volume of the reactor m³
V = Vtot / n        ; Volume of a discrete element m³
R = 4500            ; Internal recirculation, turbulence m³ h⁻¹
Cin = 0
init C[1] = 1*Q/V   ; Initial condition for the first element, h⁻¹
init C[2..n] = 0    ; Initial conditions for elements 2.. n
d/dt(C[1]) = (Q*Cin + R*C[2]–(Q+R)*C[1])/V
                    ; Balance for the first element, bound. cond.
d/dt(C[2..n–1]) = ((Q+R)*(C[i–1]–C[i]) + R*(C[i+1]–C[i]))/V
                    ; Balance for elements 2..n–1
d/dt(C[n]) = (Q+R)*(C[i–1]–C[i])/V   ; Balance for the last element, bound. cond.
fn = C[n]           ; Effluent concentration f(t) h⁻¹
```

Example 7.16: Numeric simulation of an open plug-flow reactor

The dynamic simulation of an open plug-flow reactor is difficult because the boundary conditions at the upper and lower end are obtained only with difficulty. We can manage by modeling a closed reactor which is extended depending upon turbulence number N_T above and below the system of interest (see Fig. 6.17). The larger N_T, the larger becomes the necessary extension. For the determination of the required internal recirculation R Fig. 7.19, Table 7.1, and the relevant equations are sufficient if $n > 10$.

Example 7.17: Turbulent diffusion in a cascade of stirred tank reactors

A turbulent plug-flow reactor has a length of $L = 50\,m$ and a cross section of $A = 50\,m^2$. The flow is $Q = 2500\,m^3\,h^{-1}$. An experimentally determined residence time distribution indicates that you can model this reactor with a cascade of six equal reactors.
How large is the turbulent diffusion coefficient D_T in this reactor?
From Table 7.1 a value of $N_T = 0.092$ results for $n = 6$, and from Eq. (7.32) we obtain:

$$D_T \approx N_T \cdot u \cdot L = 229\,m^2\,h^{-1} \text{ (see also Example 7.14).}$$

You want to model the reactor as a turbulent plug-flow reactor and solve the balance equation for 25 discrete nodes. *How large will you have to choose the internal recirculation R?*
The fraction of turbulence provided by the cascade is: $n = 25$, $N_{T,C} = 0.02$. The recirculation must thus provide $N_{T,R} = N_T - N_{T,C} = 0.092 - 0.02 = 0.072$. From Eq. (7.39) you obtain

$$R = N_{T,R} \cdot n \cdot Q = 4500\,m^3\,h^{-1}.$$

Figure 7.20 compares the residence time distribution $f(\tau)$ of the two models. They clearly differ, although both distributions exhibit the same variance σ^2. While in the cascade of CSTRs the small residence times τ result in an excess of variance, whereas a prolonged tailing (delayed concentration) in the reactor with internal recirculation leads to a late compensation of the variance. This difference can have large effects on the performance of the reactor: in a first-order disinfection reac-

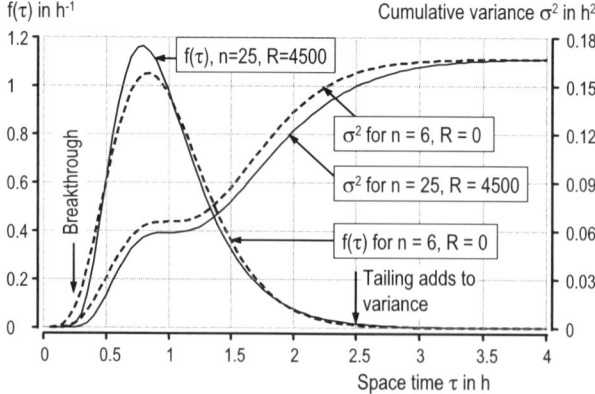

Fig. 7.20 Comparison of the residence time distribution $f(\tau)$ and the cumulative variance $\sigma^2(f(\tau))$ for the two reactor systems in Example 7.17

tion, the ideal turbulent reactor would have a degradation of $C_{out}/C_{in} = 1 \cdot 10^{-8}$, while the cascade of stirred tank reactors would only reach $C_{out}/C_{in} = 1.4 \cdot 10^{-6}$.

7.5 Reactor Combinations

For subsystems that are closed for turbulence, expected values τ_m and variances σ^2 of the residence time distribution are additive (Fig. 7.8). We can use this fact to assign the statistical characteristics of the residence time distribution to individual parts of the reactor system.

We frequently make experiments which permit at the same time to characterize several reactors in series. However, the inlet signal to a later reactor is distorted, as it is already affected by earlier reactors. If the individual subsystems are closed for turbulence, we can derive the statistical characteristics of the residence time distribution of a subsystem (expected value τ_m and variance σ^2) with the help of Eqs. (7.15) and (7.16) (see also Figs. 7.7 and 7.8):

$$\tau_{m,Reactor} = \tau_{m,Effluentsignal} - \tau_{m,Influentsignal}, \tag{7.41}$$

$$\sigma^2_{Reactor} = \sigma^2_{Effluentsignal} - \sigma^2_{Influentsignal}. \tag{7.42}$$

These statistical characteristics may then be used to find an ideal reactor that has similar properties.

7.6 RTD with Stochastic Models

Alternatively to the use of deterministic models for the derivation of the residence time distribution, we can also make use of stochastic models. We model the residence time distribution of particles in a system by modeling possible pathways of individual particles through the reactor and evaluating the results statistically. This possibility is introduced here in two case studies.

Stochastic models do not primarily serve the determination of the residence time distribution, but they are used if special problems relating to the behavior of individual particles are to be described. With the programs introduced here, we simulate the behavior of a large number of particles and follow their possible paths through the system. We then analyze in the effluent how long these particles remained in the system. Such models can easily be enhanced to include the interaction of the particles with their environment, which then allows to make predictions of system performance. Gujer and von Gunten (2003) give an example of such an application for the disinfection of drinking water.

7.6.1 Stochastic Model of a Cascade of Stirred Tank Reactors

A particle leaves a stirred tank reactor in the effluent within a small time step Δt with the probability $p = Q \cdot \Delta t / V$ (see Figs. 7.1 and 7.9). We use this relation to develop a stochastic model of a cascade of CSTRs.

In order to obtain the cumulative residence time distribution of a cascade of stirred tank reactors, we add at time $t = 0$ a large number n_P of particles to the influent of the cascade. For each time step we decide for each particle whether it is washed into the next or it remains in the same reactor. The appropriate probability amounts to:

$$p_i = \Delta t \cdot \frac{Q}{V_i} . \tag{7.43}$$

The use of Eq. (7.43) requires that Δt is small enough such that $p_i < 0.01$ or else we risk numeric inaccuracy. Such small time steps ensure that p_i remains nearly constant during the time step Δt (in reality p_i decreases exponentially as $1 - \exp(p_i \cdot t)$ and Eq. (7.43) is only a linearized form of this decrease).

Table 7.2 introduces a code that allows a large number of particles to be followed through the cascade (here $n_P = 10{,}000$). In line 7, p is computed by using Eq. (7.43). In line 10, which applies to $t = 0$, all particles are dosed into the first reactor compartment. In line 11 the particles are pushed through the cascade. In line 12 all particles which have already left the reactor are marked and in line 13 the particles in the effluent are counted and standardized to unity. The result is the cumulative frequency of the residence time $F(\tau)$. In line 14 the derivative of $F(\tau)$ is formed in order to obtain $f(\tau)$ from Eq. (7.4). While $F(\tau)$ does not differ significantly from an analytical solution with $n_P = 10{,}000$, the numeric differentiation requires a large number of particles ($n_P = 100{,}000$) or else the result must be smoothed statistically.

Table 7.2 Code for the implementation of a stochastic model of a cascade of stirred tank reactors in Berkeley Madonna (tested)

1: STARTTIME = 0	; Beginning of the simulation
2: STOPTIME = 2	; End of the simulation
3: DT = 0.001	; Time step, must be small such that $p < 1\%$
4: nR = 5	; Number of reactors
5: Vtot = 1	; Total volume m^3
6: V = Vtot / nR	; Volume of a single reactor compartment
7: p = DT*Q/V	; Probability that a particle leaves the reactor compartment in the time step DT, ; must be smaller than 0.01
8: Q = 2	; Flow rate
9: nP = 10000	; Number of particles

10: init loc[1..nP] = 1   ; Location of particles at time t (reactor comp.)
11: next loc[1..nP] = if random(0,1) < p then loc[i] + 1 else loc[i]
12: effluent[1..nP] = if loc[i] > nR then 1 else 0
                        ; Marking of the particles in the effluent
13: InEffluent = arraysum(effluent[*])/nP
                        ; Counting the particles already in the effluent
14: f = (InEffluent-delay(InEffluent,DT))/(DT)
                        ; numeric derivative, requires nP > 100,000

Example 7.18: Recirculation in a stochastic model

How can a recirculation from the last to the first compartment be inserted into the model in Table 7.2?
Due to the recirculation the probability p on line 7 increases to $p = DT*(Q+R)/V$. This may require a reduction of the time step DT (line 3).
From the last reactor there is the probability $R/(Q+R)$ that the particles are recycled into the first reactor and $1-R/(R+Q)$ that it is discharged into the effluent. Line 11 must therefore be replaced by:
next loc[1..nP] = if loc[i] < nR then if random(0,1) < p then loc[i] + 1 else loc[i] else if loc[i] = nR then if random(0,1) < p then if random(0,1) < R/(Q+R) then 1 else nR + 1 else loc[i] else loc[i]
The nested logical decisions are difficult to read but required by Berkeley Madonna.

7.6.2 Stochastic Model of Turbulence

Contrary to completely mixed reactors with infinitely large turbulence, we must follow the individual particles if turbulence is finite. The model to be used is based on a random walk of the individual particles (see Sect. 4.2.3).

In order to obtain the cumulative residence time distribution of a turbulent plug-flow reactor, we add at time $t = 0$ a large number of particles to the head end of the reactor and then follow these particles on their way through the reactor. Their path is subject on the one hand to the flow (advection) and on the other hand to the turbulence which is modeled as a random walk. If we count after each time step, the number of particles that have already left the reactor, we can derive the cumulative residence time distribution. The boundary conditions are critical: a random walk can move the individual particles against the direction of flow. Since in the closed reactor the turbulence does not reach into the influent or effluent pipe, we must in each time step examine whether the particles are located correctly.

Table 7.3 provides a possible code for the simulation of the transport of the chosen $n_P = 10,000$ particles through a turbulent plug-flow reactor. The time step Δt (line 3) must be small, such that a particle can be transported through the reactor only as a consequence of many single steps. With the following limits, approximately 100 time steps are ensured:

$$\Delta t < 0.01 \cdot \frac{L}{u} \quad \text{and} \quad \Delta t < 0.0001 \cdot \frac{L^2}{2 \cdot D_T} \tag{7.44}$$

L = length of the reactor [L]
u = average flow velocity [L T^{-1}]
D_T = turbulent diffusion coefficient [L^2 T^{-1}]

We design the random walk with Eq. (4.9) by specifying the standard deviation of the normally distributed random step (line 9). For each particle i we follow the momentary location x(i) (line 12 at time t=0 and line 13 for t>0). Line 11 first computes a new location for all particles independent of the boundary conditions. In line 13 the particles are not allowed to diffuse back into the feed line (x<0) or back into the reactor from the effluent pipe (x>L). In line 14 all particles which have already left the reactor are marked and in line 15 they are counted and normalized to unity, by dividing by n_P. The result is the cumulative distribution of the residence time F(τ). In line 16 the derivative of F(τ) results in the residence time distribution f(τ) according to Eq. (7.4). The numeric derivative requires a large number of particles, which can be simulated in Berkley Madonna, e.g., in a batch run or else the result must be smoothed.

Table 7.3 Code for the implementation of a stochastic model of a turbulent plug-flow reactor based on a random walk in Berkeley Madonna (tested).

1: STARTTIME=0	; Beginning of the simulation
2: STOPTIME=1	; End of the simulation
3: DT=0.001	; Time step must be small, see Eq. (7.44)
4: L=10	; Length of the reactor m
5: A=1	; Cross section of the reactor m^2
6: Q=20	; Flow rate m^3 d^{-1}
7: u=Q/A	; Flow velocity, advection m d^{-1}
8: D=5	; Dispersion coefficient m^2 d^{-1}
9: Sig=SQRT(2*D*DT)	; Standard deviation of the random step, Eq. (4.9) m
10: nP=10000	; Number of particles
11: xnew[1..nP]=x[i]+normal(u*DT,Sig)	
	; new location without boundary conditions m
12: init x[1..nP]=0	; initial location of particles with boundary conditions m
13: next x[1..nP]=if xnew[i]<0 then X[i] else if x[i]>L then x[i] else xnew[i]	
	; Examination of the boundary conditions
14: Effluent[1..nP]=if x[i]>L then 1 else 0	
	; Marking the particles in the effluent
15: InEffluent = arraysum(Effluent[*])/nP ; Counting the particles in the effluent	
16: f = (InEffluent-delay(InEffluent,DT))/(DT)	
	; numeric derivative, requires NP>100,000

Fig. 7.21 Comparison of
the computed cumulative
distribution of the residence
time $F(\tau)$ from the deterministic and a stochastic
model. The stochastic model
approaches the deterministic
result as the number of
particles is increased. With
10,000 particles no difference is visible anymore

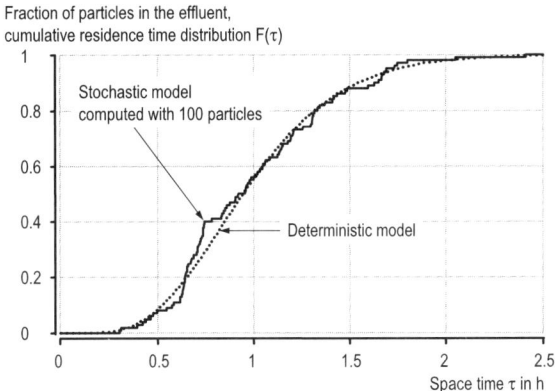

In Figure 7.21 the result of a stochastic simulation is compared with the deterministic solution. The graph is based on only 100 particles, with 10,000 particles
the difference to an analytical solution nearly disappears.

Chapter 8
Modeling of Real Reactors

The behavior of real, built reactors differs from the mathematically computed behavior of the ideal models. In reality, however, the real reactor and not the model is of interest to us. In the development of a model of a real system, our goal is to derive a mathematical construct that allows us to answer our questions as well as possible (good enough) at small expense. Observations and data from operation, together with the residence time distribution and related experiments on the one hand and the ideal reactors as components on the other hand, are the basis for the compilation of the mathematical models.

8.1 Goal

When modeling a reactor we pursue the goal of compiling at small expense a mathematical model that we can examine in detail in place of the real reactor and which gives us reliable answers to our questions.

In modeling technical systems, we achieve our goal, by combining our understanding of the behavior of ideal as well as real reactors. In doing this, we must keep in mind from the beginning the expenditure of later necessary numeric computations and for obtaining absolute values for model parameters.

While it is possible to discuss and analyze ideal reactors in all details, we must make do with real reactors with limited information. Planned reactors do not exist yet at all; built reactors must be characterized with expensive experiments. Apart from the experiments we must always rely on attentive observation, because different models can frequently cope well with experimental results, but may fail as soon as new (transformation) processes are introduced. The next section points out some aspects that require special attention.

Modeling of a real reactor is an art as well as a science. The employment of the best method is based on scientific training and experience. Thus, no generally accepted methods can be introduced here; at best it is possible to identify what is important and to introduce case studies that demonstrate a possible procedure.

8.2 Time of Mixing

The point in a reactor at which old water is mixed with younger water is called the time of mixing. In a plug-flow reactor no such mixing takes place; in a stirred tank reactor the influent is immediately mixed with all water in the system.

Figure 8.1 shows three reactor systems, which all have the same residence time distribution as presented in Fig. 8.2. The difference in the three systems lies in the time of the mixing, which obviously cannot be identified with the help of the residence time distribution. The time of mixing has a large effect on the length profile of a material which is subject to a reaction. Figure 8.3 represents exemplarily for a first-order reaction, how the concentration decreases along the three reactor chains. Since the time of the mixing cannot be obtained from simple tracer experiments, we must obtain this information from observation, or in the case of planned reactors, from our experience or expectation.

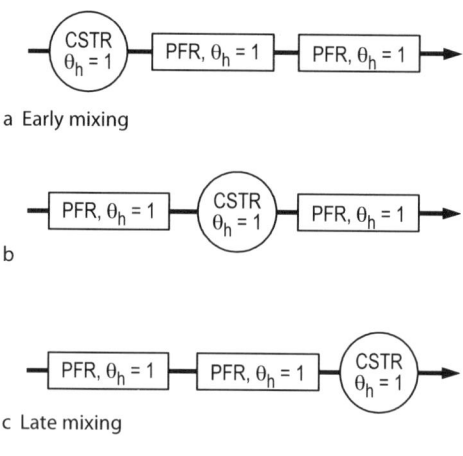

Fig. 8.1 Different combinations of one stirred tank reactor (CSTR) and two plug-flow reactors (PFR)

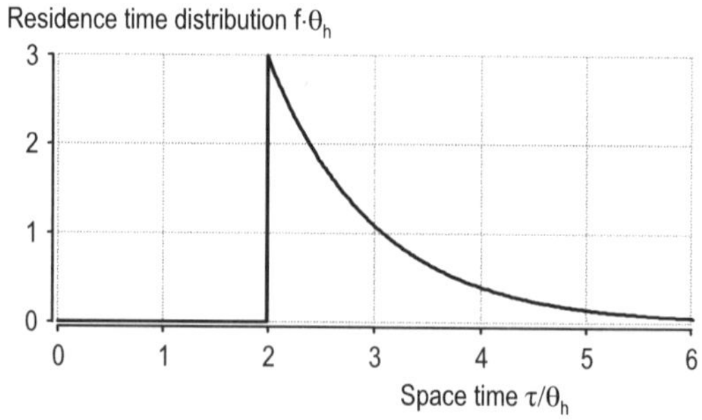

Fig. 8.2 Common distribution of the residence time for the three reactor combinations in Fig. 8.1

For first-order reactions the time of the mixing does not affect the performance of the reactor, for all other reactions and for reactions with several educts, as is the case, e. g., in biological wastewater treatment, this time is decisive.

Example 8.1: Effect of time of mixing on reactor performance

The time of mixing affects the performance of a reactor. Early mixing is beneficial for reaction orders smaller than one whereas higher reaction orders benefit from late mixing. The reactor scheme in Fig. 8.4 allows shifting the dominant mixing process along the length of the reactor.

Figure 8.5 shows the predicted effluent pollutant concentration depending on the location of the large reactor compartment. Clearly the performance of the reactor is significantly affected by reaction order and the location of mixing.

Figure 8.5 was obtained from a parameter plot (iCSTR = 1 to 21) with the following BM code:

METHOD RK4   STARTTIME = 0   STOPTIME = 5   DT = 0.001
m = 1   ; Reaction order
iCSTR = 1   ; Location of large CSTR

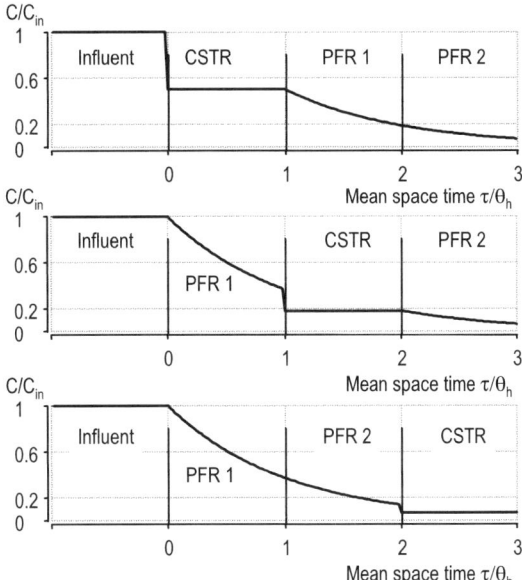

Fig. 8.3 Length profiles in the three reactor combinations in Fig. 8.1, showing the concentration of a material that is degraded in a first-order reaction

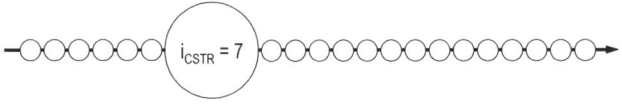

Fig. 8.4 Schematic of reactor for Example 8.1 with variable location of dominant mixing processes

Effluent pollutant concentration

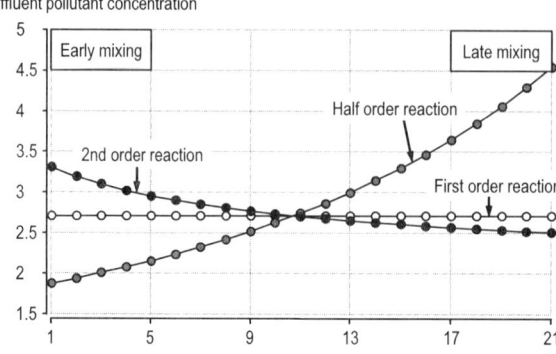

Fig. 8.5 Effluent pollutant concentration as a function of time of mixing and reaction order (Example 8.1)

---

n = 21   ; total number of compartments
Q = 1 Vtot = 1   ; Influent flow rate and total volume
V[1..n] = if i = iCSTR then Vtot/2 else Vtot/(2*(n−1))
                                      ; volume of each compartment
C0 = 100 k = 5   ; Influent concentration and rate constant
r[1..n] = k*C[i]^m   ; reaction rate
init C[1..n] = C0   ; concentration in all compartments
d/dt(C[1]) = Q*(C0−C[i])/V[i]−r[i]
d/dt(C[2..n]) = Q*(C[i−1]−C[i])/V[i]−r[i]
Cout = C[n]   ; effluent concentration

8.3 Methods for Model Identification

From descriptive statistics we know the methods of minimizing the sum of squared errors and the method of the moments for the determination of the parameters of probability distributions. Both methods are used in systems analysis.

8.3.1 Method of Moments

In the method of moments the empirically determined, statistical characteristics of the experimentally observed distribution of the residence time (their moments) are equated to the analytically well-known characteristics of the residence time of the ideal model reactors. This allows the derivation of the model parameters.

If the residence time distribution is experimentally determined, we can compute estimated values for the expected value τ_m and the variance σ^2 and, if necessary, higher moments directly from the data. In addition Eq. (7.9) (the zeroth moment) gives us the possibility to test (calibrate) the measuring device for the flow rate Q. If the inlet signal is subject to variance (i. e., not a Dirac pulse), we can obtain τ_m and σ^2 from Eqs. (7.41) and (7.42).

If the appropriate values of the model are available in function of the model parameters, we can determine the parameters from the following equations:

$$\tau_{m,\text{Experiment}}(\text{Data}) = \tau_{m,\text{Model}}(\text{Parameter})$$
$$\sigma^2_{\text{Experiment}}(\text{Data}) = \sigma^2_{\text{Model}}(\text{Parameter})$$

(8.1)

A case example is introduced in Sect. 8.4.

Advantages of the Method

This method is easily applied for the evaluation of data; conventional spreadsheet programs and a minimum of programming knowledge are sufficient. It is possible to use information from experiments where the entire course of the residence time distribution is not available, but only the statistical characteristics (the moments) of these distributions are known (Fig. 7.7).

Disadvantage of the Method

The empirical moments of the residence time distribution can only be determined, if sufficiently dense time series of tracer measurements exist and in particular, if all tracer has left the reactor before the end of the experiment. In addition, the separation of background concentrations may cause additional measuring expenditure. If the inlet signal already exhibits a large variance, the identified variance of the system may be subject to considerable uncertainty.

The moments of the residence time distribution of the mathematical model are available only for some few ideal reactors. For more complicated models and combined reactors this method can barely be applied.

Example 8.2: Volume and model of a fish pond

An unknown, constant quantity of water Q flows through a fish pond with an unknown volume V.

You would like to obtain volume and flow rate as well as an approximate characterization of mixing in the pond. You add to the influent $E_A = 200\,\text{kg}$ of a tracer (NaCl) in the form of a Dirac pulse and measure in the effluent a time series of tracer concentration. The statistical characterization of the concentration in the effluent, after subtracting the background concentration, leads to the following results:

Zeroth moment: area underneath the course of concentrations $M_0 = 2500\,\text{h g m}^{-3}$

First moment of the course of concentrations $M_1 = 30{,}000\,\text{h}^2\,\text{g m}^{-3}$

Second central moment of the course of concentration $M_2 = 600{,}000\,\text{h}^3\,\text{kg m}^{-3}$

According to Fig. 7.4 we have $M_0 = E_A/Q$ and thus $Q = 200{,}000/2500 = 800 \, m^3 h^{-1}$.
Following Eqs. (7.11) and (7.7): $\tau_m = M_1/M_0 = \theta_h = V/Q = 12 \, h$ and thus
$V = 12 \cdot 800 = 9600 \, m^3$.
According to Eq. (7.6) or (7.12): $\sigma^2 = M_2/M_0 - \tau_m^2 = 56 \, h^2$.
Modeled with a cascade of CSTRs according to Eq. (7.26) yields $n = \theta_h^2 / \sigma^2 = 1.5$
The fish pond appears to be nearly completely mixed, a fact that should be validated with visual inspection.
With one simple experiment an ill-defined system could be reasonably well characterized with the help of the first three moments. We obtained Q, V, and some information on internal mixing.

8.3.2 Adjustment of the Model to the Measurements

By adjusting the parameters of the model, the weighted sum of the squared deviation between the measurement and model prediction is minimized.

We first develop a model of the system and simulate the experiment with estimated model parameters. With the help of an optimization routine, we adapt the model parameters in such a way that the test characteristic χ^2 is minimized:

$$\chi^2 = \sum_{i=1}^{n} \left(\frac{y_{m,i} - y_i(p)}{\sigma_{m,i}} \right)^2 \tag{8.2}$$

χ^2	=	chi square, sum of the squares of the weighted deviation of the measured from the computed state variables [−]
$y_{m,i}$	=	measured value of a state variable in the real system in the ith measurement [typically $M \, L^{-3}$]
$\sigma_{m,i}$	=	standard deviation of the measuring error of $y_{m,i}$
$y_i(p)$	=	computed value of the model state which corresponds to the measurement $y_{m,i}$ in kind, location and time [typically $M \, L^{-3}$]
p	=	set of model parameters
n	=	number of data points

If the values of $\sigma_{m,i}$ are not available, we neglect the weights and set $\sigma_{m,i} = 1$. Thereby χ^2 loses its absolute meaning in the appropriate test on normal distribution of the residuals. See Chap. 12 for the topic of parameter identification.

For the search of the optimal parameters there are various strategies, which are not discussed here. It is important that the flow rate Q is included as a parameter because its measurement is frequently afflicted with a large bias error. If the different data loggers are not exactly time synchronized, their time series must be time shifted by an unknown time step, which adds additional parameters.

Advantage of the Method

This method leaves us many degrees of freedom in the organization of the experiments and is applicable to the most different systems, even if we do not know an analytical solution for the residence time distribution.

The method is applicable even if the experiment is terminated before the entire tracer has left the reactor. Accordingly the experiment can be shortened. In addition the method does not react very sensitively to a constant background concentration and can handle influent signals with a large variance of the concentration of the tracer in the influent.

Disadvantage of the Method

This method requires that either an analytic solution for the expected residence time distribution is available, or that a dynamic model of the system that allows the system to be simulated is developed. Only specialized software offers options to identify large sets of parameters by optimizing an objective function such as χ^2. Conventional spreadsheets make routines for parameter identification and optimization available: their use is, however, frequently not trivial.

8.4 Case Study

The following case study is based on simulated data. This has on the one hand the advantage that the correct parameters of the reactor model are available, on the other hand artificial data have the disadvantage that not all practical problems may develop.

Task

We want to develop a mathematical model of an activated sludge tank which can describe the conditions along the reactor with "sufficient" accuracy, but which is as simple as possible in its application. From experience we know that it is sufficient

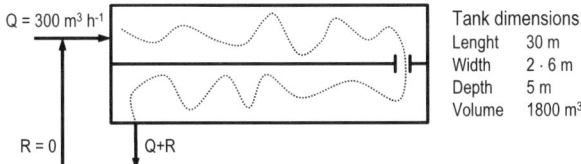

Fig. 8.6 Geometry of the activated sludge tank

to characterize the length profile of the state variables in four locations. The reactor has the dimensions shown in Fig. 8.6. It is separated into two equal parts, the connection between Sects. 1 and 2 is so narrow that no back mixing is possible.

Procedure

We will first determine the distribution of the residence time $f(\tau)$. For this we add a certain quantity of salt solution (NaCl) as pulse to the influent of the tank and then follow the conductivity in the effluent. During the experiment, the flow of water Q to the observed reactor is kept constant and the return sludge is fed into parallel process trains in order not to disturb the experiment by salt, which is recirculated. The aeration is operated normally in order to produce the typical mixing.

The "measured" conductivity in the effluent of the activated sludge tank is shown in Fig. 8.7; the data were simulated with the model of two turbulent plug-flow reactors in series and an artificial noise (measuring error) was added (see Fig. 8.8).

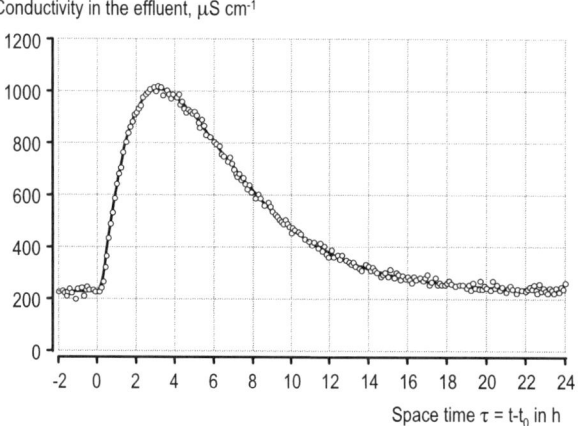

Conductivity in the effluent, $\mu S\ cm^{-1}$

Space time $\tau = t\text{-}t_0$ in h

Fig. 8.7 "Measured" and simulated conductivity in the effluent of the activated sludge tank. Data is obtained from a simulation with two turbulent plug-flow reactors in series: simulation is with only four reactors in series with internal recirculation (see Fig. 8.8)

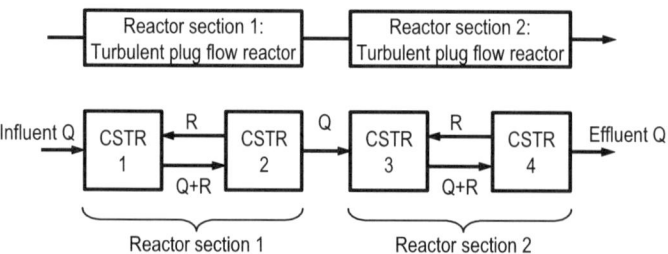

Fig. 8.8 Examples of possible reactor models

Reactor Model

Possible simple models of the tank are shown in Fig. 8.8. The turbulent plug-flow reactor must be discretized for a simulation in order to solve the nonstationary form of the underlying partial differential equations. This model supplies a length profile with high spatial resolution which is not necessary. We opt for the model with four stirred tank reactors and internal recirculation. With $R=0$ this model also includes an appropriate cascade of CSTRs.

The intended model considers the information that between the second and the first reactor section no back mixing is possible. However, no other obstacles restrain longitudinal mixing.

Method of Moments

With the method of moments we first determine the expected value for τ_m and the variance σ^2 of the residence time distribution $f(\tau)$. We can then derive the parameters of the reactor model. We can apply this method only if the experiment lasted long enough to recover the entire mass of tracer in the effluent, which is here approximately valid. It is problematic that for the conductivity an a priori unknown background concentration C_{BG} is present. The average of all data with $\tau < 0$ and $\tau > 22$ h yields $C_{BG} = 232\ \mu S\ cm^{-1}$.

With the code in Example 8.3 we obtain the expected value and the variance of the residence time distribution for the entire reactor:
$\tau_m = 5.98$ h and thus $Q = V/\tau_m = 301\ m^3\ h^{-1}$
$\sigma^2 = 15.4\ h^2$.

Since the tank is divided into two equal segments in series, we distribute the two values to the two partial reactors (σ^2 and τ_m are additive). If the turbulence added by aeration is equal in both segments:

$\tau_{m,1} = 2.99$ h, $\sigma_1^2 = 7.7\ h^2$,

$\dfrac{\sigma_1^2}{\tau_{m,1}^2} = 0.861$ and with Eq. (7.37) we obtain after iteration: $N_T = 2.14$.

We model each segment of the tank with two reactors in series. The appropriate cascade ($n=2$) has a turbulence number of $N_{T,C} = 0.391$ (see Table 7.1), thus the internal recirculation must provide for $N_{T,R} = 2.14 - 0.391 = 1.75$. Equation (7.40) yields:

$R = N_T \cdot n \cdot Q = 1.75 \cdot 2 \cdot 301 = 1053\ m^3\ h^{-1}$.

Thus, all parameter values of the model are known.

The residence time distribution computed with these parameters is compared with the measured data in Fig. 8.7. The agreement is very good, even though the original data were obtained with another model. With genuine experimental data the agreement would not be as good.

Example 8.3: Expected value and variance of the residence time distribution from data

{The following simple code computes the weighted means and the variance of the time series TS}

```
STARTTIME = 0            ; Beginning
STOPTIME = 24            ; End
DT = 0.1                 ; Time step of the data
TS = #RTD(Time)−232      ; Time series = data-background concentration
init M[0..2] = 0         ; 0th–2nd moment of the data
next M[0..2] = M[i] + DT*TS*time^i
taum = if M[0] > 0 then M[1]/M[0] else 0          ; Expected value τm
sig2 = if M[0] > 0 then M[2]/M[0]−taum^2 else 0 ; Variance σ²
{Is valid τm = taum(STOPTIME) and σ² = sig2(STOPTIME)}
```

Example 8.4: Parameters of the model of a turbulent plug-flow reactor

What parameters result from the empirical data in Fig. 8.7 for the model of a turbulent plug-flow reactor?

The flow velocity u (advection) in the reactor amounts to (Fig. 8.6):

$u = Q/A = 301 \, m^3 \, h^{-1}/30 \, m^2 = 10 \, m \, h^{-1}$.

From $N_T = D_T/(u \cdot L) = 2.14$ it follows that $D_T = 2.14 \cdot 10 \, m \, h^{-1} \cdot 30 \, m = 642 \, m^2 \, h^{-1}$.

Thus, the parameters of the model are known.

Direct Adjustment of the Model Parameters

For this procedure a simulation program must be available in order to simulate the experiment. Here it is Berkeley Madonna.

We first develop the code for the model (see Example 8.5) and then fit the simulated effluent concentration with the help of the option *Curve Fit* to the measured data. The parameters to be adjusted are:

E_A = quantity of the dosed conductivity [$\mu S \, cm^{-1} \, m^3$]
Q = influent flow of water [$m^3 \, h^{-1}$]
R = internal recirculation [$m^3 \, h^{-1}$]
C_{BG} = background concentration of the conductivity [$\mu S \, cm^{-1}$]

The results of this method lead in this example to values that are very close to the results obtained with the method of moments. This is due to the fact that the data is simulated. Typically a line fit by adjustment of the parameters is more reliable, because the results can also be evaluated if the experiment is broken off prematurely, here, e.g., already 12 h after addition of the tracer (see also Table 8.1). After the identification of the parameters the statistical characteristics of the residence time distribution, if they are of interest, may be obtained with the help of the ideal models, or from the simulated distribution.

Table 8.1 Comparison of the identified parameters

Parameter	Simulated data	Method of moments	Parameter identification		Units
			−2–24 h	−2–12 h	
E_A	6000	6036	6048	6084	$\mu S\ cm^{-1}\ m^3$
Q	300	301	299	299	$m\ h^{-3}$
R	–	1053	1275	1297	$m^3 h^{-1}$
C_{BG}	235	232	232	229	$\mu S\ cm^{-1}$
Resulting statistical characteristics of the residence time distribution					
$\tau_m = \theta_h$	6	5.97	6.00	6.00	h
σ^2	15.4	14.94	15.34	15.36	h^2

The simulated residence time distribution $f(\tau)$ for the three models which were identified in Table 8.1, can barely be distinguished by eye.

Example 8.5: Code for the direct identification of the model parameters with Berkeley Madonna

```
{Simulation of activated sludge tank in Fig. 8.8 below, tested}
METHOD RK4
STARTTIME = −2
STOPTIME = 24
DT = 0.1
TS = #RTD(Time)        ; Measured time series
Q = 299                ; Flow rate, identified, m³ h⁻¹
R = 1275               ; Internal recirculation, identified, m³ h⁻¹
Vtot = 1800            ; Total volume, fixed, m³
V = Vtot / 4           ; Volume of partial reactors m³
Ea = 6048              ; Mass of added tracer, identified, m³ µS cm⁻¹
CBG = 232              ; Background conc. of conductivity, identified, µS cm⁻¹
C0 = pulse(Ea,0,100) + CBG; Concentration in the influent
init C[1..4] = CBG     ; Initial concentration in the partial reactors
d/dt(C[1]) = (C0*Q−C[1]*(Q + R) + R*C[2])/V
                       ; Balance for 1. partial reactor
d/dt(C[2]) = ((Q + R)*(C[1]−C[2]))/V
d/dt(C[3]) = (Q*C[2]−(Q + R)*C[3] + R*C[4])/V
d/dt(C[4]) = ((Q + R)*(C[3]−C[4]))/V
Cout = C[4]            ; Effluent concentration which is fitted to TS
```

In Berkeley Madonna we now have the option to fit the output Cout to the observed data TS. For this we have to identify the parameters to be adjusted. Here they are Q, R, E_A, and C_{BG}.

Application

With the code in Example 8.5 we now have the model available to predict the behavior of the activated sludge tank. We must still consider how the aeration intensity affects the turbulence and therefore the parameter R. We assume a positive correlation which we could derive from literature data or from additional experiments and our experience.

Figure 8.9 illustrates the simulated performance of the activated sludge tank for a material which is degraded in a first-order reaction. The comparison between the plug-flow reactor with turbulence, which leads to a fastidious model, and the cascade with internal recirculation and only two reactors per reactor component shows that for many questions this simple model might be sufficient.

Example 8.6: Influence of the aeration intensity on turbulence and required internal recirculation

How can the aeration intensity be integrated into the model of the reactor in Fig. 8.8?

Example 7.14 gives the following relationship for the dispersion coefficient D_T in function of the air flow rate A [m^3 h^{-1}]:

$$D_T \sim A^{0.346} \text{ with A = air flow rate in m}^3 \text{ h}^{-1}$$

Thus, we can obtain a current turbulence number $N_{T,current}$ from:

$$N_{T,current} = N_T(t) = N_{T,Experiment} \cdot \left(\frac{A_{current}}{A_{Experiment}} \right)^{0.346}$$

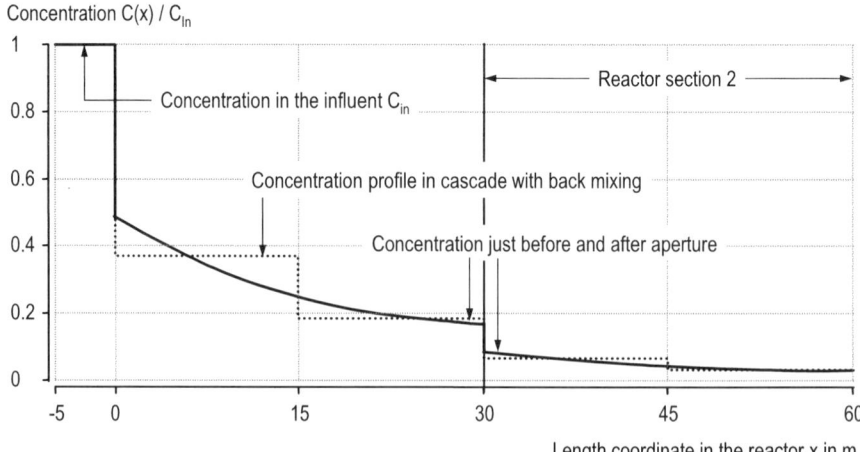

Fig 8.9 Comparison of the simulated performance of the reactor at steady state. Degradation with a first-order model is shown. *Continuous line*: two turbulent plug-flow reactors in series; *broken line*: cascade with internal recirculation

and with Eq. (7.40), $R = (N_T(t) - N_{T,C}) \cdot n \cdot Q(t)$ and Table 7.1 with $n = 2$:

$$R(t) = \left(2.14 \cdot \left(\frac{A(t)}{A_{Experiment}} \right)^{0.345} - 0.391 \right) \cdot 2 \cdot Q(t) .$$

This equation can be integrated into the code for the simulation of the activated sludge tank. However, the information on the current value of the airflow A must be available.

Thus, we have used the experiment only in order to obtain the dispersion coefficient D_T for the operating conditions during the experiment.

Chapter 9
Heterogeneous Systems

In homogeneous systems, the chemical and microbiological processes take place evenly distributed over the entire reaction space in only one phase. In heterogeneous systems these processes take place in limited regions only, within one or more phases of a multiphase system. Heterogeneous systems are very prominent in environmental science and technology.

9.1 Classification of Processes and Systems

We differentiate between homogeneous and heterogeneous processes and systems. A process is homogeneous if it proceeds in only one phase. It is heterogeneous if at least two phases are necessary for the process to proceed at its characteristic rate. A system is homogeneous if therein only one phase has importance. It is heterogeneous if its characteristic performance depends on at least two phases, which both occur in the system and are included in its description.

In homogeneous systems the characteristic transformation processes proceed distributed over the entire reaction space. In heterogeneous systems the transformation may be limited to a part of the system.

In urban water management we typically deal with heterogeneous systems, whereas the homogeneous systems are the exception for very specific cases (see Table 9.1). Whether we must model the heterogeneity depends, however, on the question that we wish to answer. Frequently we can use empirical models that do not deal explicitly with the heterogeneity (see Example 9.1).

Several phases are involved in heterogeneous processes, thus transformation rates are controlled by mass transfer of materials across boundary layers as well as local reaction rates. The detailed description of these processes is frequently rather involved.

179

Table 9.1 Examples of homogeneous and heterogeneous processes and systems

Classification	Process	
Homogeneous	Acid–base reactions, neutralization, buffer systems	
	Oxidation of Fe^{2+} to Fe^{3+} by oxygen in activated sludge systems	
	Reactions of ammonium, NH_4^+ with chlorine, Cl_2 in disinfection	
	Reactions of ozone, O_3 with dissolved pollutants	
	Oxidation of sulfite by O_2 in the characterization of an aeration	

Classification	Process	Phases
Heterogeneous	Microbial processes	Water – organisms
	Growth of biofilm	Water – organisms – substratum
	Degradation of particles	Water – particles – organisms
	Disinfection	Water – microorganisms
	Sedimentation	Water – particles
	Adsorption	Water – adsorbents
	Ion exchange	Water – Ion exchange resin
	Flocculation	Water – particles
	Precipitation processes	Water – ions – precipitation products
	Filtration	Water – filter material – particles
	Gas exchange	Water – gas

Example 9.1: Aeration, a heterogeneous system which we usually describe homogeneously

Analyzed in detail, the aeration of wastewater with small bubbles consists of two processes in series: (i) in the interaction of each bubble with the wastewater, oxygen will be transferred from the bubble to the wastewater and (ii) this oxygen is then distributed by turbulence from the gas–water interface over the entire reactor. Of the two processes, the first determines the rate of the oxygen transfer; both the surface of the boundary between the air and water, a $[m^2_{Interfacial\ Area}\ m^{-3}_{Water}]$ and a mass transfer coefficient k_l $[m\ d^{-1}]$ are of importance. These two parameters are combined into the product $k_l a$ $[d^{-1}]$ to form a first-order rate constant. We use this model, which is completely equivalent to a homogeneous reaction, where the oxygen is generated distributed over the entire volume.

Why is this simple model acceptable?

Aeration distributes thousands of small bubbles evenly over the reaction space. The oxygen transfer takes place over the surfaces of the bubbles, and rising bubbles produce turbulence that distributes the generated dissolved oxygen rapidly (in comparison to the oxygen consumption rate) and evenly over the entire volume. For bacteria the oxygen seems to be produced homogeneously.

9.2 Multiphase Systems

Many reactors and systems which we apply in water and wastewater treatment, as well as the self-purification of rivers, are based on heterogeneous systems. Since

these are frequently difficult to model, we try to describe them in terms of homogeneous processes.

Homogeneous systems are generally easier to model than heterogeneous ones. Therefore, we try first to seize a problem with a homogeneous description. Only if the question does not permit this will we differentiate between several phases and their corresponding behavior and interactions at additional expense.

Today extensive models that describe biological wastewater treatment systems and in particular the activated sludge process are used with success (Henze et al., 1987; Gujer et al., 2000). They are primarily based on homogeneous models. If we analyze, however, these systems in detail, it becomes obvious that they are based on heterogeneous processes: water, dissolved and particulate pollutants, microorganisms, precipitates and air bubbles, all refer to different phases. Deriving model equations, we connect our analysis with experience and know-how, hopefully without loss of significance of the developed models. The case studies in Sect. 9.4 will, however, demonstrate that this is not always possible.

9.2.1 Microbial Degradation of Stored Pollutants

This section requires some previous knowledge of biological wastewater treatment, especially the family of activated sludge models ASM1 to ASM3 should be known to fully appreciate these explanations. They demonstrate an approach to modeling of heterogeneous systems with the aid of homogeneous models.

Many microorganisms can store organic or mineral materials and make these storage products available for later degradation. Degradation requires that both microorganisms with their biochemical apparatus and storage materials (and possibly oxygen) are present in the reactor. For soluble materials we typically use Monod kinetics as the degradation kinetics (Table 9.2). In biological wastewater treatment we use sedimentation to concentrate the biomass X_H (the solid phase), but this does not have an effect on the concentration of the soluble material S_S. Concentrating the biomass by a factor of two results in an increase of the degradation rate by a factor of two, which appears to be reasonable.

Storage products are inside the microbial cells (Fig. 9.1). Per unit volume, sedimentation will increase the concentration of the microorganisms as well as the concentration of the storage compounds. Using Monod kinetics (a homogeneous approach, the second process in Table 9.2) would increase the degradation rate of storage products anywhere between factors of two to four depending on the ratio of storage product to biomass X_S/X_H. This is not reasonable since storage products are still exposed to the same biochemical apparatus as before the increase of their concentration. We would expect to observe an increase of the degradation rate in parallel with the increase of their concentration.

A more reasonable form for this process rate is given by the third process in Table 9.2. The ratio of stored product to biomass X_S/X_H in the kinetic expression

Fig. 9.1 Bacteria containing granules of storage products. *Left*: before concentration and *right* after concentration by sedimentation/thickening

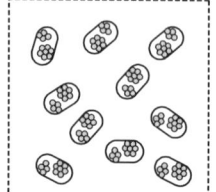 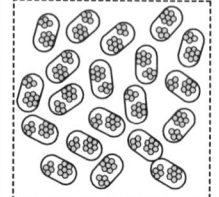

Table 9.2 A simple kinetic model to describe storage of substrate X_S by heterotrophic organisms X_H based on soluble organic substrate S_S

Process	Substrate S_S COD	Biomass X_H COD	Stored substrate X_S COD	Process rate ρ
Storage or substrate removal based on Monod kinetics	$-\dfrac{1}{Y_{Sto}}$		1	$k_{Sto} \cdot \dfrac{S_S}{K_S + S_S} \cdot X_H$
Degradation of storage products based on Monod kinetics		1	$-\dfrac{1}{Y_{Gro}}$	$\mu_{Gro} \cdot \dfrac{X_S}{K_X + X_S} \cdot X_H$
Degradation of storage products based on "heterogeneous" kinetics		1	$-\dfrac{1}{Y_{Gro}}$	$\mu_{Gro} \cdot \dfrac{X_S / X_H}{K_X + X_S / X_H} \cdot X_H$

leads to the expected behavior: doubling the concentration of the solids results in exactly twice the volumetric rate of the degradation. Thus, by choosing an adapted form of the kinetics, it is possible to describe the heterogeneous processes with a homogeneous mathematical format.

9.3 Behavior of Individual Particles

We can identify individual particles and follow their changes as they pass through a reactor, so they have an identity. Individual molecules do not have characteristics that differentiate them from other molecules of the same compound, so they do not have an identity.

Individual particles have a history that forms them. In the course of the reaction time they change their composition, shape and structure: bacteria were exposed to a certain dose of disinfectant and are damaged accordingly, precipitation products change their surface-to-volume ratio, etc. Individual molecules have no individual characteristics: either they belong to this material or else to another one.

With stochastic models (see Sect. 7.6) we can follow individual particles along their paths through a reactor. If we simulate for each particle its interaction with the environment, we can obtain statistical information about the change of the properties of these particles over time. In special cases this may provide interesting additional information (see the following examples).

Example 9.2: Disinfection of oocysts of Cryptosporidium with ozone

Many dormant forms of microorganisms (spores, cysts), must be exposed to a minimum dose of ozone before they are disinfected; this also applies to the oocysts of Cryptosporidium, a protozoa that impairs water quality. In a batch reactor this has the consequence that we observe a delayed disinfection activity or a lag phase during which no organisms are killed. Subsequently, the retarded disinfection starts. Since in a batch reactor all particles have the same history, we can easily model this system, even, if in the course of time the ozone concentration decreases. In a mixed system the individual organisms are transported along different paths through the reactor and are therefore exposed to a different history of damaging ozone. In this case we cannot easily model the effect of the lag phase. In a stochastic approach we model the possible paths of the organisms through the reactor, integrate the local ozone concentration along these paths, and estimate the resulting damage. By simulating the fate of many organisms and statistical evaluation of the results, we can analyze the consequences of the delayed disinfection.

Such a model is described in detail by Gujer and von Gunten (2003).

Example 9.3: Storage materials in bacteria

In modern biological wastewater treatment plants, especially for biological phosphorus removal, we make use of bacteria that can store polyphosphates and other materials. Here the uptake of phosphate relies on the availability of stored organic material. If an individual bacteria has been shortcut through some subsystems, it may be void of these materials and can therefore not participate in the phosphorus removal process.

Since biological processes are usually nonlinear, bacteria that contain different quantities of storage material may react quite differently. With stochastic models we can simulate the behavior of the individual bacteria and evaluate the result for the whole population statistically.

An example of such a model is given by Gujer (2002).

Example 9.4: Gas bubbles change their composition

In an aeration system the air bubbles deliver oxygen, and thereby the composition of the gas in the bubbles changes continuously. Depending upon the situation we must consider this change.

Example 9.5: Activated carbon and ion exchange resin have a history

Organic impurities adsorb on activated carbon. In the upper part of an adsorption column the activated carbon is subject to a higher charge of pollutants than in the lower part. If we now mix the activated carbon by backwashing, activated carbon

granules with the most different charges will become distributed over the entire column. *How can we model the performance of such a backwashed column?*
An ion exchange resin behaves quite similarly. The top of the column is first occupied with water with greater hardness. After mixing of the resin, e. g., by a backwash procedure, the resin is mixed and behaves entirely differently from in its original firm layering (see Sect. 9.4.5).

Following a particle individually is comparable to a system in which small batch reactors (particles) are transported, where each has its own history. Thus, in these models, batch reactors are introduced for each particle and their development is simulated. This requires for each particle a separate transport equation (one for each dimension) and a balance equation for each state variable that characterizes the particles. Thousands of differential equations may result, which are to be solved simultaneously. An example is given by Gujer and von Gunten (2003).

Levenspiel (1999) provides analytical solutions for the description of heterogeneous systems for some simple situations. Additional key words are segregation, micro- and macrofluid.

9.4 Case Studies

For the description of heterogeneous systems there are hardly any generally accepted rules. Frequently it will, however, be necessary to characterize first a usually microscopic subsystem which reaches up to a phase boundary. Subsequently, the result for the subsystem is integrated into the macroscopic balance equation, which covers the entire system.

The following case studies highlight some possibilities.

9.4.1 Transformation Processes in a Sewer

In a sewer, microorganisms are active both in the flowing wastewater itself as well as in the fixed biomass (biofilm) on the wetted sewer walls. In addition oxygen is transferred through the free surface of the flowing water. In this heterogeneous system the characteristic processes are influenced by several phases: the water phase, the fixed biomass, and the gaseous phase. A balance equation for pollutants in the wastewater must therefore consider these three phases.

The geometrical conditions in a sewer are illustrated in Fig. 9.2. The flowing wastewater stands in interaction with the biofilm, which covers the wetted surface of the sewer and is biologically very active. The wastewater composition changes along the direction of flow by degradation processes in the wastewater itself, in

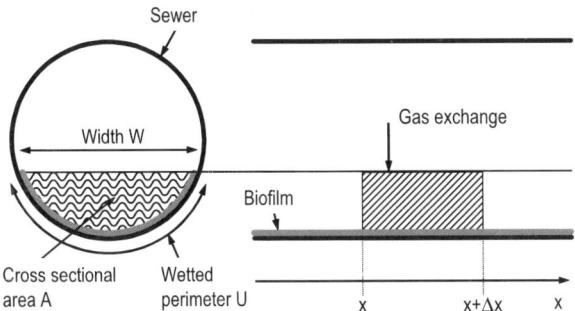

Fig. 9.2 Geometry of a sewer. *Left*: cross section. *Right*: length profile. The biofilm (the fixed biomass) covers the wetted perimeter, and oxygen is delivered from the air by reaeration

the biofilm and by the exchange of oxygen with the atmosphere. In order to model these changes, we must derive the appropriate balance equations.

Pollutants C_S are degraded both in the wastewater and exchanged with the biofilm. As a consequence of the degradation of pollutants the biofilm grows, i. e., it becomes thicker; the shear stress of the wastewater leads to erosion, which returns pollutants (suspended solids) from the biofilm to the wastewater. Oxygen S_{O2} is used both in the wastewater and in the biofilm, but delivered from air through the free water surface. The solids in the biofilm X_{BF} are quantified here as mass per unit surface $[M_{BF} \, L^{-2}]$. For a short length Δx of the sewer (Fig. 9.2) the following balances result:

$$A \cdot \Delta x \cdot \frac{\partial C_S}{\partial t} = Q \cdot \left(C_S(x) - C_S(x + \Delta x)\right) + \left(j_{Degrad,S} + j_{Erosion,S}\right) \cdot U \cdot \Delta x + r_S \cdot A \cdot \Delta x$$

$$A \cdot \Delta x \cdot \frac{\partial S_{O2}}{\partial t} = Q \cdot \left(S_{O2}(x) - S_{O2}(x + \Delta x)\right) + j_{Degrad,O2} \cdot U \cdot \Delta x$$
$$+ k_1 \cdot W \cdot \Delta x \cdot (S_{O2,sat} - S_{O2}) + r_{O2} \cdot A \cdot \Delta x$$

$$A \cdot \Delta x \cdot \frac{\partial X_{BF}}{\partial t} = \left(j_{Degrad,BF} + j_{Erosion,BF}\right) \cdot U \cdot \Delta x \, .$$

After division by the water volume $A \cdot \Delta x$ and transition from Δx to ∂x this yields:

$$\frac{\partial C_S}{\partial t} = -u \cdot \frac{\partial C_S}{\partial x} + \frac{j_{Degrad,S} + j_{Erosion,S}}{R_h} + r_S \, , \qquad (9.1)$$

$$\frac{\partial S_{O2}}{\partial t} = -u \cdot \frac{\partial S_{O2}}{\partial x} + \frac{j_{Degrad,O2}}{R_h} + \frac{k_1}{h_m} \cdot \left(S_{O2,sat} - S_{O2}\right) + r_{O2} \, , \qquad (9.2)$$

$$\frac{\partial X_{BF}}{\partial t} = j_{Degrad,BF} + j_{Erosion,BF} \, . \qquad (9.3)$$

C_S = pollutant concentration in the wastewater $[M_S L^{-3}]$
S_{O2} = concentration of dissolved oxygen in wastewater $[M_{O2} L^{-3}]$
X_{BF} = mass of solids per surface of biofilm $[M_{BF} L^{-2}]$
u = mean flow velocity of the wastewater $[L \cdot T^{-1}]$
x = flow distance, length coordinate $[L]$
r_S, r_{O2} = transformation rates of pollutant and oxygen within the flowing water $[M_i L^{-3} T^{-1}]$
$j_{Degrad.}$ = mass flux into the biofilm as a consequence of the degradation of pollutants, S pollutant, BF biofilm material, O_2 oxygen, each in $[M_i L^{-2} T^{-1}]$
$j_{Erosion}$ = erosion of biofilm material back to the wastewater $[M_i L^{-2} T^{-1}]$
k_l = mass transfer coefficient for dissolved oxygen $[L \cdot T^{-1}]$
U = wetted perimeter (see Fig. 9.2)
R_h = hydraulic radius (U/A) $[L]$
W = width of water surface (see Fig. 9.2)
h_m = mean depth (W/A) $[L]$

Equations (9.1)–(9.3) clearly show the problems that develop if we model heterogeneous systems:

- Beside the conventional transport processes (here Q and u) we need additional transport parameters in the form of mass flux (j_{Degrad}, $j_{Erosion}$, j_{O2}) over different parts of system boundaries (R_h, h_m).
- Mass transfer coefficients (k_l) become important.
- State variables (material concentrations) do not only refer to the reaction volume (here S_{O2}, C_S) but also to surfaces (here X_{BF}).
- The balance equations do not all refer to the same subsystem (here the wastewater and biofilm).
- Apart from the reaction rates r_i, which link stoichiometry and kinetics, mass transfer processes must be considered which may also relate to stoichiometry and kinetics (see Table 9.3).

Table 9.3 Possible simple representation of the mass transfer processes over the phase boundaries in the sewer example. The representation is similar to the stoichiometric matrix (Sect. 5.3) but resulting in a mass flux depending upon the state variables and a stoichiometric coefficient that contains geometrical variables, for example: $j_{Degrad,S} = -\varphi_1/R_h$

	S_{O2} $gO_2 \, m^{-3}$	C_S $gCOD \, m^{-3}$	X_{BF} $gCOD \, m^{-2}$	Process rate φ $g_i \, m^{-2} d^{-1}$
Degradation	$-\dfrac{1-Y}{R_h}$	$-\dfrac{1}{R_h}$	Y	$k_{BF} \cdot \dfrac{S_{O2}}{K_{O2}+S_{O2}} \cdot \dfrac{C_S}{K_S+C_S} \cdot X_{BF}$
Aeration	$\dfrac{1}{h_m}$			$k_l \cdot (S_{O2,sat} - S_{O2})$
Erosion		$+\dfrac{1}{R_h}$	-1	$k_{Erosion} \cdot X_{BF}^n \cdot u^m$

9.4.2 Activated Sludge Flocs

The activated sludge process is based on heterogeneous processes: microbial aggregates or flocs are in interaction with dissolved and particulate materials. Only if the pollutants are transported to the microorganisms within the flocs can the degradation processes proceed with their characteristic rate. Transport of dissolved materials is via molecular diffusion within the water phase inside the flocs.

A simple model of an activated sludge floc regards these as a sphere (Fig. 9.3) with homogeneously distributed microorganisms. The activity of these organisms is determined by the supply of nutrients, which are transported by molecular diffusion in the stagnant water inside the floc.

A balance for the material A in the thin spherical element in Fig. 9.3 has the form:

$$4 \cdot \pi \cdot r^2 \cdot \Delta r \cdot \frac{\partial C_A}{\partial t} = j_A(r) \cdot 4 \cdot \pi \cdot r^2 - j_A(r + \Delta r) \cdot 4 \cdot \pi \cdot (r + \Delta r)^2 + R_A \cdot 4 \cdot \pi \cdot r^2 \cdot \Delta r$$

r   = radius [L]
C_A = concentration of the material A $[M_A L^{-3}]$
j_A = mass flux of material A $[M_A L^{-2} T^{-1}]$
R_A = transformation rate of the material A $[M_A L^{-3} T^{-1}]$

After division by $4 \cdot \pi \cdot \Delta r$ and transition from Δr to ∂r this yields:

$$r^2 \cdot \frac{\partial C_A}{\partial t} = -\frac{\partial (r^2 \cdot j_A)}{\partial r} + r^2 \cdot R_A . \tag{9.4}$$

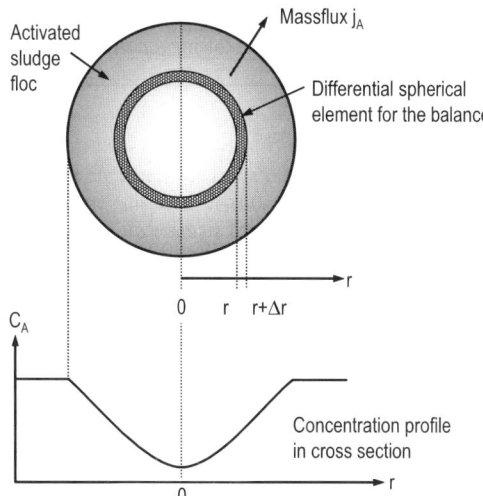

Fig. 9.3 Schematic representation of a spherical activated sludge floc, the definition of a differential spherical element, and a transverse profile of the concentration of a degradable pollutant

With Fick's first law for molecular diffusion inside the floc $j_A = -D_A \cdot \partial C_A / \partial r$ we have

$$r^2 \cdot \frac{\partial C_A}{\partial t} = \frac{\partial \left(r^2 \cdot D_A \cdot \dfrac{\partial C_A}{\partial r} \right)}{\partial r} + r^2 \cdot R_A, \tag{9.5}$$

or for the steady state and constant D_A

$$\frac{d \left(r^2 \cdot \dfrac{dC_A}{dr} \right)}{dr} = -r^2 \cdot \frac{R_A}{D_A}, \tag{9.6}$$

and using the chain rule

$$\frac{d \left(r^2 \cdot \dfrac{dC_A}{dr} \right)}{dr} = r^2 \cdot \frac{d^2 C_A}{dr^2} + 2 \cdot r \cdot \frac{dC_A}{dr} = -r^2 \cdot \frac{R_A}{D_A} \tag{9.7}$$

or

$$\frac{d^2 C_A}{dr^2} = -\frac{2}{r} \cdot \frac{dC_A}{dr} - \frac{R_A}{D_A}. \tag{9.8}$$

Equation (9.8) requires two boundary conditions for its solution:

$$\left. C_A \right|_{r=r_0} = C_A \text{ (surface)}$$

$$\left. \frac{dC_A}{dr} \right|_{r=0} = 0 \quad \text{because the concentration profile } C_A \text{ is symmatrical} \tag{9.9}$$

For $r=0$ the term $-\dfrac{2}{r} \cdot \dfrac{dC_A}{dr}$ is indefinite; in Berkeley Madonna it cannot be integrated. Therefore we define the second boundary condition not in the center of the floc ($r=0$) but at a small radius, say $r=1\,\mu m$. Thus, the integration will proceed without significant loss of accuracy.

With the help of Eqs. (9.8) and (9.9) the concentration profiles of all soluble materials involved can be predicted in function of different environmental conditions and floc diameters. The degradation performance of a single floc results from Fick's law applied to the surface (r_0). If we make the simplifying assumption that all flocs are of equal size, we can convert the activity of the activated sludge to the volume of the activated sludge tank:

$$r_A = \frac{4 \cdot \pi \cdot r_0^2 \cdot j_A (r_0)}{\dfrac{4}{3} \cdot \pi \cdot r_0^3 \cdot \gamma_{BS}} \cdot X_{BS} = \frac{3 \cdot j_A (r_0)}{r_0 \cdot \gamma_{BM}} \cdot X_{BS} = -\frac{3}{r_0} \cdot D_A \cdot \left. \frac{dC_A}{dr} \right|_{r_0} \cdot \frac{X_{BS}}{\gamma_{BS}} \tag{9.10}$$

r_A = activity of the activated sludge per reactor volume $[M_A L^{-3} T^{-1}]$
$j_A(r_0)$ = flux of material A out of the flocs $[M_A L^{-2} T^{-1}]$

r_0 =   radius of the flocs [L]
D_A =   diffusion coefficient of the material A within the flocs [$L^2 T^{-1}$]
X_{BS} =   concentration of the activated sludge per reactor volume [$M_{AS}L^{-3}$]
γ_{BS} =   concentration of the activated sludge per floc volume [$M_{AS}L^{-3}$]

Converting the floc volume to the tank volume in Eq. (9.10) allows our analysis to continue based on homogeneous models. However, the flux of material must be computed with the heterogeneous model.

An application of such a model is given by Manser et al. (2006).

Example 9.6: Computation of the concentration profiles in an activated sludge floc

The following code translates Eq. (9.8) into a code for BM. The code can easily be extended to several materials. Time is used to integrate over the radius. (tested)

METHOD RK4
STARTTIME = 1e−6   ; S_{O2}'' depends on r^{-1}, we start the computation at r = 1 μm
STOPTIME = 0.5E−3   ; Radius of the floc, 0.5 mm
DT = 1E−5   ; Integration step in space
rename time = r   ; Computation over space instead of time
DO2 = 1.8E−4   ; Diffusion coefficient for O_2 inside the floc, $m^2 d^{-1}$
SO20 = 2   ; Dissolved oxygen at the surface, g O_2 m^{-3}
KsO2 = 0.05   ; Monod saturation coefficient for O_2, g O_2 m^{-3}
rO2 = −10000*SO2/(KsO2 + SO2)   ; Oxygen consumption inside the floc gO_2 m^3 d^{-1}
init SO2 = 0.1   ; Initial value S_{O2} in the center of the floc (estimate), gO_2 m^{-3}
init SO2' = 0   ; Initial value of dSO2/dt in the center of the floc (at STARTTIME)
SO2'' = −2*SO2'/r−rO2/DO2   ; Balance equation (9.8)
jO2 = −DO2*SO2'   ; Mass flux of oxygen j_{O2}, gO_2 $m^{-2} d^{-1}$

Since the two boundary conditions in Eq. (9.9) do not refer to the same location, the numeric solution of Eq. (9.8) occurs iteratively. The initial value of S_{O2} is improved until the boundary condition at the surface can be satisfied. BM makes a module available which automates this iteration.

9.4.3 Self-purification in a Brook

In a small river or a brook, the biomass that is responsible for the self-purification forms a thin layer, a biofilm, on the surfaces of stones, sediment, and even leaves of higher plants. Nutrients and pollutants are supplied to this biomass from the

flowing water by molecular diffusion. As a consequence of self-purification the concentration of the pollutants decreases along the brook. How can we quantify this decrease?

We analyze the problem in two steps:

- first the behavior of a biofilm in direct contact with the local nutrient concentration is of interest and
- subsequently, the consequence of this behavior on the decrease of the nutrients along the brook can be analyzed.

We describe the river as a heterogeneous system in which the characteristic rate of autopurification can only be captured if both the flowing water and the fixed biofilm are described.

A Simple Biofilm Model

Figure 9.4 represents schematically the geometry and nutrient conditions in a biofilm. Since we expect gradients of the concentration of the nutrients in the biofilm over the depth of the film, we first develop the material balance equation for a small subsystem with the volume $A \cdot \Delta z$. Subsequently, we integrate this balance equation over the entire biofilm and derive the flux of material $j_{S,0}$ through the surface of the biofilm.

With the reaction rate r_S the material balance equation for dissolved pollutants S results in:

$$A \cdot \Delta z \cdot \frac{\partial S}{\partial t} = j_S(z) \cdot A - j_S(z + \Delta z) \cdot A + r_S \cdot A \cdot \Delta z . \qquad (9.11)$$

For the steady state, making the replacements $j_S(z + \Delta z) - j_S(z) = \Delta j_S \to dj_S$ and $\Delta z \to dz$ and dividing by $A \cdot dz$,

$$\frac{dj_S}{dz} = r_S . \qquad (9.12)$$

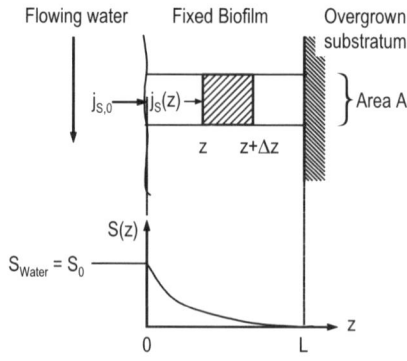

Fig. 9.4 Schematic representation of a biofilm

Within the biofilm neither advection nor turbulence are observed. The transport of soluble materials can be modeled with Fick's law for molecular diffusion

$$j_S = -D_S \cdot \frac{dS}{dz} . \tag{9.13}$$

From Eqs. (9.12) and (9.13) we can isolate the local coordinate z and separate the variables:

$$dz = \frac{dj_S}{r_S(S)} \quad \text{and} \quad dz = -\frac{D_S}{j_S} \cdot dS . \tag{9.14}$$

There remains:

$$j_S \cdot dj_S = -D_S \cdot r_S(S) \cdot dS . \tag{9.15}$$

The conditions for the solution of Eq. (9.15) are:

$r_S = -k \cdot S$:   In rivers which fulfill legal requirements, the pollutant concentrations are small. Their degradation can reliably be described with a first-order reaction.

$j_S = 0$ at $z = L$:   There is no flux of material into the impermeable substratum.

$S = 0$ at $z = L$:   Biofilms will grow ever thicker until at depth no more substrate is present and growth is stopped by lack of nutrients.

$S \approx S_0$ at $z = 0$: This approximation neglects a possible mass transfer resistance in a laminar sublayer at the surface of the biofilm.

$j_S = j_{S,0}$ at $z = L$: This is the amount of pollutant that is lost from the water in the brook to the biofilm.

Upon integration, Eq. (9.15) becomes:

$$\int_0^{j_{S,0}} j_S \cdot dj_S = -D_S \cdot k \cdot \int_0^{S_0} S \cdot dS \tag{9.16}$$

with the solution

$$\frac{j_{S,0}^2}{2} = \frac{k \cdot D_S \cdot S_0^2}{2} \quad \text{or} \quad j_{S,0} = \sqrt{k \cdot D_S} \cdot S_0 = k_f \cdot S_0 . \tag{9.17}$$

The degradation rate of the entire biofilm is first-order relative to the pollutant concentration in the river, S_0. The rate constant $k_f = (k \cdot D_S)^{0.5}$ [L T^{-1}] corresponds to the geometric mean of the diffusion coefficient D_S (a rate constant for transport) and the reaction rate constant k for the reaction inside the biofilm. Thus, transport and reaction affect the behavior of the biofilm in equal ways.

Equation (9.17) describes the interaction between the biofilm and the local pollutant concentration in the brook. The next step consists of describing the effect of the biofilm on the length profile of the pollutant concentration in the river.

A Simple Model of a River

Along the river, too, we expect a gradient of the pollutant concentration. Thus, we select again a small, differential subsystem for which we can derive the material balance. Figure 9.5 shows geometrical definitions in the river.

In the steady state longitudinal concentration gradients are rather small, therefore dispersion can be neglected and the flow rate Q becomes the dominant transport process (see Example 4.19). The material balance equation for the short length Δx of the brook has the form:

$$V \cdot \frac{\partial S}{\partial t} = Q(x) \cdot S(x) - Q(x + \Delta x) \cdot S(x + \Delta x) - j_{s,0} \cdot \Delta A \ . \tag{9.18}$$

For the steady state, after substituting Eq. (9.17) with the replacements $\Delta x \rightarrow dx$, $(Q(x + \Delta x) \cdot S(x + \Delta x)) - (Q(x) \cdot S(x)) = \Delta(Q \cdot S) \rightarrow d(Q \cdot S)$, where $(Q \cdot S)$ is the local load of pollutants, and $\Delta A \rightarrow dA$ as well as division by dA, one has

$$\frac{d(Q \cdot S)}{dA} = -j_{s,0} = -k_f \cdot S \ . \tag{9.19}$$

Dividing Eq. (9.19) by the load $Q \cdot S$ and separating variables leads to

$$\frac{d(Q \cdot S)}{Q \cdot S} = -k_f \cdot \frac{dA}{Q} \ . \tag{9.20}$$

The integral of Eq. (9.20) along the flow distance has the form

$$\int_{x=0}^{x} \frac{d(Q \cdot S)}{Q \cdot S} = -k_f \cdot \int_{x=0}^{x} \frac{dA}{Q} \ . \tag{9.21}$$

Q is constant in sections of the brook without additional influent; it may, however, increase stepwise upon joining with a side arm. If added water does not contain any pollutants, the load $Q \cdot S$ does not change. For this situation Eq. (9.21) has the solution:

$$\ln \frac{(Q \cdot S)_x}{(Q \cdot S)_{x=0}} = -k_f \cdot \sum_i \frac{A_i}{Q_i} \tag{9.22}$$

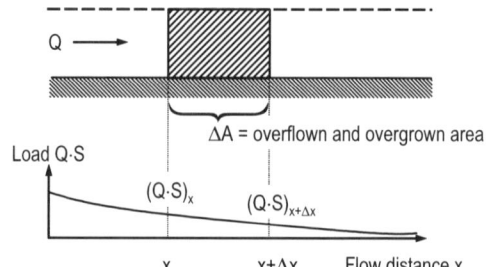

Fig. 9.5 Geometric definitions for the brook

Q_i =   constant flow of water in section i (e.g. between two side brooks) $[L^3 T^{-1}]$
A_i =   wetted surface with biofilm growth in section i (with constant Q_i) $[L^2]$
$(Q \cdot S)_x$ =   pollutant load at location x $[M_S T^{-1}]$
k_f =   self purification constant from Eq. (9.17) $[L T^{-1}]$

Application of the River Model

The following use of Eq. (9.22) is based on a real example.

A small cheese manufactory in the countryside does not have its own wastewater treatment plant. It discharges approximately 2 kgDOC d^{-1} readily degradable soluble organic material (primarily in the form of lactose) into a small brook with the geometry shown in Fig. 9.6. Discharge is evenly distributed over 24 h in order to obtain optimal dilution. Swiss regulations for receiving waters specifies that as a consequence of wastewater discharge no colonies of heterotrophic bacteria which are visible to the naked eye should develop. Based on experiments in model rivers, this demand can be translated into the requirement that the concentration of readily degradable organic compounds should be lower than 0.2 gDOC m^{-3}. *After what flow distance does the pollutant load from the cheese production not induce any bacteria colonies any more that are visible by naked eye?*

Applied to this case Eq. (9.22) becomes

$$\ln \frac{(Q \cdot S)_x}{(Q \cdot S)_{x=0}} = \ln \left(\frac{3500 \text{ m}^3 \text{d}^{-1} \cdot 0.2 \text{ gm}^{-3}}{2000 \text{ gd}^{-1}} \right) = -1.05 = -k_f \cdot \sum_i \frac{A_i}{Q_i}.$$

For the degradation of sugar, $k_f = 1.5$ m d^{-1} is a realistic value at 15°C (derived from experiments in model rivers). Thus, for the flow distance at which visible colonies of heterotrophic bacteria might develop, one has

$$\sum_i \frac{A_i}{Q_i} = \frac{1.05}{1.5 \text{ md}^{-1}} = 0.70 \text{ dm}^{-1}.$$

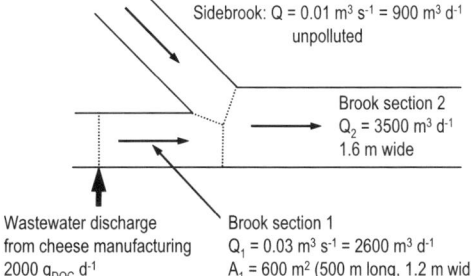

Fig. 9.6 Geometric conditions for the wastewater discharge into the brook

Sidebrook: Q = 0.01 m³ s⁻¹ = 900 m³ d⁻¹ unpolluted

Brook section 2
Q_2 = 3500 m³ d⁻¹
1.6 m wide

Wastewater discharge from cheese manufacturing 2000 g$_{DOC}$ d⁻¹

Brook section 1
Q_1 = 0.03 m³ s⁻¹ = 2600 m³ d⁻¹
A_1 = 600 m² (500 m long, 1.2 m wide)

For the first section this yields

$$\frac{A_1}{Q_1} = \frac{600 \text{ m}^2}{2600 \text{ m}^3\text{d}^{-1}} = 0.23 \text{ dm}^{-1},$$

and for the second section

$$\frac{A_2}{Q_2} = 0.70 - 0.23 = 0.47 \text{ dm}^{-1}.$$

With $Q_2 = 3500 \text{ m}^3\text{d}^{-1}$ this yields $A_2 = 0.47 \cdot 3500 = 1650 \text{ m}^2$. With a mean width of the second section of 1.6 m the polluted length of this section is $L_2 = 1650 \text{ m}^2/1.6 \text{ m} = 1030 \text{ m}$. The total flow distance that is polluted becomes $L_1 + L_2 = 500 + 1030 \text{ m} = 1530 \text{ m}$. The disturbance of the brook is thus visible over a surprisingly long flow distance. Over this flow distance the initial concentration of 0.76 gDOC m^{-3} is reduced to the permissible 0.2 gDOC m^{-3}, partly by self-purification and partly by dilution. In the summer season, when macrophytes provide additional leaf surfaces for biofilm growth, the required flow distance for self-purification may be shorter.

Because Eq. (9.21) was integrated for loads (Q·S), the dilution by influent from side brooks is already considered in the resulting equation. The necessary flow distance is hardly shortened, however, by the dilution. Without dilution the brook would be 1.2 m wide and the required flow distance would be 1520 m. Fast degradation with higher concentration compensates dilution.

Example 9.7: Self purification in a trickling filter

A trickling filter corresponds to a river with a very small flow depth (0.2 mm) and a very large overgrown surface that is responsible for the degradation of the pollutants.
Can the behavior of the pollutants in a river (Eq. (9.22)) be transferred to a trickling filter?
In a trickling filter that is designed for the degradation of organic materials (BOD_5), a surface area of $A = 10 \text{ m}^2$ must be provided per inhabitant. With a wastewater production per inhabitant of $Q = 0.3 \text{ m}^3 \text{ d}^{-1}$ and a self-purification coefficient of $k_f = 1.5 \text{ m d}^{-1}$, the following treatment would result:

$$\ln \frac{\text{Load}_{\text{Effluent}}}{\text{Load}_{\text{Influent}}} = -1.5 \text{ md}^{-1} \cdot \frac{10 \text{ m}^2}{0.3 \text{ m}^3\text{d}^1} = -5.0.$$

This corresponds to an efficiency of $\eta_{BOD5} = 1 - \exp(-5.0) = 99.3\%$, which is clearly too high.
Obviously we cannot transfer the self-purification model without closer inspection to a trickling filter. Here we must consider the role of oxygen and particulate material as well as maximum degradation rates (saturation) of the microorganisms in the biofilm model. They all reduce the self-purification coefficient k_f.

9.4.4 Gas Exchange in a Stirred Tank Reactor

Volatile materials can be stripped from water to the atmosphere by injecting small air bubbles into a reactor: Rising bubbles take up more and more of the material and transfer it to the environment when leaving the reactor.

In order to quantify the removal from a reactor, we must:

- first examine the behavior of an individual bubble and
- subsequently analyze the effect of many bubbles on the reactor.

Analogous to the stripping of volatile materials is the transfer of oxygen (or ozone) from air to water. The models which will be deduced here provide the correct result with the correct choice of the boundary conditions.

A Model for the Rising Bubble

Bubbles with a diameter of 3–5 mm have a typical rising velocity in water of $0.3\,\mathrm{m\,s^{-1}}$. Thus, they remain for approximately 10 s in a reactor with a depth of 3 m. If the mean hydraulic residence time of the stirred tank reactor is $\theta_h > 1000\,\mathrm{s}$, we can assume that over the residence of a single bubble the concentration in the reactor barely changes.

In Fig. 9.7 a rising bubble is drawn. Inside the bubble the gas circulates as indicated due to wall friction. The gas bubble is surrounded by a laminar boundary layer in which water packages shift from top to bottom. These packages seem to a large extent to remain intact. In the short time of contact of these packages with the bubble, material diffuses from the water package into the gas bubble (or in

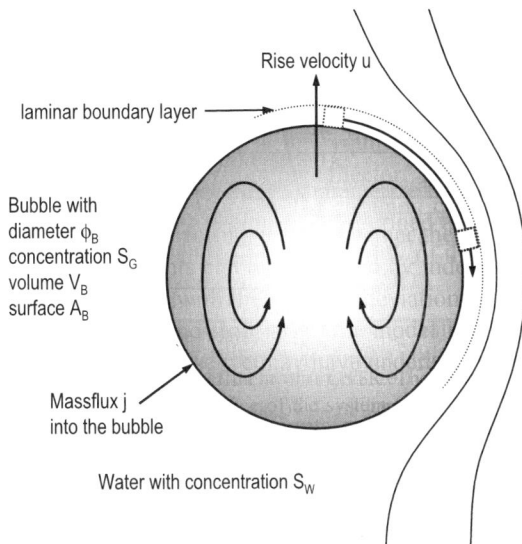

Rise velocity u

laminar boundary layer

Bubble with
diameter ϕ_B
concentration S_G
volume V_B
surface A_B

Massflux j
into the bubble

Water with concentration S_W

Fig. 9.7 Schematic representation of a rising bubble

Fig. 9.8 Evolution of the concentration profile over the gas–water interface and concentration conditions at the interface

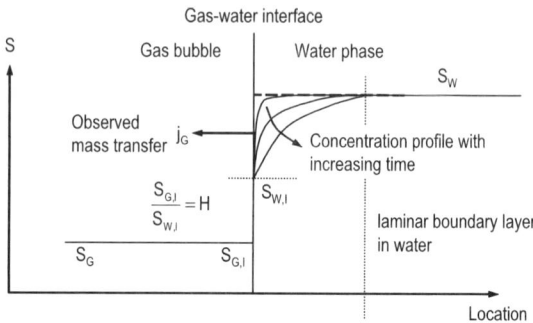

reverse) which is a nonstationary process. At the lower end of the bubble these packages are mixed into the water phase again.

Figure 9.8 shows the concentration profile of a material that diffuses from the water into the gas. Beginning with the concentration in the bulk water (S_W), material diffuses over the phase boundary, thus the concentration in the water package which moves along the bubble decreases with increasing time. At the gas–water interface the material concentration is controlled by a thermodynamic equilibrium: the Henry coefficient H indicates the relationship between the concentration in the gas $S_{G,I}$ and directly at the surface in the water $S_{W,I}$:

$$H = \frac{S_{G,I}}{S_{W,I}} \qquad\qquad (9.23)$$

H   = Henry coefficient, here dimensionless [–] or [$L^3_{water} L^{-3}_{gas}$]
$S_{G,I}$ = concentration of the gas at the phase boundary [$M_S L^{-3}_{air}$]
$S_{W,I}$ = concentration of the material on the water side of the phase boundary [$M_S L^{-3}_{water}$]

Example 9.8: Easily and hardly soluble materials

An easily soluble material has a small Henry coefficient. That leads to the fact that, in contact with a certain concentration in the gas, the compound reaches a high concentration in the water.
A barely soluble material has a large Henry coefficient and thus the material reaches only a low concentration in the water.
Examples of Henry coefficients are:

Oxygen O_2 H = 30   hardly soluble
Carbon dioxide CO_2 H = 1   well soluble, reacts weakly with the water
Ammonia NH_3 H = 0.0006   very well soluble, hydrolyzes immediately in water

---

The mass transfer from water into the gas phase increases as the difference $S_{W,I} - S_W$ becomes larger. In addition the diffusion coefficient D_W of the material in the water and the contact time of the water packages with the surface of the bubble affect mass transfer. Bird et al. (1960) derive the following equation for this situation:

$$j_W = -j_G = \sqrt{\frac{4 \cdot D_W \cdot u}{\pi \cdot \phi_B}} \cdot \left(\frac{S_{G,I}}{H} - S_W \right) = k_1 \cdot \left(\frac{S_{G,I}}{H} - S_W \right)$$ or

$$j_G = -k_1 \cdot \left(\frac{S_G}{H} - S_W \right) \text{ with } k_1 = \sqrt{\frac{4 \cdot D_W \cdot u}{\pi \cdot \phi_B}} \qquad (9.24)$$

j_W, j_G = mass flux over the interface [$M_i\, L^{-2}\, T^{-1}$]
u = rise velocity of bubbles (frequently ≈ 0.3 m s^{-1}) [$L\, T^{-1}$]
ϕ_B = diameter of the bubble [L]
k_1 = mass transfer coefficient [$L\, T^{-1}$]

A material balance for an individual gas bubble (neglecting pressure gradients along the ascent) results in:

$$V_B \cdot \frac{dS_G}{dt} = j_G \cdot F_B = -k_1 \cdot \left(\frac{S_G}{H} - S_W \right) \cdot A_B \qquad (9.25)$$

V_B = volume of the gas bubble [L^3]
A_B = surface area of the gas bubble [L^2]
k_1 = mass transfer coefficient over the interface [$L\, T^{-1}$]

With the assumption that a newly formed gas bubble enters the reactor at time $t = 0$ with a concentration $S_{G,0}$ of the volatile compound, we obtain after separation of variables:

$$\int_{S_{G,0}}^{S_G} \frac{dS_G'}{S_G' - H \cdot S_W} = -\frac{k_1}{H} \cdot \frac{A_B}{V_B} \cdot \int_0^\tau dt \,,$$

with the solution:

$$S_G(\tau) = H \cdot S_W \cdot \left[1 - \exp\left(-\frac{k_1}{H} \cdot \frac{A_B}{V_B} \cdot \tau \right) \right] + S_{G,0} \cdot \exp\left(-\frac{k_1}{H} \cdot \frac{A_B}{V_B} \cdot \tau \right) \qquad (9.26)$$

τ = time since the bubble entered the reactor [T].

Starting from $S_{G,0} = 0$, the increase of the concentration S_G on the course of a rising gas bubble is illustrated in Fig. 9.9. The concentration of the volatile material in the bubble strives asymptotically towards the value of $S_G = S_W \cdot H$, which is called equilibrium or saturation concentration, because no concentration gradients over the interface exist any more. The rate of approach to the asymptote depends on the mass transfer coefficient k_1, the size of the bubble ϕ_B and the Henry coefficient. Small bubbles have a favorable surface-to-volume ratio, which results in a rapid approach to equilibrium.

With very small Henry coefficients ($H \ll 1$) the mass transfer resistance on the gas side must be considered, too. Here we cannot deal with this problem in more detail.

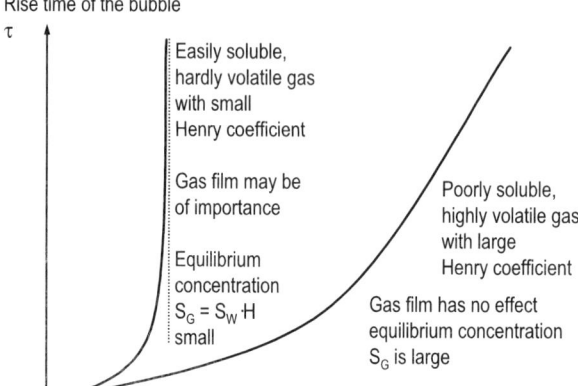

Rise time of the bubble

Fig. 9.9 Increase of the concentration S_G of a volatile material in a rising gas bubble: the situation for a very volatile, barely soluble material ($H \gg 1$) and a barely volatile easily soluble material ($H \ll 1$)

Example 9.9: Rise of a gas bubble to equilibrium

How far must a gas bubble rise until it reaches 99% of the equilibrium between gas and water?
The bubble reaches 99% of equilibrium when (see Eq. (9.26)):

$$\frac{S_G(\tau)}{H \cdot S_W} = 1 - \exp\left(-\frac{k_l}{H} \cdot \frac{A_B}{V_B} \cdot \tau\right) > 0.99 .$$

For spheres one has $\dfrac{A_B}{V_B} = \dfrac{6}{\phi_B}$ (where ϕ_B is the diameter) and thus

$$\tau = \ln(1 - 0.99) \cdot \frac{H \cdot \phi_B}{k_l \cdot 6} .$$

Equation (9.24) has been validated for the estimation of k_l in clean water for bubble diameters of 3–5 mm:

$$k_l = \sqrt{\frac{4 \cdot D_i \cdot u}{\pi \cdot \phi_B}} \quad u = 0.3 \text{ m s}^{-1} = 26000 \text{ m d}^{-1}, \ \phi_B = 0.003 \text{ m} .$$

For different compounds the results are

Gas	H_i	D_i	k_l	τ (99%)	H (99%)
Oxygen	30	$2.09 \cdot 10^{-4} \text{ m}^2 \text{d}^{-1}$	48 m d^{-1}	124 s	37.2 m
Carbon dioxide	1	$1.65 \cdot 10^{-4} \text{ m}^2 \text{d}^{-1}$	43 m d^{-1}	4.62 s	1.38 m
Ammonia	0.0006	$1.69 \cdot 10^{-4} \text{ m}^2 \text{d}^{-1}$	43 m d^{-1}	0.0028 s	0.8 mm

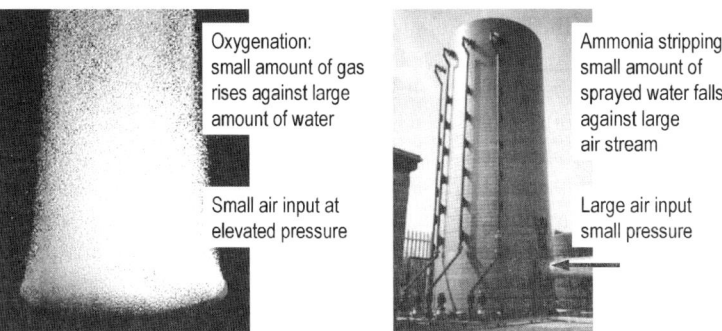

Fig. 9.10 *Left*: fine bubble aerator for oxygen transfer into water, $H_{O2} = 30$; the bubbles rise through the water. *Right*: a *stripping tower* for the removal of ammonia, $H_{NH3} = 0.006$; here drops are falling through a stream of air

While an air bubble must rise for about 37 m in order to reach equilibrium with oxygen, a rise of only 0.8 mm suffices for ammonia. Thus, for ammonia the time of the formation of the bubble is sufficient to reach equilibrium with the surrounding water.

These numbers show that it makes sense to build a reactor for oxygen transfer rather deep, while stripping of ammonia does not improve with increasing reactor depth, but uses more energy for the compression of air. Ammonia is typically stripped from water in so-called stripping towers. Here the water is sprayed in small droplets into an air flow in order to generate short transport distances on the water side. Air can then be pulled through this spray in large amounts with fans at small energy expenditure. Figure 9.10 compares these two systems.

The Model for the Reactor

The stripping of volatile materials from a stirred tank reactor is to be modeled (see Fig. 9.11). Many bubbles rise in the reactor simultaneously with increasing content of volatile material from bottom to top (assuming $S_{G,0} = 0$). We draw the system borders outside of the reactor. Thus, we must only consider the gas and the water flow over the system boundary. Only mass transfer but no reaction is considered (although this could easily be added). Since many bubbles coexist in the reactor, we average over all bubbles:

V_G = volume of all bubbles in the reactor = total volume of gas in the reactor $[L^3]$

A_G = total surface of all gas bubbles in the reactor $[L^2]$

Q_G = flow rate of the gas $[L^3_{gas}\ T^{-1}]$

V = volume of the water in the system $[L^3_{water}]$

a = A_G/V = specific surface of the gas bubbles per reactor volume
        $[L^2_{gas}\ L^3_{water}] = [L^{-1}]$

θ_G = mean residence time of a gas bubble in the reactor
        = τ at exit from the reactor $[T]$

Fig. 9.11 Schematic representation of the stirred tank reactor, from which volatile materials are to be stripped

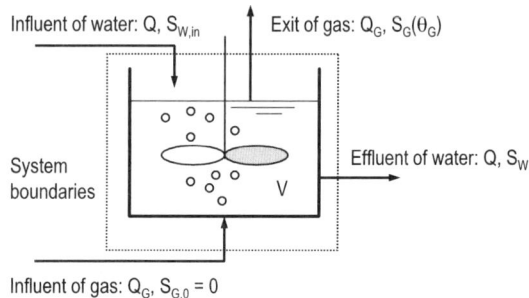

With these definitions one has

$$\frac{A_B}{V_B} = \frac{A_G}{V_G} = \frac{a \cdot V}{V_G} \quad \text{and} \quad \theta_G = \frac{V_G}{Q_G} \quad \text{analogous to Eq. (7.7).} \tag{9.27}$$

For the material balance equation for a volatile material in the stirred tank reactor without reaction we obtain from Eq. (9.26)

$$V \cdot \frac{dS_W}{dt} = Q \cdot (S_{W,in} - S_W) - Q_G \cdot S_G(\theta_G)$$

$$= Q \cdot (S_{W,in} - S_W) - Q_G \cdot H \cdot S_W \cdot \left[1 - \exp\left(-\frac{k_1}{H} \cdot \frac{A_B}{V_B} \cdot \theta_G \right) \right]. \tag{9.28}$$

$Q_G \cdot S_G(\theta_G)$ corresponds to the removal of the volatile material from the reactor into the gas stream. For the steady state substitution of Eq. (9.27) leads to

$$\frac{S_W}{S_{W,in}} = \frac{1}{1 + H \cdot \dfrac{Q_G}{Q} \cdot \left[1 - \exp\left(-\dfrac{k_1}{H} \cdot \dfrac{a \cdot V}{Q_G} \right) \right]}, \tag{9.29}$$

which can be simplified for two cases:

Case 1: H ≪ 1

When H is very small, i. e., the material is well soluble in water (which is partially valid for CO_2 and definitely valid for NH_3)

$$\exp\left(-\frac{k_1}{H} \cdot \frac{a \cdot V}{Q_G} \right) \text{ tends towards 0, so}$$

$$\frac{S_W}{S_{W,in}} = \frac{1}{1 + H \cdot \dfrac{Q_G}{Q}}. \tag{9.30}$$

The rising gas bubbles rapidly approach their equilibrium with water. The gas is saturated and an increased depth of the reactor does not improve performance. θ_G does not influence the result. However, an increased gas flow rate (Q_G) would improve the performance.

Case 2: H >> 1

When H is very large, i. e., the material is hardly soluble in water (which is true for O_2, H_2, N_2, and CH_4 etc.):

$$\exp\left(-\frac{k_1}{H}\cdot\frac{a\cdot V}{Q_G}\right) \text{ tends towards 1.}$$

Expanding $\exp(x)=1+x$ for small x ,

$$\frac{S_W}{S_{W,in}} = \frac{1}{1+\dfrac{k_1\cdot a\cdot V}{Q}}. \tag{9.31}$$

Here reactor performance can be improved primarily by increasing $a\cdot V=A_G$, the surface of the bubbles in the reactor by:

- increasing the gas flow rate to increase a, because more bubbles are simultaneously in the reactor;
- decreasing the bubble size to increase their surface-to-volume ratio;
- building a deeper reactor such that the bubbles remain inside the reactor for longer.

An increase of the reactor volume V would not improve the situation because at constant gas flow rate this would decrease the specific bubble surface, a, by an equal amount.

Thus, with simple assumptions, statements can be made on the behavior of this two-phase system which are very material specific and substantial.

Example 9.10: Stripping perchlorethylene (PER) from groundwater

A groundwater contains 0.1 gPER m^{-3} from a previous accident. The concentration is to be reduced by a factor of 100. Groundwater is pumped at a rate of $Q=0.1\,m^3\,s^{-1}$.

The Henry coefficient of PER amounts to $H_{PER}=1.06$; the diffusion coefficient $D_{PER}=1.1\cdot10^{-9}\,m^2\,s^{-1}$.

How much clean air has to be blown through the groundwater, if a stirred tank reactor is used which is approximately 2 m deep? The aerators supply gas bubbles with a diameter of 3 mm.

$$u=0.3\,m\,s^{-1} \quad \text{thus} \quad \theta_G=H/u=2/0.3=6.7\,s$$

Eq. (9.24) $k_l = (4 \cdot D_{PER} \cdot u / (\pi \cdot \phi_B))^{0.5} = 0.00037 \text{ m s}^{-1}$

$$\frac{a \cdot V}{Q_G} = \frac{A_G}{Q_G} = \frac{V_G \cdot A_B}{Q_G \cdot V_B} = \theta_G \cdot \frac{6}{\phi_B} = 6.7 \cdot \frac{6}{0.003} = 13'400 \text{ m s}^{-1}$$

Substituted into Eq. (9.29) yields for $S_W/S_{W,in} = 0.01$ the ratio $Q_G/Q = 94$.

Thus, approximately $10 \text{ m}^3 \text{ s}^{-1}$ of air must be blown into the reactor, which means a large expenditure of energy. Injecting the air at a depth of 2 m is not efficient; a smaller injecting depth could save energy. With a depth of 1 m only 10% more air would be required, leading to significant savings of energy.

How does the situation improve if a cascade of six equal CSTRs were used? (see also Example 6.9)

With six reactors in series the required performance per reactor is only: $S_{W,i}/S_{W,i-1}$ $= 0.01^{1/6} = 0.46$. The required air volume per reactor is then $Q_G/Q = 1.1$. For all six reactors only $Q_{G,total} = 0.66 \text{ m}^3 \text{ s}^{-1}$ of air are necessary. The energy requirement could be further reduced if the reactors were less than 2 m deep.

9.4.5 Adsorption in an Activated Carbon Column

In Europe the production of drinking water frequently relies on granulated activated carbon in the form of filter columns. Organic materials diffuse from the water into the activated carbon and adsorb onto its very large internal surface. Adsorption is characterized by an equilibrium that is given in the form of an isotherm. In the course of a filter run, the loading of the activated carbon with organic materials increases until no further materials can adsorb, and an increasing residual concentration breaks through the filter.

Here a model for the dynamic behavior of this heterogeneous system is to be developed.

Description of the Processes

Figure 9.12 shows an adsorption column filled with activated carbon. The operation is similar to a rapid sand filter; the water flows from top to bottom of the filter bed. The activated carbon is granular, with a grain size in the range 0.5–3 mm, typically 2 mm. Due to the adsorption the pollutant concentration decreases in the direction of flow, and in response the activated carbon becomes loaded with the adsorbed pollutants. With increasing loading the adsorption efficiency decreases, an adsorption front migrates downwards, and the residual concentration in the effluent increases in the course of time. Finally, the whole activated carbon is loaded with pollutants and the column cannot adsorb any additional materials; the activated carbon must be regenerated.

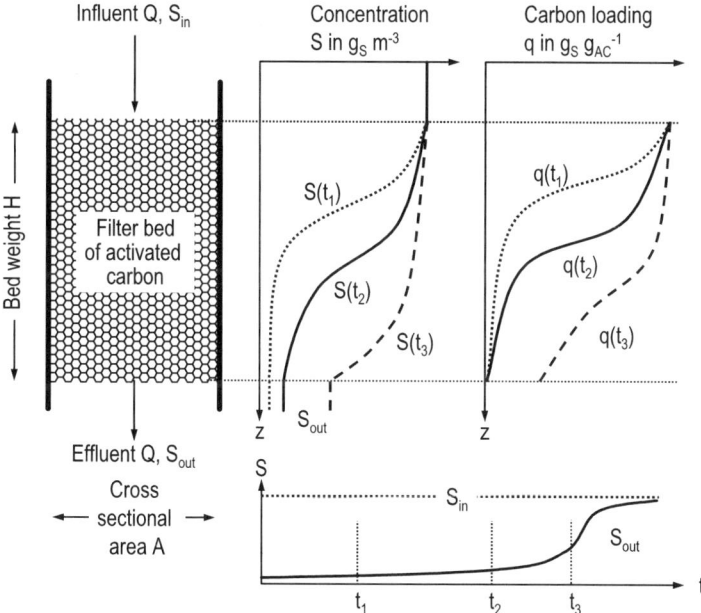

Fig. 9.12 Operating stages of an activated carbon filter with increasing time from t_1 to t_3

Isotherm

A frequently used model for the relationship between the loading of the activated carbon and the concentration of the pollutants in the water is the Langmuir isotherm. It has the following form:

$$q = q_{max} \cdot \frac{S_{equ}}{K_S + S_{equ}} \qquad (9.32)$$

q = loading of the activated carbon with pollutants $[M_S\ M_{AC}^{-1}]$
q_{max} = maximal possible loading $[M_S\ M_{AC}^{-1}]$
K_S = saturation concentration, with this concentration half of the maximum loading is reached $[M_S L^{-3}]$
S_{equ} = pollutant concentration which is in equilibrium with the loading q $[M_S\ L^{-3}]$

Typical values for the two parameters of the Langmuir isotherm are:
q_{max} = 0.15–0.35 gDOC g^{-1} activated carbon
K_S = 5–50 gDOC m^{-3}

A frequent application of Eq. (9.32) is based on its inverse form:

$$S_{equ} = K_S \cdot \frac{q}{q_{max} - q} . \qquad (9.33)$$

If the pollutant concentration S is larger than S_{equ}, there is a tendency for the activated carbon to adsorb more pollutants; if it is smaller, pollutants may start to desorb (become redissolved).

Example 9.11: Maximum loading of an activated carbon

Activated carbon, as used in the drinking water processing, has an inner surface of approximately $1500\ m^2\ g_{AC}^{-1}$ onto which organic material can adsorb. Carbon atoms in the graphitic structure of activated carbon have a separation of 3–4 Å $(1\ \text{Å} = 10^{-10}\ m)$. Thus, each C atom occupies a surface of approx. $10\ \text{Å}^2$. This results in the case of full occupation of the internal surface with adsorbed materials:

$$q_{max} = \frac{1500\ m^2 g_{AC}^{-1}}{10 \cdot 10^{-20}\ m^2\ /\ \text{C Atom adsorbed}} \cdot \frac{12\ g_{AC}\ /\ \text{Mol C}}{6 \cdot 10^{23}\ \text{C Atoms}\ /\ \text{Mol C}} = 0.3 \frac{g\ C_{adsorbed}}{g\ C_{AC}}$$

This corresponds to the maximal possible loading of activated carbon, as can be determined experimentally.

Balance Equations

In order to describe the time-dependent development of an adsorption column, we must derive the balance equations for the pollutants in the flowing water and the adsorbed materials on the fixed activated carbon.

For the adsorption column schematically shown in Fig. 9.13 the balance for the dissolved material S in the flowing water has the following form. With constant influent load on the column, the dispersion in the column can be neglected as a transport process, so

$$\Delta z \cdot A \cdot \varepsilon \cdot \frac{\partial S}{\partial t} = Q \cdot S(z) - Q \cdot S(z + \Delta z) - a \cdot j_{ads} \cdot \Delta z \cdot A \ . \tag{9.34}$$

For a mass q of adsorbed materials on the activated carbon one has

$$\Delta z \cdot A \cdot \rho_{AC} \cdot \frac{\partial q}{\partial t} = a \cdot j_{ads} \cdot \Delta z \cdot A \tag{9.35}$$

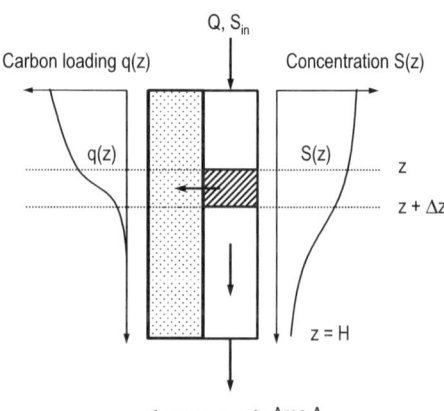

Fig. 9.13 *Left*: schematic representation of an adsorption column with loading of activated carbon.
Right: pollutant concentration in the water phase

A, Q, S, z as defined in Fig. 9.13

ε = porosity of the adsorbent column (fraction of flowing water in the cross section) [–]

a = external surface of the activated carbon grains per column volume [$L^2 L^{-3}$]

j_{ads} = mass flux of material from the water to the activated carbon [$M_S L^{-2} T^{-1}$]

ρ_{AC} = density (packed density) of the activated carbon [$M_{AC} L^{-3}_{Reactor}$]

The flux of material that is adsorbed can be approximated with the following simple model:

$$j_{ads} = k_1 \cdot \left(S - S_{equ}\right) = k_1 \cdot \left(S - K_S \cdot \frac{q}{q_{max} - q}\right) \tag{9.36}$$

k_1 = mass transfer coefficient which characterizes the flux of material through the laminar boundary layer around the activated carbon grain and the transport inside the grain [$L T^{-1}$]

Equation (9.36) expresses that the flux of material goes toward the activated carbon if the pollutant concentration S is larger than the equilibrium concentration S_{equ}, which indicates that there is still adsorption capacity left. k_1 depends on the flow velocity in the column, the grain size, and the temperature. Typical values are in the range $k_1 = 1–2.5–5$ m d^{-1}.

With the replacements $\Delta x \to \partial x$, $\Delta S \to \partial S$, $\Delta q \to \partial q$ and the substitution of Eq. (9.36) one obtains

$$\frac{\partial S}{\partial t} = -\frac{Q}{A \cdot \varepsilon} \cdot \frac{\partial S}{\partial z} - \frac{k_1 \cdot a}{\varepsilon} \cdot \left(S - S_{equ}\right), \tag{9.37}$$

$$\frac{\partial q}{\partial t} = \frac{k_1 \cdot a}{\rho_{AC}} \cdot \left(S - S_{equ}\right), \tag{9.38}$$

$$S_{equ} = K_S \cdot \frac{q}{q_{max} - q}. \tag{9.33}$$

With Eqs. (9.37) and (9.38) we have developed the model for the adsorption column. These equations must be solved numerically. We discretize the column in the form of a cascade of stirred tank reactors. Equations (9.37) and (9.38) must be adapted to

$$\frac{dS_i}{dt} = \frac{Q}{A \cdot \varepsilon} \cdot \frac{n}{H} \cdot \left(S_{i-1} - S_i\right) - \frac{k_1 \cdot a}{\varepsilon} \cdot \left(S_i - S_{equ,i}\right), \tag{9.39}$$

$$\frac{dq_i}{dt} = \frac{k_1 \cdot a}{\rho_{AC}} \cdot \left(S_i - S_{equ,i}\right), \tag{9.40}$$

$$S_{equ,i} = K_S \cdot \frac{q_i}{q_{max} - q_i}. \tag{9.41}$$

n = number of discrete nodes (stirred tank reactors) [–]
S_i = concentration in the ith node (CSTR) [$M_S L^{-3}$]
q_i = carbon loading at the ith node (CSTR) [$M_S M^{-1}_{AC}$]

Example 9.12: Specific surface of activated carbon grain in a fixed bed

While the specific internal surface of the activated carbon on which the pollutants adsorb is independent of the grain size, the phase boundary between the grain surface and water depends on the grain size. *What is the relationship between the grain diameter ϕ and the specific external surface per reactor volume a?*

In a control volume V we find the following number of spherical grains:

$n = \dfrac{(1-\varepsilon)\cdot V}{\dfrac{\pi\cdot\phi^3}{6}}$, where ε is the porosity or free water fraction. These grains have the

following external surface: $A = n\cdot\pi\cdot\phi^2$.

The specific surface is: $a = \dfrac{A}{V} = \dfrac{6\cdot(1-\varepsilon)}{\phi}$.

Typical values are: $\varepsilon = 0.40$ and $\phi = 0.002$ m, and thus $a = 1800\,m^2\,m^{-3}_{Reactor}$.

Example 9.13: Code for the implementation of the model of the adsorption column

The following code implements the model of the adsorption column. The parameter values are selected in a realistic range (tested)

```
METHOD RK4          ; Integration with fourth-order Runge–Kutta
STARTTIME = 0        ; Beginning
STOPTIME=150         ; End of simulation, d
DT = 0.0001          ; Time step, d
DTout = 1            ; Time step for outputs, d
Q0 = 2000            ; Influent, m³ d⁻¹
S0 = 2               ; Influent concentration, adsorbable, g_DOC m⁻³
kla = 5000           ; rate constant for adsorption, d⁻¹
A = 10               ; Cross section of the column, m²
eps = 0.40           ; Porosity of the carbon bed, –
rho = 420000         ; space density of the carbon, g_AC m⁻³_Reactor
qmax = 0.3           ; maximum loading, g_DOC g⁻¹_AC
KS = 10              ; Saturation concentration, g_DOC m⁻³
H = 3                ; Height of the column, m
n = 30               ; Number of discrete nodes, –
D = n*Q0/(H*A*eps)   ; Transport rate per element, d⁻¹
init S[1..n] = 0     ; Concentration profile, g_DOC m⁻³
init q[1..n] = 0     ; Loading of the carbon, g_DOC g⁻¹_AC
d/dt(S[1]) = D*(S0–S[1])–kla*(S[i]–Sequ[i])/eps
                     ; Eq. (9.39) considering boundary condition
```

d/dt(S[2..n]) = D*(S[i−1]−S[i])−kla*(S[i]−Sequ[i])/eps   ; Eq. (9.39)
d/dt(q[1..n]) = kla*(S[i]−Sequ[i])/rho   ; Eq. (9.40)
Sout = S[n]   ; Effluent concentration, g_{DOC} m^{-3}
sequ[1..n] = q[i]*KS/(qmax−q[i])
    ; Equilibrium concentration, Eq. (9.41), g_{DOC} m^{-3}
adsorbed = arraysum(q[*])*H*A*rho/n   ; Sum of adsorbed materials, g_{DOC}

Dynamic Behavior of the Activated Carbon Column

With the help of the code in Example 9.13, which is based on realistic parameters for adsorption in water treatment, we can simulate the development of the performance (DOC removal) and the loading of the carbon over time. Figure 9.14 shows how in the course of time a continuous front of the pollutant concentration moves through the column. At the same time the loading of the activated carbon in the top of the column increases. After 200 days the adsorptive capacity of the column is exhausted, and the activated carbon must be replaced and regenerated.

Figure 9.15 shows that, long before the exhaustion of the adsorption capacity, the pollutants will break through the filter. After about 120 days the adsorption performance decreases and the carbon should be replaced. Since the regeneration of the activated carbon is expensive, one tries to operate, e.g., two columns in

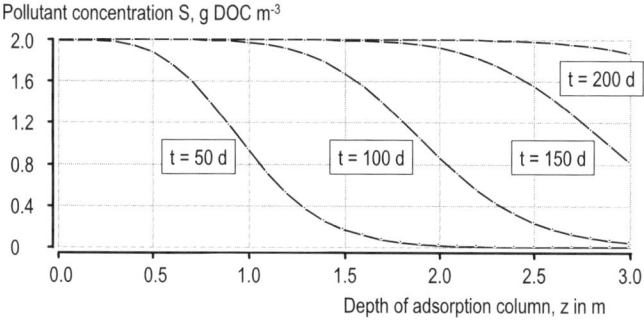

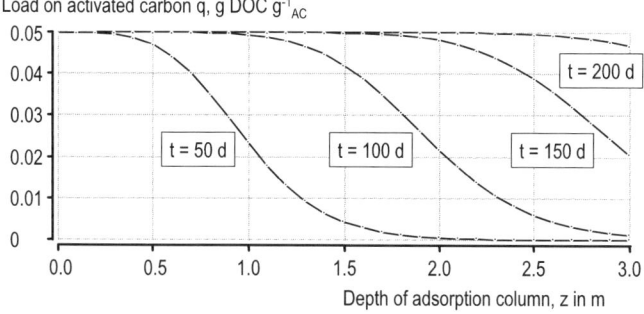

Fig. 9.14 *Above*: course of the pollutant concentration S over the depth of the adsorption column. *Below*: Associated loading of the activated carbon

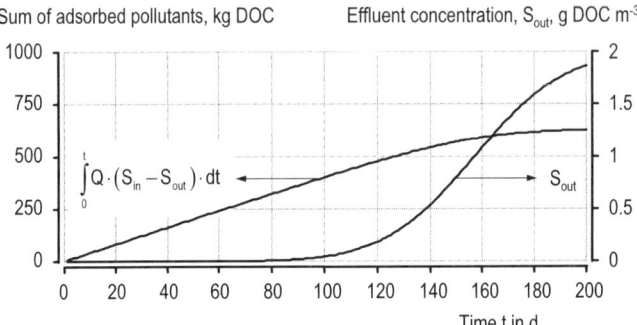

Fig. 9.15 Time course of the effluent concentration of the pollutants S_{out} and the total mass of adsorbed pollutants

series. This allows the operation of a column until the adsorption capacity is exhausted. Once the second column breaks through, it will become the first and the first column is refilled with fresh, regenerated coal and becomes the second filter.

Backwashing of an Activated Carbon Column

If suspended solids reach the activated carbon column, they may be retained in the filter, and the head loss will increase over time. A large head loss can only be reduced by either reduction of the hydraulic load or backwashing of the activated carbon. Backwashing may cause complete mixing of the activated carbon filter bed with very negative effects on filter performance, because now coal which is already loaded arrives in the proximity of the effluent, leading to an increase in the equilibrium concentration S_{equ}. Figure 9.16 shows this result, if a backwashing procedure becomes necessary every 50 days. Even after the first backwashing procedure the performance deteriorates and no longer meets expectations. Obvi-

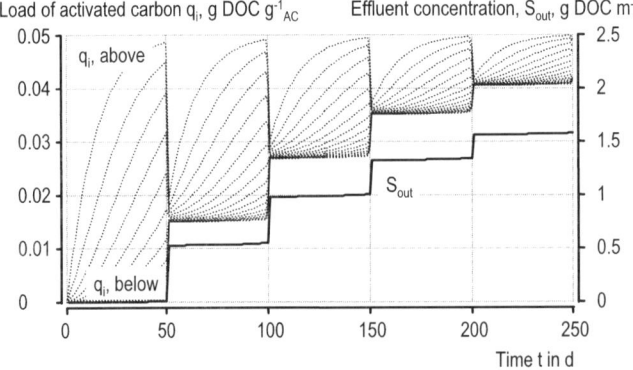

Fig. 9.16 Progress of an adsorption cycle, if the activated carbon bed is completely mixed every 50 days as consequence of backwashing

ously efficient adsorption requires good pretreatment of the water so that only the smallest quantities of particulate materials reach the activated carbon.

In the simulation that led to Figure 9.16, the loading of the activated carbon after each backwashing process was averaged over the entire column, i. e., all activated carbon grains had the same loading after backwashing. In reality the individual carbon grains have their own history and thus are differently loaded before backwashing. The performance of the resulting random mixture of differently loaded activated carbon grains after backwashing may deviate from our prediction based on a homogeneous filter bed with averaged loading. The expenditure required to follow each grain individually is, however, large and would require stochastic models.

Chapter 10
Dynamic Behavior of Reactors

The dynamic, time-dependent behavior of reactors depends equally on the variation of the load and the hydraulic details of the reactor (transport and mixing) as on the transformation processes. Many effects are counterintuitive, and accordingly we cannot capture them without careful considerations and computations.

The discussion of how different kinds of dynamic loads and degradation processes affect the performance of the different ideal reactors will be based on case studies. The practical treatment of such problems requires the use of numeric integration, and therefore appropriate software is necessary.

Example 10.1: Dynamic load of a completely mixed and a plug-flow reactor

Two reactors of equal size, one a CSTR, one a plug-flow reactor with turbulence, are both loaded with wastewater, which contains a material that is subject to degradation in a first order-reaction. As a consequence of a rain event, the flow of water is doubled. Since the pollutant is discharged independent of the rain event, its load remains constant, but its concentration is halved by dilution.
What effect does a rain event have on the effluent concentration?
Figure 10.1 shows the computed paths of the concentrations. Initially (in the steady state) the plug-flow reactor performs better, as its hydraulics are more favorable. With the beginning of the rain, after 1 h conditions change. Over time, the rain event has a lasting but quite different effect on both systems.
The interaction of load, transport, and reaction is complex and cannot be predicted without careful considerations.

Influent m^3 h^{-1}, concentration g m^{-3}

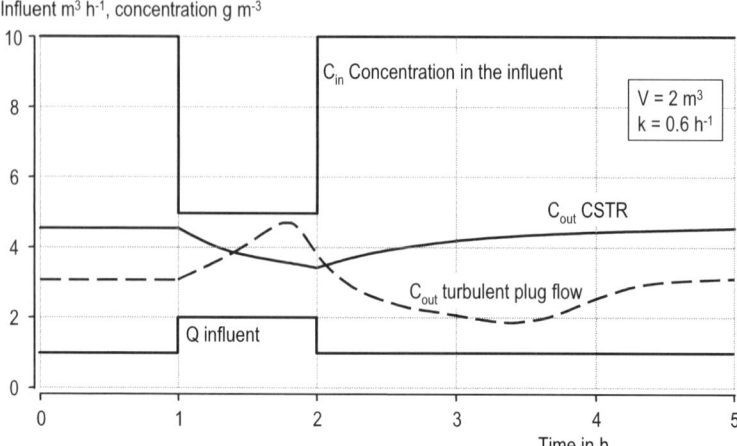

Fig. 10.1 Time course of the inlet and effluent concentration during a rain event. The load Q·C$_{in}$ remains constant. The effluent concentrations refer to a stirred tank reactor and a turbulent plug flow reactor, respectively. The material is subject to a slow first-order degradation process

10.1 Causes of the Dynamics

Both in water supply and in urban drainage, we do not load the systems with temporally constant disturbancies. Accordingly these systems must be arranged in such a way that they can deal with varying loads. Whereas in water treatment this is frequently done via storage in reservoirs, wastewater treatment plants are exposed to extraordinarily large and rapid load fluctuations

There are many effects that can lead to dynamic, time-dependent phenomena in the behavior of urban water systems and treatment plants. We differentiate between the following cases:

- *Case 1.* Variation of the disturbances (the loads) is caused by an individual event that is frequently of stochastic nature, i.e., we can neither predict the time nor the extent exactly: rain events, accidents, the fire brigade, and incorrect control manipulations, etc.
- *Case 2.* The disturbances are subject to periodic variations. We are dealing with foreseeable events that repeat themselves regularly in very similar form: diurnal variation as a consequence of the societal and industrial activities; weekly, seasonal, and yearly variations: production cycles, etc.
- *Case 3.* The load is subject to a long-term trend: growth or decrease of the population, economic growth or recession, change in societal habits or industrial production processes, reduction of the specific water consumption, etc.
- *Case 4.* The plants are operated intermittently or cyclically. Batch processes, SBR systems (sequencing batch reactor), intermittent aeration, binary controllers, etc.

- *Case 5.* Disturbances or operational parameters are subject to smaller, overlaid, stochastic processes that we cannot or do not want to capture in detail; these are considered as noise.
- *Case 6.* The different transport processes of waves (celerity) and materials (flow velocity) in sewers can contribute to the emergence of dynamic loads: sewers, primary clarifiers, etc. (see Fig. 10.1 and Example 10.2).

Typically, these different causes (cases 1–6) are superimposed and frequently must be included together into the prediction of the behavior of a planned system.

Primarily the transformation rates couple the model equations (material balance equations). They frequently do not represent linear relationships so that analytic solutions of these equations can only rarely be found. If qualitative considerations are not sufficient to answer the questions posed, the differential equations must be solved numerically. Depending upon the problem, different simulation programs are available which either allow some predefined systems to be dealt with or else require the relevant model equations to be derived and coded. The well-known simulation programs for water supply systems and storm water management, EPANET and SWMM, are examples of the first group, *Berkeley Madonna* belongs to the second group.

The forward integration of ordinary differential equations requires the initial conditions to be specified, while for partial differential equations the boundary conditions for all state variables must also be given. If these conditions do not result from the assignment or the problem at hand, it is often reasonable to begin the simulation from a steady state for average operating conditions. A simple, but sometimes costly (in computer time) possibility to obtain the steady state of a system is relaxation. Here we start with best estimates for the initial and boundary conditions and use constant disturbances in the forward integration of the original set of dynamic material balance equations until the results become constant and thus independent of the initial conditions. If the disturbances are subject to periodic variations, then a *cyclic steady state* can be defined, which is found by integration through several cycles until the results for an entire period do not depend on the initial conditions any more.

Example 10.2: Behavior of an idealized primary clarifier during a rain event

Figure 10.2 presents a primary clarifier that we consider to be an ideal plug-flow reactor. During a rain event a pollutant is produced with a constant load (see also Example 10.1), however, its concentration is diluted by the additional rain water (an example of such a material is ammonium, which originates in households but not in rainwater). The effluent flow reacts immediately to the increased influent; the pollutant concentration, however, will first remain at the old undiluted level. Only after the entire volume of the clarifier has been exchanged with diluted water will the lower concentration reach the effluent. The result is a strongly variable load (and a potential overloading) of the following biological treatment process, although in the influent to the plant the load remained constant.

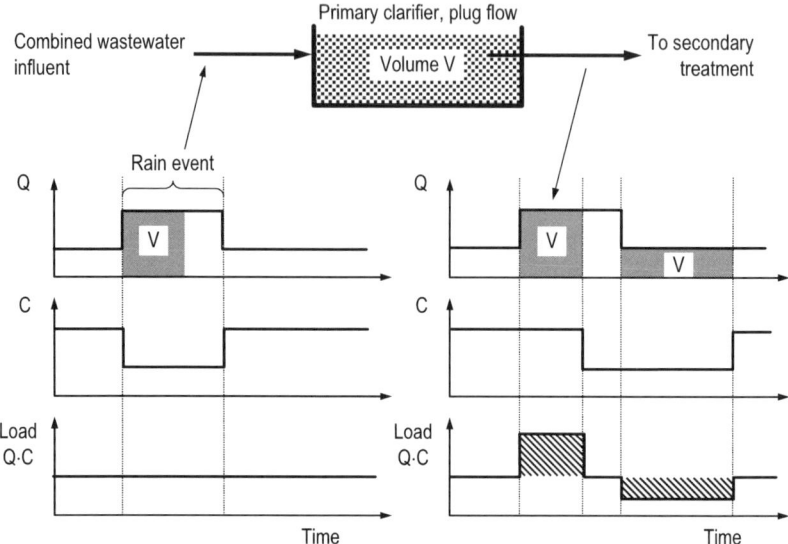

Fig. 10.2 Influent and effluent of an idealized primary clarifier during a rain event. Due to the hydraulic conditions the constant load in the influent becomes a strongly variable load in the effluent (or influent to the following biological treatment plant)

10.2 Adjustment to Step Changes in Load

Intuitively we assume that a mixed reactor with a hydraulic residence time of 24 h will hardly react to load variations of 1 h duration, if the change of the load is not very large. This assumption is wrong. The better the treatment performance of an ideally mixed reactor, the faster its reaction to variations in load and the faster it will reach a new steady state.

The following balance equation describes a material subject to a first-order reaction in a stirred tank reactor:

$$\frac{dS}{dt} = \frac{Q}{V} \cdot (S_{in} - S) - k \cdot S .$$

With the dilution rate $D = Q/V$ we obtain

$$\frac{dS}{dt} = D \cdot (S_{in} - S) - k \cdot S . \tag{10.1}$$

For time periods t_1 to t with temporally constant load (Q, D, S_{in}) Eq. (10.1) can be integrated

$$\int_{S(t_1)}^{S(t)} \frac{dS'}{D \cdot S_{in} - (k + D) \cdot S'} = \int_{t_1}^{t} dt'$$

with the solution

$$t - t_1 = -\frac{1}{D+k} \cdot \ln\left(\frac{D \cdot S_{in} - (D+k) \cdot S(t)}{D \cdot S_{in} - (D+k) \cdot S(t_1)}\right)$$

or

$$S(t) = \underbrace{S(t_1) \cdot \exp\left(-(D+k) \cdot (t - t_1)\right)}_{\text{Becoming independent of initial conditions}} + \underbrace{S_{in} \cdot \frac{D}{D+k}}_{\text{New steady state}} \cdot \underbrace{\left[1 - \exp\left(-(D+k) \cdot (t - t_1)\right)\right]}_{\text{Asymptotic approach to new steady state}}.$$

(10.2)

Both the decay of the initial condition $S(t_1)$ and the adjustment to the new steady state occur with the rate $D+k$. If the pollutant is to be degraded efficiently in the plant, then $k \gg D$ applies. Thus, the adjustment rate is controlled by the reaction (k) and not by the hydraulic residence time ($\theta_h = 1/D$).

Figure 10.3 shows results from Eq. (10.2) for different reaction rates k. Starting from steady state at time $t < 0$, the concentration is doubled for $t \geq 0$ but the flow is kept constant. For materials with slow degradation ($k \approx 0$) the new steady state is reached after about 2 hrs (equivalent to four hydraulic residence times $\theta_h = 1/D = 0.5$ h). With increasing degradation (k increases) the effluent adapts ever faster to the new steady state. With $k = 15$ d^{-1} only $D/(D+k) = 12\%$ of the inlet concentration remain at steady state and the new steady state is reached after approximately 0.2 h.

The derivative of Eq. (10.2) with respect to time for the situation in Fig. 10.3 results for $t_1 = 0$ in

$$\left.\frac{dS}{dt}\right|_{t=0} = D \cdot \left(S_{in}(t \geq 0) - S_{in}(t < 0)\right) = D \cdot \Delta S_{in}\big|_{t=0},$$

which is independent of the reaction rate k; i. e., immediately after the concentration change at time $t = 0$ the concentration in the effluent increases in all cases at the same rate. With rapid degradation the new steady state changes only slightly and is therefore reached again very rapidly.

With a fast reaction (large degradation) a small increase of the effluent concentration (ΔS_{out}) results. The additional amount of material stored in the reactor ($\Delta M = \Delta S_{out} \cdot V$) is also small, but is fed into the system with a large change of the disturbance in the influent ($Q \cdot \Delta S_{in}$). The smaller ΔS_{out} is, the more quickly the new steady state is reached again.

In plug-flow reactors the advection outweighs the mixing; the turbulence number of N_T is small. Accordingly the hydraulic residence time becomes ever more important with decreasing turbulence. In the ideal plug-flow reactor, the duration of the adjustment to the new steady state is exclusively fixed by the hydraulic residence time (Fig. 10.4), independent of the reaction rate.

Mixing, advection, and degradation together determine the response of a reactor to load variations. The larger the mixing and the faster the reaction, the sooner a variation in load becomes apparent in the effluent of a system.

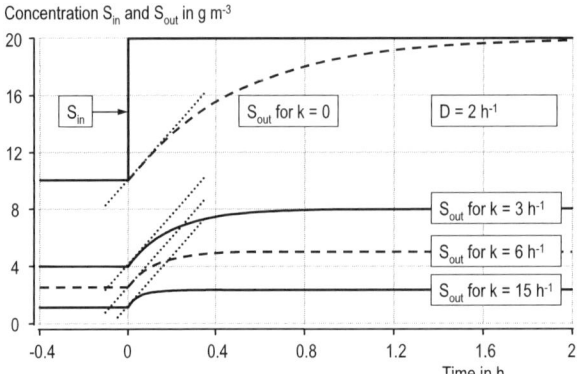

Fig. 10.3 Change of the effluent concentration of a material in a stirred tank reactor after a concentration step change in the influent at t = 0. The material is degraded in a first-order reaction. Traces starting from steady state are given for different reaction rates k

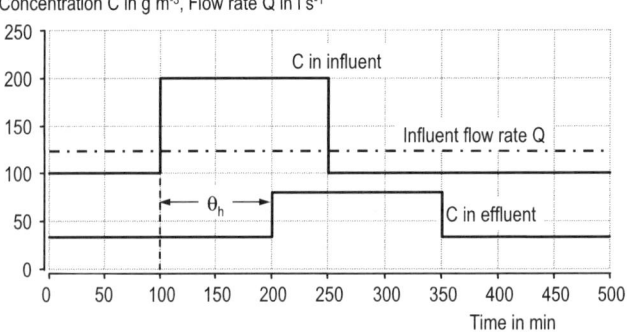

Fig. 10.4 Behavior of the concentration in the effluent of a plug flow reactor during and after a concentration step change in the influent at t = 0. The behavior is controlled by the hydraulic residence time (time shift) and the reaction (concentration level)

Example 10.3: Reaction of a nitrifying activated sludge plant to the diurnal variation of the ammonium load

Figure 10.5 indicates the fast reaction of the ammonium concentration in a completely mixed activated sludge tank after a step change of the load in the influent. The performance of the nitrification is no longer sufficient, and the ammonium accumulates in the tank. For an increase of the concentration in the basin from $0.5\,\mathrm{g\,m^{-3}}$ during the night to approximately $4\,\mathrm{g\,m^{-3}}$ at the peak time, only a short time is necessary. The inlet concentration, which increases from 6 to $22\,\mathrm{g\,m^{-3}}$, fills this reservoir quite rapidly. The concentration in the effluent from the stirred tank reactor reacts directly to the change in the influent composition.

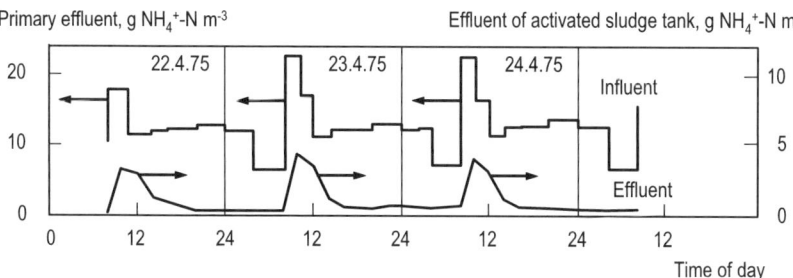

Fig. 10.5 Development of the ammonium concentration in the influent and effluent of a completely mixed, nitrifying activated sludge tank. Results from pilot experiments for the design of the wastewater treatment plant Werdhölzli in Zurich (Gujer 1976)

10.3 Periodic Load Variation

Cyclic load variations are of paramount importance in wastewater treatment. Daily and weekly load patterns affect biological processes; yearly variation of temperature is decisive for the performance of microorganisms and the rate of chemical reactions as well as physical processes.

Any kind of load pattern can be expressed approximately with the help of Fourier transformation as a sum of periodic oscillations. An understanding of the behavior of reactor systems under periodic load variations is thus of special interest.

Here we will discuss the situation in which the flow rate remains constant but the pollutant is subject to sinusoidal concentration variations in the influent according to:

$$S_{in} = S_{m,in} \cdot \left(1 + A_{in} \cdot \sin(2 \cdot \pi \cdot f \cdot t)\right)$$

$$(10.3)$$

$S_{in}(t)$ = concentration in the influent $[M\ L^{-3}]$
$S_{m,in}$ = mean (average) value of S_{in} $[M\ L^{-3}]$
A_{in} = relative amplitude of the sinusoidal variation, $A_{in} < 1$ [−]
f = frequency of the variation $[T^{-1}]$
t = time [T]

The variable flow rate does not lead to results which are different in principle. The degradation in each case is first order, according to $r_S = -k \cdot S$.

10.3.1 Stirred Tank Reactor

With a sinusoidally varying inlet concentration and constant inlet flow rate Q, the material balance for the stirred tank reactor results in:

$$\frac{dS}{dt} = \frac{1}{\theta_h} \cdot \left(S_{m,in} \cdot \left(1 + A_{in} \cdot \sin(2 \cdot \pi \cdot f \cdot t)\right) - S\right) - k \cdot S \tag{10.4}$$

$\theta_h = V / Q =$ hydraulic residence time [T]

Equation (10.4) has an asymptotic solution (independent of initial conditions, cyclic steady state) of the form:

$$S = S_{out} = S_{m,in} \cdot \left(\frac{1}{1 + k \cdot \theta_h} + A_{out} \cdot \sin(2 \cdot \pi \cdot f \cdot t + \chi)\right), \tag{10.5}$$

$$\frac{A_{out}}{A_{in}} = \frac{1}{\sqrt{\left(1 + k \cdot \theta_h\right)^2 + \left(2 \cdot \pi \cdot f \cdot \theta_h\right)^2}} . \tag{10.6}$$

$$\chi = -\text{arctg}\left(\frac{2 \cdot \pi \cdot f \cdot \theta_h}{1 + k \cdot \theta_h}\right) \tag{10.7}$$

$A_{out} =$ relative amplitude of the concentration in the effluent [–]
χ = phase shift relative to the influent [–]

Equation (10.5) describes the variation of the concentration in the effluent. The average value $S_{m,out}$ is reduced by the factor $1/(1+k\cdot\theta_h)$ as we know from the steady-state result. This average value is overlaid with a sinusoidal variation, with attenuated amplitude A_{out} and a phase shift χ relative to the variation in the influent. Equation (10.6) describes the attenuation of the variation, which is composed of a portion caused by degradation $(1/(1+k\cdot\theta_h))$, and a second part $(2\cdot\pi\cdot f\cdot\theta_h)$ that stems from mixing of the influent into the reactor. Equation (10.7) permits the computation of the phase shift, which is always smaller than $\pi/2$.

Figure 10.6 shows the resulting attenuation of the effluent concentration for different frequencies of the disturbance. The effluent concentrations of the three systems are compared in Fig. 10.7. With constant k it is primarily the dimensionless product $f\cdot\theta_h$ that determines the attenuation. The high-frequency disturbance (Fig. 10.6 *top*, $f\cdot\theta_h = 5$) is nearly completely filtered out. Here it would be possible to obtain the system performance from the average influent concentration to $S \approx S_{m,in}/(1+k\cdot\theta_h)$. The situation with a low frequency of disturbance (Fig. 10.6, *bottom*, $f\cdot\theta_h = 0.05$) is entirely different. The variation is so slow that attenuation does not become effective in the comparatively small reactor volume. Here the performance is determined by the instantaneous steady state, $S(t) \approx S_{in}(t)/(1+k\cdot\theta_h)$, and a static model would be sufficient. The situations in between, with a moderate frequency of the disturbance ($f\cdot\theta_h = 0.5$), can only be described by Eq. (10.5) which considers the full dynamics of the system.

Example 10.4: Interpretation of a dynamic mass balance

The influent Q of a stirred tank reactor is constant. It contains two materials, whose identical concentration varies harmoniously. One material is degraded in a

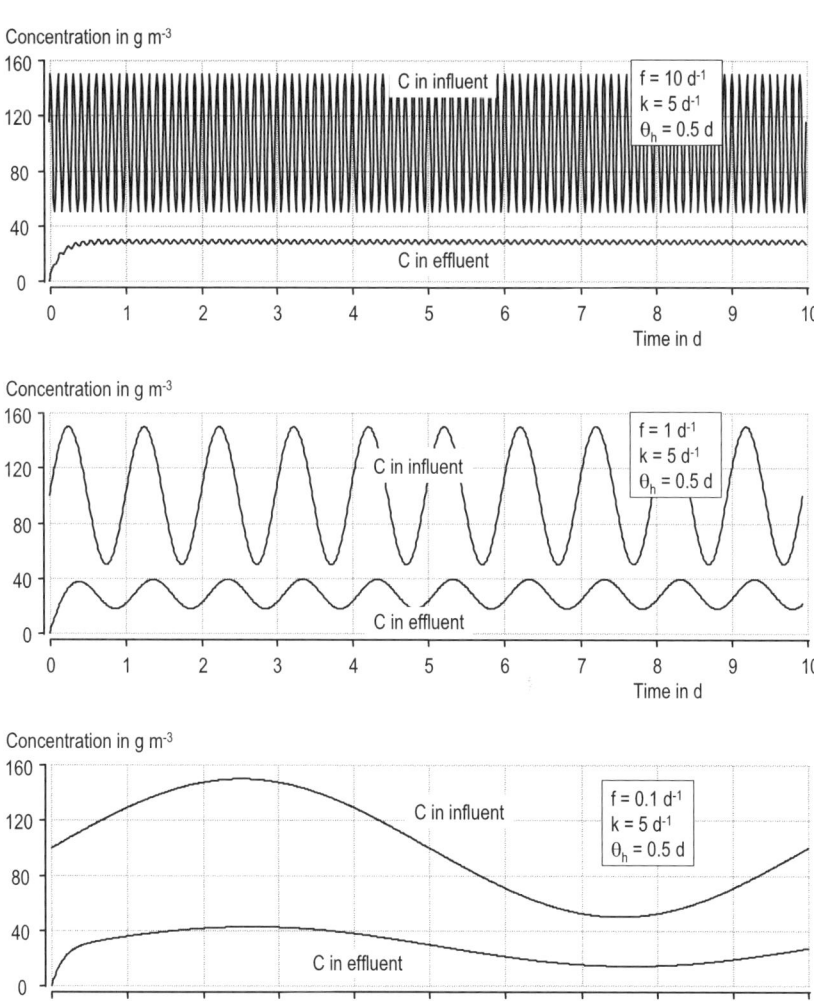

Fig. 10.6 Sinusoidal varying influent and effluent concentrations of a material in a stirred tank reactor. The reaction is first order. *Top*: high-frequency disturbance ($f \cdot \theta_h = 5$). *Middle*: the moderate frequency of the disturbance is comparable to the inverses of the hydraulic residence time ($f \cdot \theta_h = 0.5$). *Bottom*: very slow variation ($f \cdot \theta_h = 0.05$)

first-order reaction, the other in a zero-order reaction. The results (here simulated) of an experiment are shown in Fig. 10.8. The balance for the stirred tank reactor has the form:

$$\frac{dC}{dt} = \frac{Q}{V} \cdot (C_{in} - C) + r \quad \text{with} \quad r = -k \cdot C \quad \text{or} \quad r = -k_0 \, .$$

Questions:

1. Which material is degraded in a first-order reaction, and which in a zeroth-order reaction?

At the intersections of the concentration curves $C_{in} = C$ or $C_{in} - C = 0$ and thus $dC/dt = r$. At points 1 and 2 the slope dC/dt has the same gradient, which suggests to a zeroth-order reaction for material 1. At the points 3 and 4 the gradient is reducing with the concentration, which points to a first-order reaction.

2. How large are the two reaction rates?

These result from the gradients of C_1 and C_2 at the points 1–4.

3. What is the hydraulic residence time of the stirred tank reactor?

This can be obtained from the phase shift between points 5 and 6 or 7 and 8 with the help of Eq. (10.7). Alternatively the ratio of the amplitudes gives an indication.

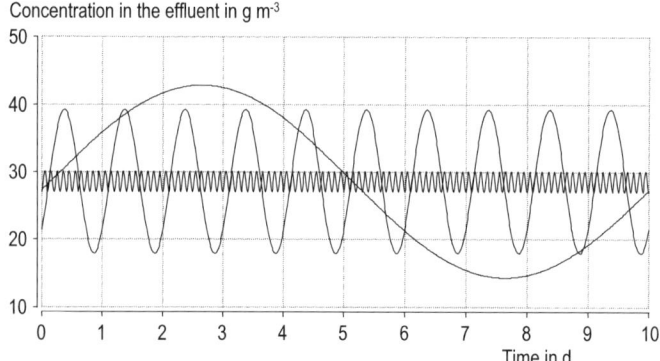

Fig. 10.7 Comparison of the effluent concentration of the three systems shown in Fig. 10.6

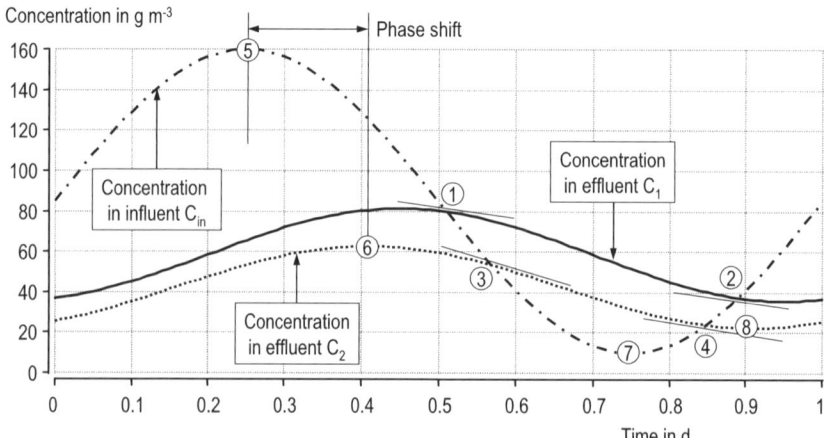

Fig. 10.8 Results of a dynamic experiment with two materials: see Example 10.4 for details

10.3.2 Cascade of Stirred Tank Reactors

Equations (10.5)–(10.7) can be applied to a cascade of n equal stirred tank reactors in series. With $\theta_h = V_{tot}/Q$ the hydraulic residence time of the individual reactor results in θ_h/n. Since in the effluent of the individual reactors a sinusoidal concentration variation arises again and again, the mentioned equations are valid for all reactors individually, the phase shift is additive, and the attenuation results from the product of the dampening for individual reactors. Summarized for the entire cascade the result is

$$S = S_{m,in} \cdot \left(\left(\frac{1}{1 + k \cdot \theta_h / n} \right)^n + A_{out} \cdot \sin(2 \cdot \pi \cdot f \cdot t + \chi) \right)$$

(10.8)

$$\frac{A_{out}}{A_{in}} = \left(\frac{1}{\sqrt{\left(1 + k \cdot \theta_h / n\right)^2 + \left(2 \cdot \pi \cdot f \cdot \theta_h / n\right)^2}} \right)^n ,$$

(10.9)

$$\chi = -n \cdot \text{arctg} \left(\frac{2 \cdot \pi \cdot f \cdot \theta_h / n}{1 + k \cdot \theta_h / n} \right).$$

(10.10)

Figure 10.9 shows the variation of the effluent concentration for different reactors. The load variation is clearly visible in the effluent of the stirred tank reactor ($\pm 10\%$). It disappears in the cascade of six equal stirred tank reactors ($<1\%$) and reappears with decreasing longitudinal mixing of the reactor ($n = 200$) until it is fully developed again in the ideal plug-flow reactor.

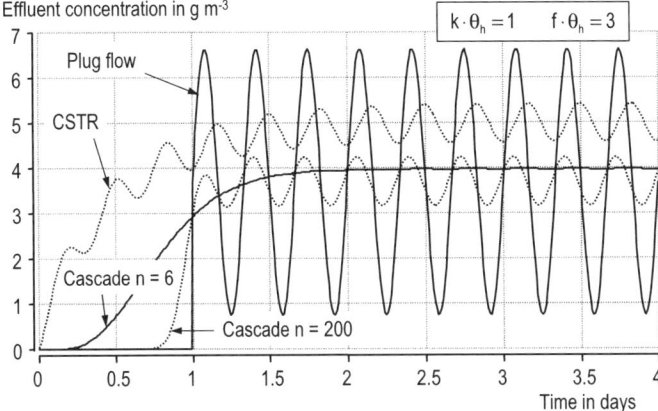

Fig. 10.9 Comparison of the effluent cncentration for different reactor systems

10.3.3 Plug-Flow Reactor

The effluent concentration in the ideal plug-flow reactor results from the time delay θ_h and the degradation; hydraulic attenuation (mixing) is not expected

$$S_{out} = S_{in}(t - \theta_h) \cdot \exp(-k \cdot \theta_h).$$

The following reduction of the amplitude of the concentration variation in the effluent results:

$$A_{out} = A_{in} \cdot \exp(-k \cdot \theta_h). \tag{10.11}$$

The ideal plug-flow reactor is not a suitable model for the description of systems in which any attenuation of the variation takes place by longitudinal mixing. Figure 10.9 demonstrates that even a small amount of dispersion (n = 200) can cause a significant dampening. The expected attenuation depends on the reaction, hydraulic residence time, and dispersion in the reactor. We can illustrate these dependencies in the so-called Bode diagram.

10.3.4 Bode Diagram

In a Bode diagram we characterize the response of a system to a disturbance with a harmonious oscillation (sinusoidal load variation). Of particular interest is the attenuation of the amplitude of input disturbances. Figure 10.10 shows a Bode diagram for a stirred tank reactor. The axes refer to a dimensionless frequency $(f \cdot \theta_h)$ and the attenuation of the amplitude (A_{out}/A_{in}) of a harmonic disturbance of the reactor. The heavy line corresponds to Eq. (10.6) in a double-logarithmic coordinate system. A dimensionless reaction rate $k \cdot \theta_h$ remains in Eq. (10.6) in contrast to Fig. 10.10 which is specific to the reaction rate $k \cdot \theta_h = 3$. Some characteristic

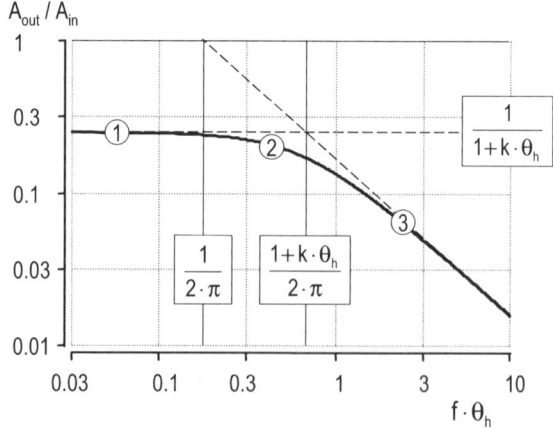

Fig. 10.10 Bode diagram for a stirred tank reactor, drawn for the dimensionless reaction rate of $k \cdot \theta_h = 3$. The points 1–3 refer to the three pictures in Figure 10.6, *top*, *middle*, and *bottom*

variables (frequencies, attenuations) result from the structure of Eq. (10.6). We read the diagram as follows:

- Disturbances with small frequencies $f \cdot \theta_h \ll (1 + k \cdot \theta_h)/(2 \cdot \pi)$ result in a constant attenuation, which is the result of degradation. This situation corresponds to point 1 in Fig. 10.10 or Fig. 10.6 *bottom*. Such a system is always in a momentary steady state. We apply static models.
- Disturbances with high frequencies $f \cdot \theta_h \gg (1 + k \cdot \theta_h)/(2 \cdot \pi)$ are subject to strong attenuation. This situation corresponds to point 3 in Fig. 10.10 or Fig. 10.6, *above*. The system is always in an approximate steady state relative to the mean load. The variable disturbance is filtered out and does not have an influence.
- For disturbances with moderate frequencies $f \cdot \theta_h \approx (1 + k \cdot \theta_h)/(2 \cdot \pi)$, attenuation by dispersion as well as by reaction must be considered. This situation corresponds to point 2 in Fig. 10.10 or Fig. 10.6, *middle*.

Figure 10.11 characterizes the behavior of three different reactor systems. With small frequencies, it becomes evident that the performance of a stirred tank reactor is inferior to that of plug-flow systems. With high frequencies the cascade of stirred tank reactors exhibits better damping than the stirred tank reactor whereas we cannot expect any damping from the ideal plug-flow reactor, since it neglects any possible dispersion. The same picture also arises from Figs. 10.7 and 10.9.

Application of the Bode Diagram: Case Studies

In the Bode diagram we can introduce a further dimension, the reaction rate k. Figure 10.12 shows the influence of an increasing first-order degradation rate. The

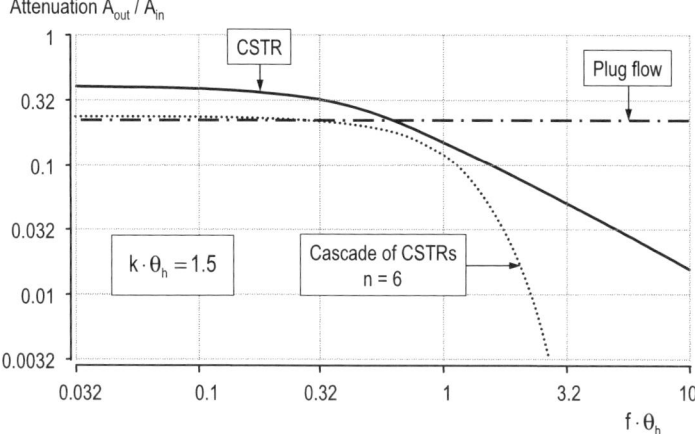

Fig. 10.11 Comparison of the attenuation of harmonic disturbances in different systems. The Bode diagram is based on the Eqs. (10.6), (10.9) and (10.11)

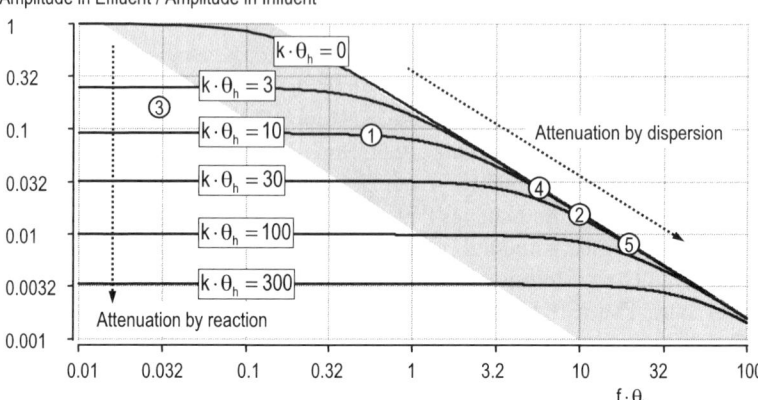

Fig. 10.12 Bode diagram for a completely mixed system (stirred tank reactor). Apart from the frequency also a variable, the dimensionless reaction rate $k \cdot \theta_h$ is introduced. The points 1–5 refer to the cases 1–5 in the text

better the degradation performance of the system, the higher the frequency that reaches the effluent becomes. In the grey range of Fig. 10.12 stationary models are not sufficient for accurate predictions; only dynamic models can provide reliable results.

The following case examples show how the diagram can be used. In order to realize the relevance of the individual cases, some basic knowledge of biological wastewater treatment may be necessary.

Case 1: How can the nitrification performance of an activated sludge system sub-ject to a diurnal load variation be predicted (see also Example 10.3)? Nitrifying activated sludge systems are subject to a strong diurnal load variation ($f = 1 \; d^{-1}$). The mean hydraulic residence time typically amounts to $\theta_h = 0.5 \; d$. From the typ-ical efficiency of the plant (>90%), we compute the degradation rate for ammo-nium to be $k \approx 20 \; d^{-1}$. The data pair $k \cdot \theta_h = 10$ and $f \cdot \theta_h = 0.5$ is indicated in Fig. 10.12 by point 1, which lies in the middle of the grey range. Therefore nitrification per-formance can only be predicted reliably with the help of dynamic simulation.

Case 2: Does the variation of the concentration of nitrifying bacteria in the acti-vated sludge have to be considered in the evaluation of diurnal variation in case 1? The concentration of nitrifying bacteria depends on their maximum growth rate $\mu = 0.5 \; d^{-1}$ (15°C). The residence time of the organisms in the system corresponds to the solids retention time $\theta_x = 10 \; d$. The data pair $k \cdot \theta = \mu \cdot \theta_x = 5$ and $f \cdot \theta_x = 10$ is indi-cated in Fig. 10.12 by point 2, which lies in the range where the amplitude (= variation) is strongly attenuated. Even with an average concentration, the pre-diction will be exact. A dynamic simulation of the nitrifier concentration is not required but is typically included in the models.

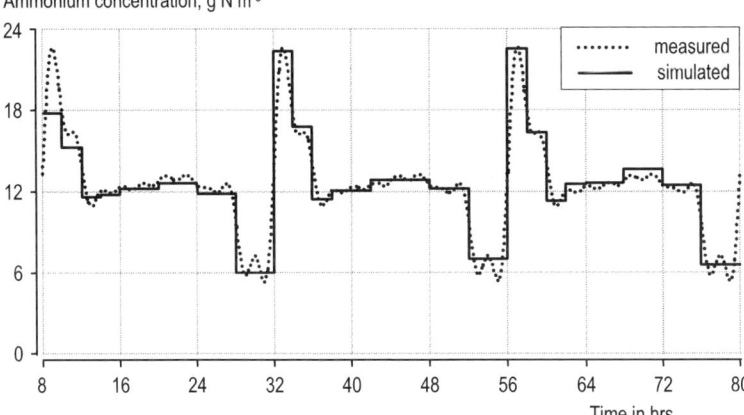

Fig. 10.13 Comparison of a measured and a simulated ammonium concentration curve. Measurements are available in 2 and 4 h composite samples. The simulation is based on a discrete Fourier transformation using the frequencies of $1-10\,d^{-1}$. Compare also Fig. 10.5

Case 3: How does the yearly temperature variation ($f = 1/365\,d^{-1}$) have to be included in the considerations? Similarly to case 2, we obtain $\mu \cdot \theta_x = 5$ and $f \cdot \theta_x = 0.03$. This data pair is indicated in Fig. 10.12 by point 3. The temperature can be considered at its instantaneous value (winter or summer conditions). These two situations do not influence each other, as the temporal distance is too large.

Case 4: Which frequency of sampling in the influent is necessary in order to be able to predict the variation of the ammonium concentration in the effluent of a nitrifying activated sludge system? Similar to case 1 the hydraulic residence time amounts to $\theta_h = 0.5\,d$ and the relevant degradation rate is $k = 20\,d^{-1}$. We are looking for the frequency which is to a large extent absorbed in the reactor. A further increase of the sampling frequency would then not lead to information which can be recognized in the effluent. Figure 10.5 shows that the amplitude of the ammonium concentration in the influent amounts to approximately $10\,gN\,m^{-3}$. If we want to make a prediction that is more accurate than $0.5\,gN\,m^{-3}$ in the effluent, the amplitudes in the influent must be reduced to below $A_{out}/A_{in} = 0.5/10 = 0.05$. In Fig. 10.12, point 4 is situated at $k \cdot \theta_h = 10$ and $A_{out}/A_{in} = 0.03$, which leaves $f \cdot \theta_h = 5$ for the residual degree of freedom. With $\theta_h = 0.5\,d$ we obtain the required sampling frequency of $f = 10\,d^{-1}$. Thus, the analysis of 2 h composite samples ($f = 12\,d^{-1}$) is sufficient for our problem setting. That corresponds to the frequency which has been used in order to obtain Fig. 10.5.

With the help of a Fourier transformation a diurnal variation is modeled with a set of harmonic oscillations. Here, too, we must ask ourselves what is the largest frequency to be taken up in the model. The aforementioned considerations apply here as well. Figure 10.13 shows the result of the synthesis of the inlet concentration in Fig. 10.5 with the help of the frequencies with $f \leq 10\,d^{-1}$. A further refinement does not seem to achieve much improvement.

Case 5: Can an anaerobic digester with a hydraulic residence time of $\theta h = 20\,d$ be fed once a day (f = 1\,d^{-1}) or is continuous feeding required? The degradation rate of the rate determining solids at 33°C amounts to $k = 0.2\,d^{-1}$. The data pair $k \cdot \theta_h = 4$ and $f \cdot \theta_h = 20$ is indicated in Fig. 10.12 by point 5. The variation as a consequence of a periodic loading is strongly reduced in the reactor, so the extra effort for a constant feeding over 24 h/d is therefore not worthwhile. Despite periodic filling the reactor performance can be evaluated with static models as if it were continuously fed.

10.3.5 Stochastic Processes

Stochastic load variations, like periodic variations, are also attenuated. The more complete the degradation, the more random concentration variations in the influent can still be recognized in the effluent.

Figure 10.14 shows the concentration in the influent to a completely mixed system, as a purely harmonic variation (dotted line) and when overlaid with random variations (noise, continuous line). *Can the stochastic processes (noise) still be recognized in the effluent or are they absorbed by the system?*

The noise that is superimposed onto the harmonic variation introduces high-frequency (fast-changing) disturbances into this system. If the degradation performance is small (Fig. 10.14, *center*), the noise is absorbed; visually we judge the remaining noise in comparison to the absolute value of the effluent concentration: We can barely recognize any residual noise, and the effluent concentration appears to be controlled primarily by the harmonic contribution to the variation. In the Bode diagram shown in Fig. 10.12 attenuation by dispersion outweighs dampening by the reaction for the data pair $\theta_h \cdot k = 0$ and $\theta_h \cdot f = 1$. Dispersion is effective in filtering high frequencies. With increasing degradation (Fig. 10.14, *lower*) attenuation by reaction becomes dominant ($\theta_h \cdot k = 100$ and $\theta_h \cdot f = 1$). The effluent concentration still shows low as well as high frequencies in its concentration variation.

This example shows that sampling is more difficult in the effluent of a system that has good degradation performance than in systems with poor degradation and therefore enhanced mixing. Point or grab samples hardly allow reliable information to be obtained in well-performing plants.

10.3.6 Dynamic Operation of Plants

Technical plants can be operated cyclically, intermittently, or unstably. Examples are the sequencing batch reactor (SBR) systems, the intermittent aeration of denitrifying activated sludge plants or unintended oscillations caused by unsuitable automatic control loops or oscillations introduced by binary controllers.

Influent concentration C_{in}, g m^{-3}

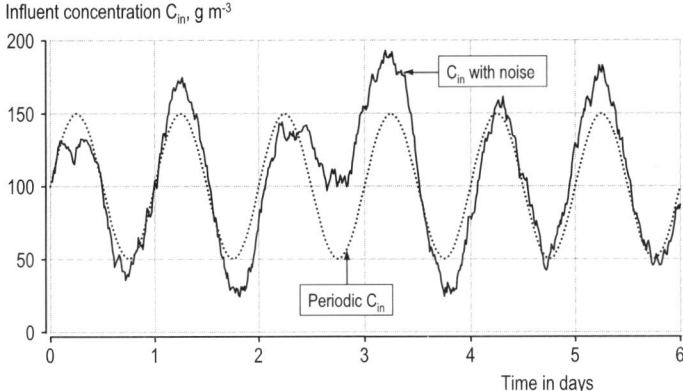

Effluent concentration, g m^{-3}

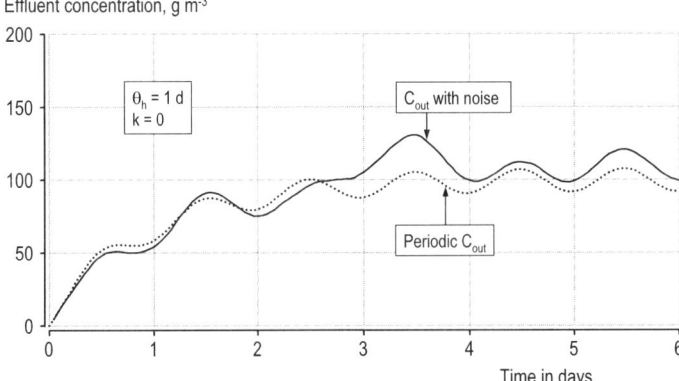

Effluent concentration, g m^{-3}

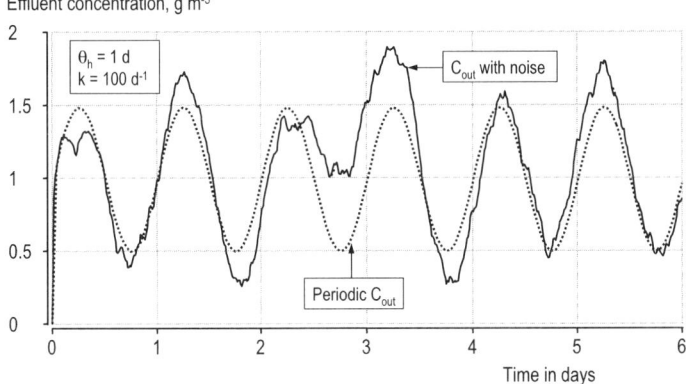

Fig. 10.14 Concentration course in the influent and in the effluent of a CSTR. The stochastic processes (noise) are absorbed with poor degradation performance of the system (*center*), while they are not filtered, if the system has good degradation performance (*below*)

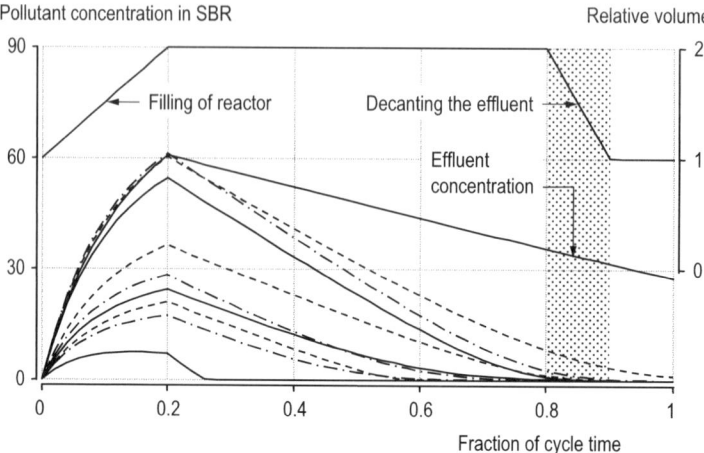

Fig. 10.15 Traces of the concentration in different cycles in the operation of an SBR

Sequencing Batch Reactor

An SBR is a reactor in which we obtain internal mixing at the earliest possible time and afterwards make time available for reaction. In the design of an SBR we must ensure that the available reaction time for all occurring loads is sufficient to reach the goal. The load variation expresses itself in particular in the time required to reach the full extent of the reaction. Figure 10.15 shows traces of the reduction of a pollutant concentration in a simulated SBR. The degradation is based on Monod kinetics. Figure 10.15 indicates that, depending upon the original load (and temperature), breakthrough of the pollutant into the effluent becomes possible.

The SBR is designed as a technical system. The choice of the operating parameters (duration and sequence of a cycle, fraction of the exchanged water volume, type of the wastewater, etc.) results in large dynamics in the reactor, which in a state-of-the-art design can, however, barely be observed in the effluent.

Intermittent Aeration of Activated Sludge Systems

Many activated sludge systems that must nitrify have (at least in the summer) excess capacity, which can be used for denitrification. A rather simple option to use these reserves is intermittent aeration. If the ammonium concentration is low, aeration is switched off and denitrification starts. If the ammonium concentration is increased, the aeration is started again, nitrification sets in, and denitrification ceases. The result is highly dynamic operation, which is exemplarily shown in Fig. 10.16.

Concentration, gN m^{-3}

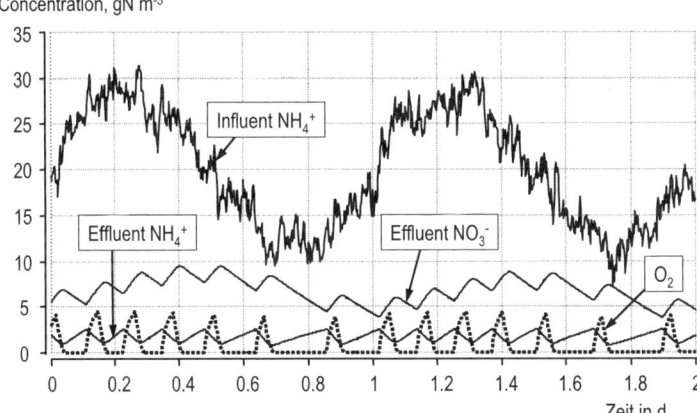

Zeit in d

Fig. 10.16 Concentration curves for ammonium, nitrate and oxygen in an intermittently aerated activated sludge system. The aeration is controlled such that the ammonium concentration remains within the range of 1.5–2.5 gN m^{-3}

Automatic Process Control

An unwanted part of the dynamics produced by technical interferences is the consequence of faulty control loops in the automatic operation of plants. If a measuring system causes large delays, automatic control loops can become unstable and begin to swing. This case will be discussed in Chap. 13.

10.4 Discussion of Time Constants

The time-dependent behavior of systems is controlled by a set of time constants, which we can evaluate based on the rate of individual processes. The comparison of such time constants permits the evaluation of which processes have a large and which have only a slight influence on system behavior.

10.4.1 The Residence Time of Individual Materials

Water and other materials have a characteristic residence time within systems that depends on transport (effluent) and reaction (consumption). From this characteristic time we learn how fast the related state variables may respond to a change of operating conditions.

A material is degraded in a first-order reaction in a fully mixed system (a batch reactor or stirred tank reactor). The following question is to be answered:

How long does a particle (a molecule) of this material remain in the system?

Particles can enter a system either in the influent or due to production (transformation process). They leave the system either in the effluent or by degradation. Similarly to the determination of the hydraulic residence time distribution $f(\tau)$ (Chap. 7), we base the derivations on an experiment: at time $t=0$ we add an amount of material M_0 to the fully mixed system and follow how this material disappears over time from the system. The result is a probability density function of the time of disappearing either in the effluent or by reaction. The expected value of this probability distribution corresponds to the mean residence time of a particle.

The balance equation for the material has the form:

$$\frac{dM}{dt} = V \cdot \frac{dC}{dt} = -Q \cdot C - k \cdot C \cdot V \,,$$

$$\frac{dC}{dt} = -\frac{Q}{V} \cdot C - k \cdot C = -(D_h + k) \cdot C \,. \tag{10.12}$$

M = mass of the material in the system [M]
Q = effluent from the system $[L^3 T^{-1}]$
C = concentration of the material in the system $[M\,L^{-3}]$
V = ideally mixed volume of the system $[L^3]$
D_h = hydraulic dilution rate $[T^{-1}]$, the inverse of the hydraulic residence time
k = the first-order degradation rate constant $[T^{-1}]$

The solution of Eq. (10.12) has the form:

$$\frac{C}{C_0} = \exp\left(-(D_h + k) \cdot t\right). \tag{10.13}$$

Equation (10.13) corresponds to an exponential distribution, with the expected value equal to the mean residence time of the material i:

$$\theta_i = \frac{1}{D_h + k_i} = \frac{V}{Q + k \cdot V} = \frac{\theta_h}{1 + k \cdot \theta_h} \,. \tag{10.14}$$

Applied to water, with a concentration according to its density ρ_w and using the fact that water is a conservative material ($k=0$), Eq. (1.14) yields

$$\theta_w = \frac{1}{D_h} = \frac{V}{Q} = \theta_h \,.$$

For a batch reactor with $Q=0$ we obtain $\theta_h = \infty$. However, for materials that are degraded, a finite mean residence time of $\theta_i = 1/k$ results.

Equation (10.14) applies to materials that are degraded in a first-order reaction. For other degradation kinetics we can derive a similar equation for the steady state. The residence time of a particle of a material corresponds to the quantity of the material in the system divided by the sum of all losses by degradation and effluent. For a stirred tank reactor this leads to

$$\theta_i = \frac{V \cdot C_i}{Q \cdot C_i + |r_{i,Degradation} \cdot V|}$$

$$\theta_i = \frac{\theta_h}{1 + \dfrac{|r_{i,Degradation}|}{C_i} \cdot \theta_h} . \tag{10.15}$$

Applied to any system and type of reaction we obtain:

$$\theta_i = \frac{\int\limits_V C_i \cdot dV'}{Q \cdot C_{out} - \int\limits_V r_{i,Degradation} \cdot dV'} . \tag{10.16}$$

Equation (10.16) is not frequently applied.

Example 10.5: Mean residence time of a degradable material in a stirred tank reactor

In a CSTR with a hydraulic residence time of $\theta_h = 0.5$ d a material i is degraded in a first-order reaction. The effluent concentration amounts to 5% of the inlet concentration. The reactor is in the steady state.
What is the average residence time of the material i in the reactor?
For a stirred tank reactor in the steady state one has

$$\frac{C_i}{C_{i,in}} = \frac{1}{1 + k_i \cdot \theta_h} = \frac{1}{1 + k_i / D_h} .$$

From $C_i/C_{i,in} = 0.05$ it follows that $k \cdot \theta_h = 19$ and thus with Eq. (10.14)
$\theta_i = \theta_h/(1 + k \cdot \theta_h) = 0.025$ d.
The reaction reduces the residence time of the degraded material in comparison to the water by a factor of 20.

Example 10.6: Residence time of a material in a batch reactor

For the batch reactor with $D_h = 0$ Eq. (10.14) yields $\theta_i = 1/k$. For the reaction in Example 10.5 we compute $k = 38$ d^{-1} and thus $\theta_i = 0.026$ d. The difference between θ_i in the CSTR and in the batch reactor results from the reduced loss as a consequence of the missing effluent.

Example 10.7: Residence time of oxygen in an activated sludge tank

An activated sludge tank has a hydraulic residence time of $\theta_h = 0.33$ d. It is operated with an oxygen concentration of $2 \, gO_2 \, m^{-3}$ and the oxygen consumption amounts to $450 \, gO_2 \, m^{-3} \, d^{-1}$. *What is the residence time of the dissolved oxygen in this tank?*

From Eq. (10.15) we obtain $\theta_{O2} = 0.33/(1 + 0.33 \cdot 450/2) = 0.0044 \, d = 6 \, min$.

Thus, the oxygen is renewed every 6 min despite the long hydraulic residence time of the wastewater of 8 h. Or in other words, if the aeration were to stop, the residual oxygen would be consumed in about 6 min (depending on kinetics). This surprising result is obtained due to the large degradation constant of $k_{O2} = r_{O2}/C_{O2} = 225 \, d^{-1}$.

Example 10.8: Time to reach the steady state

Equation (10.2) describes the approach of a CSTR after a step change of the concentration in the influent to its new steady state, if the degradation takes place in a first-order reaction. Presented in simplified form ($t_1 = 0$) this yields

$$S = \underbrace{S_0 \cdot \exp\left(-(D_h + k) \cdot t\right)}_{\text{Decay of initial condition } S_0} + \underbrace{S_{in} \cdot \frac{D_h}{D_h + k}}_{\text{New steady state}} \cdot \underbrace{\left[1 - \exp\left(-(D_h + k) \cdot t\right)\right]}_{\text{Approach to new steady state}}.$$

Rewritten based on the residence time of the material i in a stirred tank reactor, this reads

$$S_i = S_{i,0} \cdot \exp(-t / \theta_i) + S_{i,in} \cdot \frac{\theta_i}{\theta_h} \cdot \left[1 - \exp(-t / \theta_i)\right].$$

This equation illustrates the importance of the residence time of the materials and identifies at the same time θ_i as a decisive time constant for the approximation to a new steady state. The smaller θ_i, the faster is the adjustment to the new steady state (see Fig. 10.3).

10.4.2 Different Time Constants

Different variables, in particular rates and velocities, can be transformed into time constants, which give us a uniform basis for their comparison. Examples are:

τ_m, θ_h, V/Q	Hydraulic residence time
	Hydraulic transport
C_i/r_i, $1/k$ for first order reactions	Reaction time
θ_i, residence time of a material	Reaction and transport
$(2 \cdot \pi \cdot f)^{-1}$ (frequency of the disturbance)	Variation time, diurnal variation, etc.
L/u (velocity of flow u)	Determining time for the advection
$L^2 \cdot (2 \cdot D_i)^{-1}$ (molecular diffusion)	Necessary time for transport over a
$L^2 \cdot (2 \cdot D_T)^{-1}$ (turbulent dispersion)	distance L, see Eq. (4.17)

Examples of the use of time constants are:

- The turbulence number $(N_T = u \cdot L/D_T = (u/L) \cdot (L^2/D_1))$ relates two time constants: L/u, the time necessary for a particle to flow through a reactor with length L due to the flow velocity u (advection), and L^2/D_T, the time necessary so that turbulent transport can be observed over the whole length of the reactor.
- The Bode diagram (Fig. 10.12) relates the reaction time $(C_i/r_i, 1/k)$ to the hydraulic residence time θ_h and the time of load variation $(2 \cdot \pi \cdot f)^{-1}$, and permits us to derive which kind of disturbances are attenuated or absorbed in a system.
- The residence time of a material θ_i provides us with information about how fast a system approaches a new steady state (Example 10.8).
- In biological wastewater treatment we determine the product of the growth rate μ ($1/\mu$ corresponds to a reaction time) and the solids retention time θ_X (θ_X corresponds to the residence time of a material) as a safety factor which tells us whether we can expect a system to reach a certain goal of treatment (e. g., full nitrification).
- etc.

Frequently the comparison of time constants permits us to simplify mathematical models, e. g., if we can decide to describe some subprocesses as stationary, or even to describe whole systems as being in the steady state for a limited time period, a variable disturbance may be averaged over a longer time period without loss of information.

Example 10.9: Consideration of the temperature variation in biological treatment processes

In a biological wastewater treatment plant the biomass is renewed approximately every 10 days (SRT). The temperature of a local wastewater varies by approximately ±1.5°C around the daily average. In the winter the daily average is approximately 10°C, and in the summer it is approximately 18°C. It is well known that the temperature is a determining variable for the growth of the microorganisms, and in particular that their activity increases by approximately 10% per 1°C. *How do we have to consider the two temperature variations?*
1. Diurnal variation: The daily variation of the temperature of about 3°C affects the activity of the organisms by approximately $(1.1)^3 - 1 = 35\%$. The composition of the biomass can only adapt itself to a new situation over about 10 days (SRT). Clearly the diurnal variation of the temperature must be considered if predictions are to be made for the performance of the plant in the context of diurnal variation.
2. Yearly variation: Every year the biomass renews itself several times; it is not necessary to simulate the temperature variation over an entire year, and the analysis of some stationary operating conditions in the different seasons may be sufficient. It may, however, be necessary to overlay the seasonal conditions with the diurnal variation of the temperature.

Example 10.10: Learning is subject to different time constants

You are trying to understand this text and all that remains is a big confusion about integrals, masses, trigonometric functions, residence time distributions, reactors and reactions, stoichiometry and systems, conservation laws, and turbulence. This, too, is a problem of time constants: learning, understanding, digesting, storing, practising, and applying all have different time constants – with motivation you will succeed in balancing them against each other.

10.5 Nonstationary Effluent in Sewers

Intuitively we assume that sewers or rivers react similarly to a plug-flow reactor and to a plug-flow reactor with turbulence. Since surface waves in such channels move approximately 1.6 times faster than the water and the materials (celerity versus flow velocity), important differences in the behavior of the reactors with constant cross section (pressure pipes, plug-flow reactor) and channels with a free surface result.

The following derivation is only relevant for open, prismatic channels that are hydraulically not in a steady state, i. e., they have temporally variable effluent.

The geometric conditions for a section of a rectangular channel with a free surface and a constant width B are defined in Fig. 10.17. Conditions in a partially filled circular sewer pipe are similar, but mathematically more complicated to derive. The general balance equation (3.11)

$$\frac{\partial}{\partial t} \int_{V(t)} C_i \cdot dV' = \oint_{A(t)} j_{n,i} \cdot dA' + \int_{V(t)} r_i \cdot dV'$$

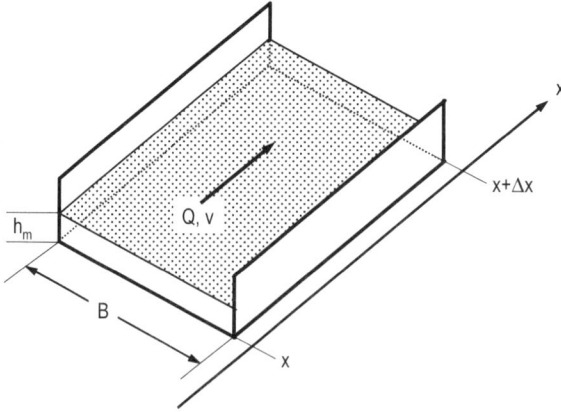

Fig. 10.17 Geometric definitions in a channel with a free water surface: B = surface width [L], h_m = mean depth [L], Δx = length of the section [L]

applied to the water in the channel with $C_i = \rho_W = const.$ and $r_W = 0$ has the form:

$$\frac{\partial V}{\partial t} = \frac{\partial B \cdot h_m \cdot \Delta x}{\partial t} = Q_{in} - Q_{out} = A(x) \cdot v(x) - A(x + \Delta x) \cdot v(x + \Delta x)$$

A = wet cross section $h_m \cdot B$ [L^2].

After division by $B \cdot \Delta x$ and replacement of Δx by ∂x, this yields

$$\frac{\partial h_m}{\partial t} = -\frac{\partial h_m \cdot v}{\partial x} = -v \cdot \frac{\partial h_m}{\partial x} - h_m \cdot \frac{\partial v}{\partial x}. \tag{10.17}$$

Equation (10.17) corresponds to the continuity equation (conservation law) for the water. The equation by Manning–Strickler for the velocity of flow in a channel with free water surface has the form:

$$v = k_{St} \cdot R_h^{2/3} \cdot J_E^{1/2}$$

v = mean flow velocity [m s^{-1}]
k_{St} = roughness factor after Strickler (inverse of Manning's n) [m$^{1/3}$ s^{-1}]
$R_h = \dfrac{B \cdot h_m}{B + 2 \cdot h_m}$ = hydraulic radius [m]
J_E = Energy gradient (friction slope) [−]

For small depths of water $h_m \ll B$, approximately

$$R_h \approx h_m,$$

and thus one obtains

$$\frac{\partial v}{\partial x} = \frac{\partial v}{\partial h_m} \cdot \frac{\partial h_m}{\partial x} \approx \frac{\partial k_{St} \cdot h_m^{2/3} \cdot J_E^{1/2}}{\partial h_m} \cdot \frac{\partial h_m}{\partial x} = \frac{2}{3} \cdot k_{St} \cdot h_m^{-1/3} \cdot J_E^{1/2} \cdot \frac{\partial h_m}{\partial x} = \frac{2}{3} \cdot \frac{v}{h_m} \cdot \frac{\partial h_m}{\partial x}.$$

After substitution into Eq. (10.17) the result is

$$\frac{\partial h_m}{\partial t} \approx -\frac{5}{3} \cdot v \cdot \frac{\partial h_m}{\partial x}, \tag{10.18}$$

which indicates that a wave (dh_m/dt) moves with celerity $c \approx 5/3 \cdot v$ faster through the channel than the flow velocity v. A material cloud is transported with the velocity of flow v (overlaid with dispersion). From this result we derive the fact that, with increasing flow distance, the discharge wave travels ahead of the cloud of pollutants that it initially contained. The observed load of material in the channel is thereby deformed unexpectedly. An example of such a wave is shown in Fig. 10.18, where it is assumed that the polluted water is discharged as a slug into a sewer at time $t = 0$. For a more detailed discussion and some experimental result see Huisman et al. (2000).

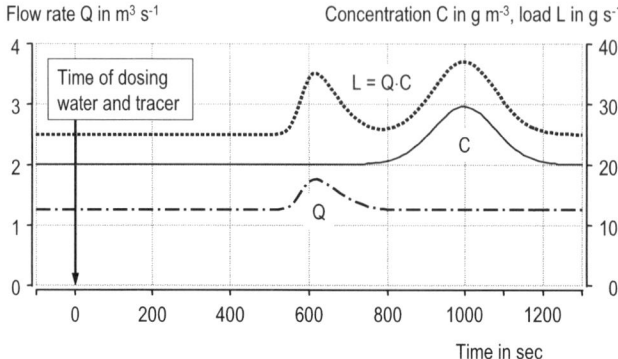

Fig. 10.18 Traces of the effluent, concentration, and load of a material in the discharge of a sewer with $L = 1000$ m. At time $t = 0$ a Dirac pulse of water is added to the influent with a material concentration of 100 g m^{-3}

Example 10.11: First flush in a combined sewer in the context of a rain event

During rain events only lightly polluted rain water drains into combined sewers, which already contain undiluted wastewater from the previous dry period. In the sewer, the information that more influent arrived in the system travels with the celerity $c = 5/3 \cdot v$ through the system, whereas the diluted wastewater travels with the flow velocity. Thus, when the effluent wave reaches the influent of the treatment plant, concentrated wastewater from the previous dry weather is discharged at an increased flow rate, causing an increase of the pollutant load to the plant. The load of the plant $(Q \cdot C)$ increases substantially, even without significant additional pollutants arriving in the system.

Example 10.12: Sampling after an accident

Due to an accident a larger quantity of water loaded with a colorless poisonous material is discharged into the sewer. You hurry to the treatment plant and wait for the wastewater quantity to increase. Now you begin to divert the water into an emergency storage tank and simultaneously take a series of samples. As the wave has passed, you stop sampling and return back to normal operation of the treatment plant since the emergency storage volume is now filled.

In the laboratory you are disappointed that the samples do not contain any poison but nevertheless the biological sludge is severely affected.

You have sampled the water wave, which consisted of the wastewater that was in the sewer before the accident occurred. The actual poisonous wave arrived only after the water wave had already passed the plant, and therefore you missed it.

If you want to divert poisonous material into an emergency tank, you should rely on the relevant plans on how to use these tanks. Using discharge information may not provide you with the information needed to make optimal use of emergency storage volumes, especially if the pollutants are discharged from a single source, over a short time, and far away.

Chapter 11
Measurement and Measurement Uncertainty

Measurements together with observation are the basis for documenting, quantifying, describing, and understanding experiments, systems, and their variables. However, the measuring process is subject to uncertainties and errors that influence the results and must be considered in the interpretation of measured values. We distinguish gross, random, and systematic deviations of measurements.

A measurement is an experimental process in which we establish the value of a variable relative to a calibrated scale.

Example 11.1: Basic units for calibration of measuring systems

A well-known basic calibration for measuring systems is initially provided by the international prototype meter for a unit length or the international prototype kilogram for a unit mass (both stored in Paris).
However, each chemical analysis or each physical measurement refers its result to a suitable calibration scale, which may be available with anything between poor to excellent accuracy and precision.

11.1 Definitions from Descriptive Statistics

Results of measurements are random variables that are subject to probability density distributions. A single realization of a measurement can have any value in the range of this probability distribution. From repeated measurements we derive empirical density distributions, which we then characterize statistically.
The true value of a variable to be measured is not known; we can approximate it, but we never know it for sure.

11.1.1 Analytical Characterization of the Distribution of Measured Values

Individual values x_i of a random variable X are subject to a certain probability density distribution $f(x)$ which is partially characterized by the expected value μ_x and the variance σ_x^2. The following relations apply (see also Chap. 7):

$$\mu_x = \int_{-\infty}^{\infty} x \cdot f(x) \cdot dx \tag{11.1}$$

$$\sigma_x^2 = \int_{-\infty}^{\infty} (x - \mu_x)^2 \cdot f(x) \cdot dx = \int_{-\infty}^{\infty} x^2 \cdot f(x) \cdot dx - \mu_x^2 \tag{11.2}$$

$f(x) =$ probability density function of the random variable X at the position x
μ_x = expected value of X
σ_x = standard deviation of X
σ_x^2 = variance of X

If X is normally distributed, we write:

$$X \sim N(\mu_x, \sigma_x). \tag{11.3}$$

The normal distribution has the form:

$$f_N(x) = \frac{1}{\sqrt{2 \cdot \pi \cdot \sigma_x^2}} \cdot \exp\left(-\frac{(x - \mu_x)^2}{2 \cdot \sigma_x^2}\right). \tag{11.4}$$

For many measured variables negative values are not permissible, whereas the normal distribution covers the whole range of the real numbers. Therefore in place of a normal distribution frequently a log-normal distribution is used, i. e., we assume that the logarithms of a random variable are normally distributed.

The log-normal distribution has the following probability density function:

$$f_{LN}(x) = \frac{1}{x \cdot \sigma_{\ln x} \cdot \sqrt{2 \cdot \pi}} \cdot \exp\left(-\frac{(\ln x - \mu_{\ln x})^2}{2 \cdot \sigma_{\ln x}^2}\right) \quad \text{for} \quad x > 0, \tag{11.5}$$

where $\mu_{\ln x}$ and $\sigma_{\ln x}$ are the mean and standard deviation of the variable's natural logarithm, respectively. The expected value and variance are:

$$\mu_x = \exp\left(\frac{2 \cdot \mu_{\ln x} + \sigma_{\ln x}^2}{2}\right) \quad \text{and} \quad \mu_{\ln x} = \ln\left(\frac{\mu_x^2}{\sqrt{\sigma_x^2 + \mu_x^2}}\right), \tag{11.6}$$

$$\sigma_x^2 = \exp\left(2 \cdot \mu_{\ln x} + 2 \cdot \sigma_{\ln x}^2\right) - \exp\left(2 \cdot \mu_{\ln x} + \sigma_{\ln x}^2\right) \quad \text{and} \quad \sigma_{\ln x}^2 = \ln\left(1 + \frac{\sigma_x^2}{\mu_x^2}\right). \tag{11.7}$$

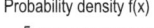

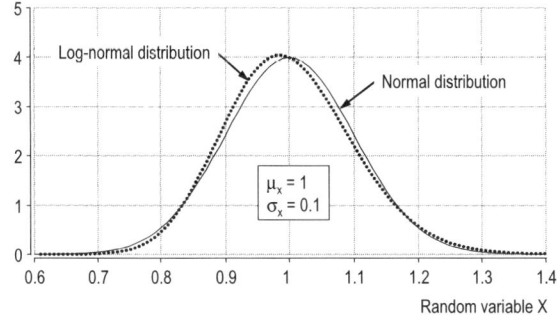

Fig. 11.1 Comparison of a normal and a log-normal distribution with a coefficient of variation $CV = \sigma_x / \mu_x = 0.1$

For small coefficients of variation $(CV) = \sigma/\mu < 0.1$ the two distributions are nearly identical, with equal characteristics $\mu_x = \exp(\mu_{lnx})$ and $\sigma_x = \exp(\sigma_{lnx})$ (Fig. 11.1).

Example 11.2: Log-normal distribution of a random variable in Berkeley Madonna

You would like to obtain a log-normally distributed random variable X that has an expected value of $\mu_X = 50$ and a standard deviation of $\sigma_X = 15$. Berkeley Madonna only provides you with a normally distributed random variable (tested)

```
STARTTIME = 1              ; Beginning of simulation
STOPTIME = 100             ; End
DT = 1                     ; Timestep, provides 100 values of x
mueX = 50                  ; Expected value of X
sigX = 15                  ; Standard deviation of X
mue_lnx = logn(mueX^2/(sigX^2 + mueX^2)^0.5)
                           ; Transformation of mue value, Eq. (11.6)
sig_lnx = (logn(1 + sigX^2/mueX^2)^0.5)
                           ; Transformation of sig value, Eq. (11.7)
X = exp(normal(mue_lnx,sig_lnx))
                           ; Generation of log-normally distributed variable
```

11.1.2 Empirical Characterization of Measured Values

If a series of experiments yields multiple results (data points) for the exact same object to be measured (replicates), we can approximately determine the distribution of a random measurement variable. The arithmetic mean m_x is used as an approximation of the expected value μ_x, and the empirical variance s^2 replaces the variance σ^2. The empirical values are computed with the following equations from discrete individual realizations x_i of the random variable X:

Arithmetic mean: $\displaystyle m_x = \frac{1}{n} \cdot \sum_{i=1}^{n} x_i$ (11.8)

Empirical variance: $\displaystyle s_x^2 = \frac{1}{n-1} \cdot \sum_{i=1}^{n} (x_i - m_x)^2 = \frac{1}{n-1} \cdot \left(\sum_{i=1}^{n} x_i^2 - n \cdot m_x^2 \right)$ (11.9)

Standard deviation: $\displaystyle s_x = \sqrt{s_x^2}$ (11.10)

s_x = empirical standard deviation or short standard deviation.

   If the number of available measured values n is small, then the results obtained from Eqs. (11.8) and (11.9) are poor (uncertain) estimates of the unknown true values μ_x and σ_x. The deviation Δx of the estimated value m_x from the true value μ_x corresponds for a specific number n of data x_i again to a random variable which for large n is normally distributed with the characteristics:

$\Delta x = m_x - \mu_x$

$\displaystyle m_{\Delta x} = 0 \text{ and } s_{\Delta x}^2 = s_{m_x}^2 = \frac{s_x^2}{n}$ (11.11)

s_{mx} = standard deviation of the average or standard error of the average

   The expected value μ_x and average value m_x are yardsticks for the position of a distribution. The standard deviation σ_x, s_x, and variance s_x^2 are indicators of dispersion, i. e., yardsticks for the dispersion of variables around the expected or average value. The term dispersion is mathematically not accurately defined, but signifies the deviation of a realization of a random variable from its expected value.

Example 11.3: Empirical characterization of a series of measurements

In order to establish the reproducibility of an analytical method to determine the ammonium concentration of a wastewater, you analyze a sample ten times. The results in $gN\ m^{-3}$ are:

14.7, 14.5, 15.1, 13.6, 14.6, 15.0, 15.6, 16.6, 13.5, 15.6

Using the above equations we obtain:

$m_x = 14.9$ $s_x^2 = 0.87$ $s_x = 0.93$ $CV = s_x / m_x = 6.3\%$ $s_{mx} = 0.30$

These values have been obtained from a generator of normally distributed random numbers with $X \sim N(15,1)$, meaning that the true values $\mu_x = 15$ and $\sigma_x = 1$ are known. The estimated values m_x and s_x are close to true values. The standard deviation of the average value s_{mx} is clearly larger than the deviation $m_x - \mu_x$. It should not have surprised us, if the resulting average value m_x would be 14.5 or 15.4.

11.2 Measuring Systems

Each measurement is obtained with the aid of a measuring system, which includes the object under test that is to be measured, the base of reference that is used for calibration, the measuring instrument that indicates the measured value, and the measurement procedure that describes the details of the measurement process.

Figure 11.2 shows schematically a measuring system. The calibration represents the reproducible yardstick for the measured value (the measurand). The measurement process affects the object under test, and all elements of the system (measuring instrument, item under test, and calibration) are subject to disturbances, which lead to measurement uncertainty.

The following terms are frequently used in this context:

Measurand

A measurand is the physical quantity to be measured. The goal of the measurement process is to obtain the value of the measurand. The measurand may also be described as the true value of the measured property.

Accuracy of a measurement

Accuracy is a qualitative concept which describes the deviation of a measuring result from the measurand (the true but unknown value).

Reproducibility

Reproducibility characterizes the closeness of the agreement between individual measurements of the same measurand, if they have been realized under different conditions (other laboratory, other person, other instrument, new calibration solution, etc.)

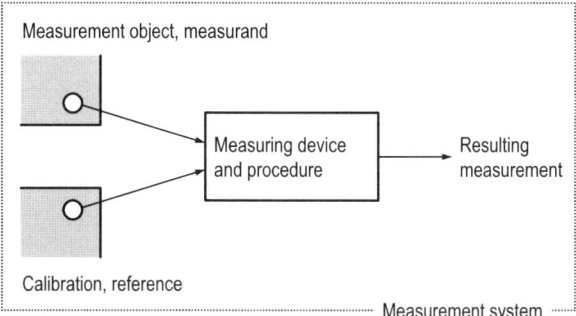

Fig. 11.2   Diagram of a measuring system (Hemmi and Profos, 1997)

Repeatability

Repeatability designates the proximity of the results if a measurement is repeated several times within a short period of time with the exact same procedures and by the same person.

Precision

Precision is the closeness of agreement between independent test results obtained under stipulated conditions (ISO 3534-1). Reproducibility and repeatability are the extremes of the possible conditions. A measure for the closeness is, e. g., the empirical standard deviation of the measured values.

Today increasingly automated measuring systems, electrodes or sensors make data with high temporal resolution available for process control systems. Such measuring systems frequently give the impression of reliability, accuracy, and precision which are far from being ensured in reality.

Measurement error

The error of measurement is the result of the measurement minus the measurand.

Random error

The random error is the result of a measurement minus the mean that would result from an infinite number of measurements of the same measurand under repeatability conditions.

Systematic error

The systematic error is the mean that would result from an infinite number of measurements of the same measurand under repeatability conditions minus the measurand.

Gross measurement error

Gross measurement errors are due to false readings, error in computations, wrong manipulations, etc.

Correction and correction factors

A correction is the value algebraically added to the uncorrected result of a measurement to compensate for systematic error. A correction factor is multiplied with the uncorrected value to compensate for systematic error.

11.3 Measuring Uncertainty

*With a measurement we try to capture the measurand with the aid of an approxi-
mation. The recommendation of the Comité International des Poids et Mesures is
to provide in place of the measurand the best estimate of this value.
A measurement is complete only when accompanied by a quantitative statement of
its uncertainty.*

Results of measurements deviate from the measurand due to various measuring
errors (Fig. 11.3):

- *gross errors*
- *random errors*
- *bias or systematic errors*

11.3.1 Gross Measurement Errors

*Gross measurement errors result from mistakes, false considerations, operating
errors, and calculation errors. It makes little sense to include measurements which
are subject to gross errors into the final analysis of data.*

The Chauvenet criterion (Coleman and Steele, 1998) is a simple method for the
identification of potential outliers. The criterion applies to *normally distributed,
random measurement errors*. It has the form:

$$\tau_i = \frac{|x_i - m_x|}{s_x} \tag{11.12}$$

with $m_x = \dfrac{1}{n} \cdot \sum_{i=1}^{n} x_i$ and $s_x^2 = \dfrac{1}{n-1} \cdot \sum_{i=1}^{n} (x_i - m_x)^2$.

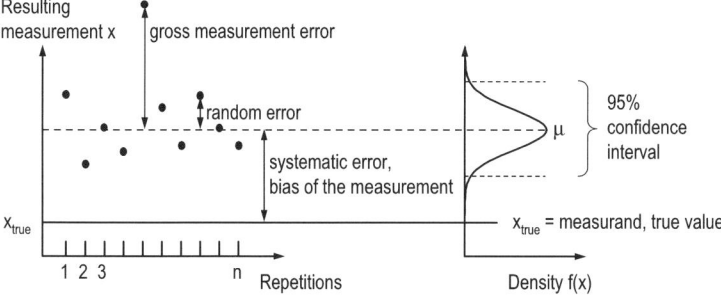

Fig. 11.3 Gross, random and systematic measurement errors (after Thomann, 2002)

Fig. 11.4 Diagram of the ranges of
the outliers criterion after Chauvenet

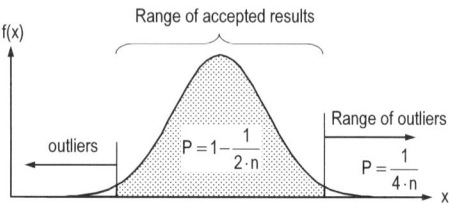

Table 11.1 Critical values τ_{krit} for the Chauvenet criterion in function of the number of measured values n

n	τ_{krit}	n	τ_{krit}	n	τ_{krit}
3	1.38	10	1.96	100	2.8
4	1.53	15	2.13	250	3.1
5	1.64	20	2.24	500	3.3
6	1.73	25	2.33	1000	3.6
8	1.86	50	2.57		

A measured value is rejected as an outlier if $\tau_i > \tau_{krit}$ (Fig. 11.4), where

$$\tau_{krit} = \Phi\left(\frac{4 \cdot n - 1}{4 \cdot n}\right) \text{ with } \Phi = \text{cumulative, standardized normal distribution.}$$

Examples of τ_{krit} as a function of the number of measured values n are given in Table 11.1.

There are arguments in the literature that the Chauvenet criterion should not be used in series, meaning that is should not be applied once the data set is reduced by eliminating positively identified outliers. However, this argument is open to debate. The Chauvenet criterion does not have a rigorous statistical basis but approaches for a single outlier the rigorously defined and more powerful Peirce criterion (Ross, 2003), which provides the possibility to test for several outliers.

Example 11.4: Application of the Chauvenet criterion

You want to identify the repeatability of your analytical procedure to measure the ammonium concentration in a wastewater. For this you repeat the analysis several times (here n = 6 is used but in reality n should be larger). The results are:

i	x_i in $gN\,m^{-3}$	τ_i for n = 6	τ_i for n = 5
1	17.6	0.25	0.54
2	17.6	0.25	0.54
3	18.9	1.83	– eliminated in the first step
4	17.9	0.01	0.50
5	18.0	0.01	0.86
6	17.7	0.13	0.21

For n = 6 one has $m_x = 17.95\,\mathrm{gN\,m^{-3}}$ $\qquad$ $s_x^2 = 0.243\,\mathrm{g^2\,N\,m^{-6}}$

$\qquad\qquad\qquad s_x = 0.493\,\mathrm{g\,N\,m^{-3}}$ $\qquad$ $\tau_{krit} = 1.73$

For i = 3 we obtain $\tau_i > \tau_{krit}$. This value is eliminated from the series as an outlier.

For n = 5 one has $m_x = 17.76\,\mathrm{gN\,m^{-3}}$ $\qquad$ $s_x^2 = 0.078\,\mathrm{g^2\,N\,m^{-6}}$

$\qquad\qquad\qquad s_x = 0.28\,\mathrm{g\,N\,m^{-3}}$ $\qquad$ $\tau_{krit} = 1.64$

All remaining measured values fulfill the Chauvenet criterion. The computed values for m_x and s_x correspond to the best estimated values. The result exhibits a clearly smaller variance after elimination of the outlier. As indicated in the text, the second round of testing is open to debate and not recommended here. For the available results it did not have an influence.

11.3.2 Random Measurement Error

Even carefully conducted measurements are subject to random measurement error. These result from influences that we cannot control. Thus, measured values correspond to random variables.

If the results of a measurement is based on a measurement procedure that is subject to many small random influences, the random measurement errors are frequently normally distributed (central limit theorem). In this case, the arithmetic mean m_x of a series of measurements x_i corresponds to the best estimated value of the measurand. The measured values are normally distributed random variables:

$$X \sim N(\mu_x, \sigma_x^2). \tag{11.13}$$

The best estimates for the expected value μ_x and the variance σ_x^2 or the standard deviation σ_x can be obtained from Eqs. (11.8) and (11.9). These estimated values are again normally distributed random variables, $m_x \sim N(\mu_x, \sigma_{m_x}^2)$. The variance of the arithmetic mean characterizes the statistical properties of the remaining random measurement error, if several measured values are available. It decreases as the number of available results n increases (Eq. (11.11)):

$$s_{m_x}^2 = \frac{1}{n \cdot (n-1)} \cdot \sum_{i=1}^{n} (x_i - m_x)^2 = \frac{s_x^2}{n}, \tag{11.14}$$

where S_{mx} is the empirical standard error for the mean of the measurement results.

If we use the empirical values m_x and s_x (as derived from Eqs. (11.8) and (11.14)) as estimates of the mean μ_x and the standard deviation σ_x of a series of measurements, they too are uncertain. Accordingly we compute the confidence interval of m_x with the help of the t-distribution with $n-1$ degrees of freedom (see also Example 11.6).

Example 11.5: Characterization of random measurement error

How large is the random measurement error in the series of measurements of the ammonium concentrations in Example 11.4.

The best estimated value of the measured variable is the arithmetic mean $m_x = 17.76$ gN m^{-3}.

The expected value of the measurement error is $E(x_i - m_x) = 0$.

The estimated value of the variance of the measurement error is $s_x^2 = 0.078$ g^2 m^{-6}.

The estimated value for the standard deviation of an individual measurement error is $s_x = 0.28$ g N m^{-3}.

The variance of the best estimated value m_x is $s_{mx}^2 = s_x^2 / 5 = 0.0156$ g^{2}m^{-6}.

The standard deviation of the best estimated value m_x is $s_{mx} = 0.125$ g N m^{-3}.

Example 11.6: 95% confidence interval for the measurand of the ammonium concentration

The 95% confidence interval indicates a range that contains with 95% probability the true average value of a series of measurements with normally distributed measurement errors.

It holds that $P\left(m_x - t_{0.975}^v \cdot \sqrt{\dfrac{s_x^2}{n}} \leq \mu_x \leq m_x + t_{0.975}^v \cdot \sqrt{\dfrac{s_x^2}{n}} \right) = 0.95$.

The number of degrees of freedom is $v = n-1$, here four, and thus from an appropriate table we find $t_{0.975}^4 = 2.78$. Thus, the 95% confidence interval has the width $17.76 \pm 0.125 \cdot 2.78 = 17.41$ to 18.11 gN m^3.

Example 11.7: Measuring error of the COD and TSS analysis

In a variation of the COD analysis, a certain quantity of dichromate is added to the sample. Some of the dichromate is then reduced by the organics; the remaining dichromate is titrated with Fe^{2+}. The COD is obtained from the difference of the amount of dichromate at the beginning and the end of the analysis. If the COD is very small, this difference is obtained from two numbers of nearly equal size.

Standard Methods (1995) give a coefficient of variation of 4.8% for samples with an average COD of 208 mg/l ($CV = s_x/m_x$).

How large is the coefficient of variation for a sample with an expected COD of 20 mg/l? Assume that the added quantity and the titration of the residual dichromate are subject to a constant variance of the measuring error. How large is the 95% confidence interval?

If we determine the total suspended solids TSS, we must subtract the initial weight of a filter paper from the final weight, which now includes the dried solids. *How do the absolute and the relative measuring error change in function of the TSS concentration? How can we control these errors?*

Example 11.8: Number of measurements

You are responsible for the design of the extension of an activated sludge system. After a pre-design you expect an average sludge loading rate of $B_{TS} = 0.1\,kgBOD_5\,kg^{-1}TSS\,d^{-1}$ and an activated sludge concentration of $TS_{AS} = 3\,kgTSS\,m^{-3}$. The plant is to treat the wastewater of about 100,000 pe. You expect the cost of the aeration tank to be approximately € 600 per m^3 additional volume.

You carefully examined all measuring systems and sampling points and are convinced that only random variations of the measured values (measuring errors and random real variations) remain, whereas all systematic deviations are eliminated.

Each additional sample, analysis, and associated evaluation costs € 150. The already existing 30 measured values for the load of the BOD_5 point to a normal distribution with a prognosis value for the future of $L_{BOD} = 4000\,kgBOD_5\,d^{-1}$ and a standard deviation of $s_{LBOD} = 1200\,kgBOD_5\,d^{-1}$. There is no recognizable weekly or seasonal cycle in the existing data.

How large will you have to design the activated sludge tank, if you do not want to perform additional measurements but want to achieve the planned sludge loading rate with at least 95% probability?

We compute the average BOD_5 load that will not be exceeded with a probability of 95% with the help of the t distribution:

The standard error of the average BOD_5 load is: $s_{\bar{F}} = \sqrt{\dfrac{s_F^2}{n}} = 220\,kgBOD_5\,d^{-1}$

With 29 degrees of freedom $t_{95\%}^{v} = 1.70$ the load becomes
$L_{BOD} = 4000 + 1.70 \cdot 220 = 4374\,kgBOD_5\,d^{-1}$.

$B_{TS} = L_{BOD}/(V_{AS} \cdot TS_{AS})$ results in $V_{AS} = 14,580\,m^3$ for the necessary volume of the tank. 1250 m^3 are to be built as a consequence of the uncertainty, with cost consequences of approximately € 750,000.-.

How many additional measurements will you obtain in order to find an economic solution for your client?

With an additional 70 samples ($n = 100$), costs of € 10,500 arise for analysis and interpretation. The uncertainty of the average value is thereby reduced to $s_{LBOD} = 120\,kgBOD_5\,d^{-1}$. If the average value remains unchanged, then this improved data situation results in 99 degrees of freedom and a reduction of the volume from 14,580 m^3 to 14,000 m^3 with expected savings of approximately € 350,000. At the same time, however, the construction will be delayed by several months and it is not sure that a reduction really results.

Time series cause large apparent random variation

We must differentiate between measurement errors (random variables) and the natural variability (a real dispersion of the measurands) of measured values. If we determine the time series of the load of a pollutant in a wastewater, then the disper-

Water distribution in 1000 m³ d⁻¹

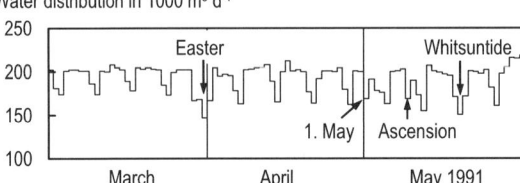

Fig. 11.5 Example of a time series: the distribution of drinking water in the city of Zurich, daily amounts over three months. The weekly cycle is clearly visible. The existing measuring errors are negligibly small in comparison to the real variation of the measured variable. Easter, the 1st of May, Ascension, and Whitsuntide can be identified as holidays

sion of the real values and the dispersion of the measurement errors overlay. If we analyze time series of real values, then periodic correlations frequently result, e. g., in the course of a week (see Fig. 11.5) or seasonally. We can model such periodic variations, identify them, and then use them to improve the estimation of the expected values (see Sect. 14.9). Truly random measuring errors are not subject to periodic processes; their influence can only be reduced by improving the measuring procedures and methods or enhancing the number of measurements.

11.3.3 Systematic Measurement Errors, Bias

Systematic measurement errors or bias have a deterministic, reproducible character that results from a fixed but possibly unknown cause–effect relationship. This relationship leads to a directed deviation of the measured value from the measurand, which can frequently be corrected based on additional expenditure.
Bias leads to correlated residuals and underestimated uncertainty of parameters when model parameters are identified from data. Today we do not have readily available statistical methods to deal with these problems (see Sect. 12.4.2).

We differentiate systematic measurement errors which are temporally constant from those that vary with time and are therefore subject to drift. In addition, the measurement error can be constant or depend on the measured value. Systematic errors cannot be recognized or reduced by simple repetition of the measurement but require additional experiments for their identification.

Example 11.9: Systematic measurement error as a consequence of the sampling procedure

You want to determine the load of the pollutants in the influent of a wastewater treatment plant. The sampling equipment draws a constant volume of wastewater at regular time intervals into the sampling bottle. You analyze the sample and multiply the result by the integrated flow rate.

Constant infiltration of unpolluted groundwater into the sewer leads to the fact that the diurnal variation of the flow of wastewater $Q(t)$ is positively correlated with the variation of pollutant concentration $C(t)$.

The daily load of the pollutants L_P results from $L_P = \int_{24h} Q \cdot C_P \cdot dt$. The sampling integrates only the pollutant concentration over time according to $m_C = \dfrac{\int_{24h} C \cdot dt}{24h}$.

Given a positive correlation between $C(t)$ and $Q(t)$, the resulting $L_P = m_Q \cdot m_C$ is systematically too small. Only flow-proportional sampling can lead to the desired result. This requires an extra effort.

With additional expenditure, systematic measurement errors may (sometimes) be identified and be corrected. A redundant information basis that is compiled in an independent way is important. Examples are:

- Careful error analysis of the measuring process and examination of the individual working procedures.
- Alternative, independent, and possibly more complex, more expensive but more reliable measuring methods that duplicate already available measurements.
- Use of a priori knowledge (e. g., mass conservation) and supplementing measuring programs, such that this knowledge can be used for uncovering systematic errors (see Sect. 11.5).
- Comparison of measured data with model prediction (here caution is required, because frequently data are collected to uncover unknown facts that are not yet captured in existing models).
- Calibration of all measuring devices.
- Identification of error models based on a priori knowledge.

Despite all of these partially very economical possibilities, many time series of measurements are grossly affected by systematic errors. The bias of flow measurements (Fig. 11.6) is of special importance. Uncorrected, this jeopardizes the bases for the extensions of many plants (Example 11.10).

Example 11.10: Systematic errors in the bases for design

As a young, inexperienced engineer you must compile the bases for the adaptation of a treatment plant to new, more stringent discharge requirements. You evaluate the measurements from the routine monitoring program of the existing treatment plant and estimate the measured pollutant load. The plant operator is convinced that he can rely on his measurements. You believe him and do not notice that the flow measurements are systematically 25% too low (see Fig. 11.6).
What are the consequences of your trust in experience?

Fig. 11.6 Observed systematic measurement error of the flow measurement in 18 full scale wastewater treatment plants in Germany (data by Port, 1994)

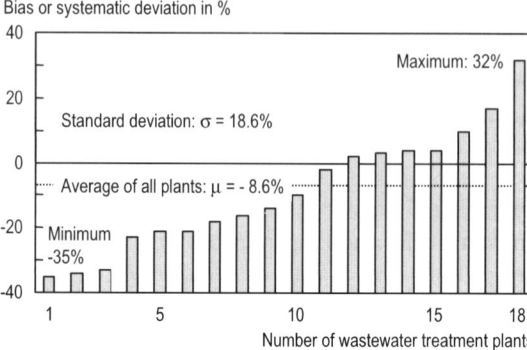

The pollutant load is proportional to the flow rate. The volume of the biological reactors is proportional to the pollutant load, as is sludge handling. The size of the sedimentation basins and the possible filtration is dependent on the quantity of water. They, too, will be 25% too small. The entire plant will be designed 25% too small and this will soon become evident once the new plant is in operation.

For the operator the relative numbers are frequently sufficient. He can use his experience from yesterday and transfer it to today based on the relative information.

Example 11.11: Systematic measurement errors raised the price of wastewater

On 22 November 2004 the *Tagesanzeiger* reported:

Cheaper Wastewater

Wiesendangen – the municipality can lower the wastewater fees substantially. Instead of SFr. 3.60 the citizens will in future pay only SFr. 2.40 per cubic meter of fresh water. In the year 2000, the municipality shut down its own treatment plant and was connected to the modern wastewater treatment plant of the city of Winterthur. Initially exact measuring of the wastewater quantity was not successful. Now it turned out that Wiesendangen delivers less wastewater than was assumed.

11.4 Case Example: COD Measurement (Standard Curve)

Figure 11.7 shows a (simulated) calibration curve for the determination of the COD. It is more extensive than is applied in practice, but it indicates individual kinds of measurement errors.

A standard solution, with well-known COD (based on the theoretical oxygen requirement TOD, Sect 5.7.3) serves as reference for the calibration of the measurement. We assume that errors in the production of the calibration solutions are small

Fig. 11.7 Calibration line for the measurement of COD with systematic and random measurement errors (simulated)

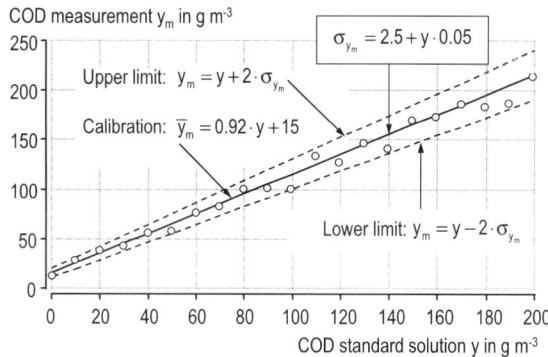

in comparison to the measurement of the COD. The regression line through the measured values does not pass through the intercept, and its slope differs from unity. This points to systematic measurement errors of the following size:

$$\Delta y_{m,systematic} = \overline{y}_m - y = -0.08 \cdot y + 15 . \tag{11.15}$$

These deviations are deterministic and can be corrected with the following relationship:

$$y = \frac{y_m - 15}{0.92} . \tag{11.16}$$

Nevertheless random deviations remain; here they increase with the measured value. With the assumption that they are normally distributed, one obtains for the corrected measurements

$$y_m = y + \Delta y_{random} ,$$

where Δy_{random} is a normally distributed random variable $N(0, \sigma_{ym})$ and the value of the standard deviation of this variable depends on y:

$$\sigma_{ym} = 2.5 + 0.05 \cdot y .$$

If we determine a COD value in the range around $100 \, g \, m^{-3}$, then the individual measurement has a standard deviation of $7.5 \, gCOD \, m^{-3}$. If we want to reduce this standard error to $5 \, gCOD \, m^{-3}$, we must make two or three parallel measurements and in addition eliminate all systematic measurement errors with Eq. (11.16).

11.5 Identifying an Error Model

Using a priori knowledge we can frequently identify an error model that can be used to eliminate systematic errors from data.

Case Example

With the aid of an oxygen electrode we follow the decrease of oxygen in an unaerated batch reactor which contains active microorganisms (Fig. 11.8, *left*). We want to derive the kinetic parameters for the use of oxygen which fit to the following model:

$$\frac{dS_{O2}}{dt} = r_{m,O2} \cdot \frac{S_{O2}}{K_{O2} + S_{O2}} \tag{11.17}$$

S_{O2} = oxygen concentration $[gO_2\ m^{-3}]$
$r_{m,O2}$ = max. oxygen consumption of the microorganisms $[gO_2\ m^{-3}\ d^{-1}]$
K_{O2} = Monod saturation concentration $[gO_2\ m^{-3}]$

The parameter of interest is K_{O2}, which influences the results primarily in the range of very low oxygen concentration ($S_{O2} < 1\ gO_2\ m^{-3}$). Visually inspecting the data reveals that the observed oxygen concentration does not approach $0\ gO_2\ m^{-3}$, as we would expect, but that it levels off at about $0.2\ gO_2\ m^{-3}$. A fit of the model to the original data results in a high K_{O2} value and residuals (model – data) which show autocorrelation (Fig. 11.9 for autocorrelation see Sect. 14.10). Adding another parameter to the model by subtracting a constant offset (to be estimated)

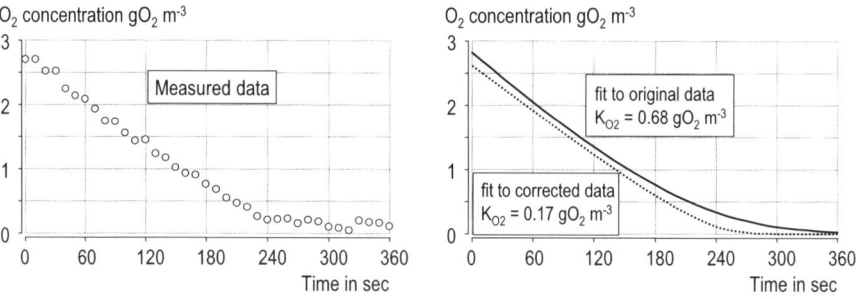

Fig. 11.8 *Left*: original data of measured oxygen concentration, one reading every 10 sec. *Right*: line fit to original data and line fit to data corrected for a constant offset (subtraction of $0.15\ gO_2\ m^{-3}$ from original data)

Fig. 11.9 Comparison of the residuals after fitting the model of Eq. (11.17) to the data. The original data leads to autocorrelated residuals whereas the corrected data results in a random sequence

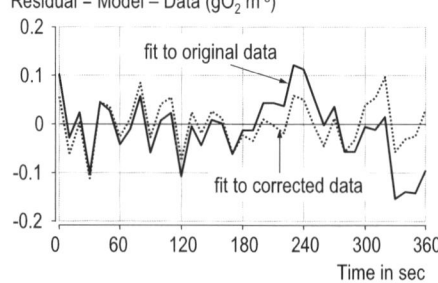

results in a much reduced K_{O2} value, and the sum of the squared errors is reduced from 0.16 to $0.08\,g^2\ m^{-6}$. We can have much more confidence in the corrected data: (i) they now asymptotically approach $0\,gO_2\ m^{-3}$ as expected, (ii) the result is in the range of our experience, and (iii) the sum of the squared errors is halved with still 56 degrees of freedom. In addition the residuals (model minus measurement) of the corrected data are randomly distributed around $0\,gO_2\ m^{-3}$, whereas the residuals from the original data show autocorrelation (Fig. 11.9). The corrected data provides a much better result, which we can accept with better confidence. See also Chap. 12.

Using the a priori knowledge that oxygen asymptotically approaches zero in our experiment allowed us here to identify an error model: a constant offset had to be subtracted from the measurements in order to obtain reliable information. There is good evidence that our model actually improves the data.

11.6 Uncovering Systematic Measurement Errors

Uncovering systematic measuring error requires redundant measurements that permit us to produce contradictions.

Example 11.12: Calibration with redundant information

The calibration of the COD analysis in Fig. 11.7 is based on redundant information. On the one hand a value for the COD results from the analysis due to the well-known composition and properties of the analytical chemicals used. On the other hand standard solutions contain a well-known quantity of an oxidizable organic material from which again the COD is known.
From the comparison of the two results Eq. (11.15) was derived, which permits us to quantify and correct systematic errors.

We can uncover measuring errors, if we procure redundant information. In addition we can use mass balances and conservation laws (see Sect. 5.7). Chemical elements in a system have neither sources nor sinks. In systems with microbiological processes, phosphorus and iron are particularly suitable, because these elements do not escape into the atmosphere (unlike, for instance, carbon as CO_2 or nitrogen with denitrification as N_2). If we capture all mass flows of phosphorus and iron with independent measurements, we can examine the measurements over a mass balance without the reaction term.

Figure 11.10 shows a simple scheme of a biological waste water treatment plant that is operated with simultaneous precipitation of phosphorus. Since the iron salts are delivered separately to the plant (and paid for), we have good control of the quantity of these salts which are dosed into the plant. The largest part is incorporated together with the phosphorus into the excess sludge. If we neglect

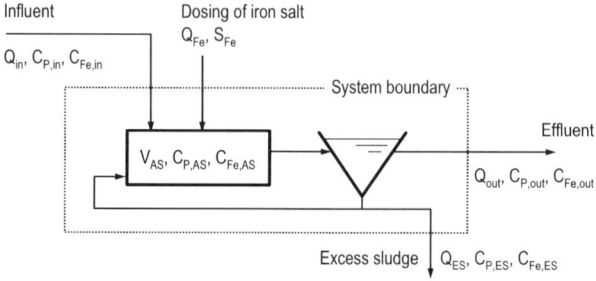

Fig. 11.10 Definitions for the derivation of the phosphorus and iron balance around a biological waste water treatment plant

storage of these elements in the secondary clarifier, the balances for the two elements P and Fe have the form:

$$V_{AS} \cdot \frac{dC_{P,AS}}{dt} = Q_{in} \cdot C_{P,in} - Q_{out} \cdot C_{P,out} - Q_{ES} \cdot C_{P,ES}$$

$$V_{AS} \cdot \frac{dC_{Fe,AS}}{dt} = Q_{in} \cdot C_{Fe,in} + Q_{Fe} \cdot S_{Fe} - Q_{out} \cdot C_{Fe,out} - Q_{ES} \cdot C_{Fe,ES} \qquad (11.18)$$

Over a long time period (many solids retention times) the accumulation term in the balance equations will add up to zero. Thus, two algebraic equations for the daily load (Q·C) result, which we can use for the examination of measurements:

$$\sum_{in} (Q_{in} \cdot C_{P,in}) = \sum_{out} (Q_{out} \cdot C_{P,out} + Q_{P,ES} \cdot C_{P,ES})$$

$$\sum_{in} (Q_{in} \cdot C_{Fe,in} + Q_{Fe} \cdot S_{Fe}) = \sum_{out} (Q_{out} \cdot C_{Fe,out} + Q_{ES} \cdot C_{Fe,ES}) \qquad (11.19)$$

In the application of these equations we must ensure that the change of the amount of material in the system is only a very small fraction of the total effluents. All concentrations must be determined in flow proportional samples. If the input deviates systematically from the output, this is an indication for systematic measuring errors. Since the dominant fraction of iron salts is subject to controlled dosing, while phosphorus arrives with the wastewater flow, a careful analysis of both balances can provide us with an indication of where we should look for the source of error: with flow rate, with the excess sludge or with analytical procedures (see also Thomann, 2002).

Example 11.13: Phosphorus balance for the identification of systematic measurement errors

A treatment plant operator provides the following annual average values from the operation of his biological treatment stage with simultaneous precipitation:

Influent measured Q_{in} = 10,000 m³ d⁻¹
                                  $C_{P,in}$ = 3 gP m⁻³, from flow proportional
                                                sampling

Effluent computed from $Q_{in} - Q_{ES}$	Q_{out}	=	$9900 \, m^3 \, d^{-1}$
measured	$S_{P,out}$	=	$0.3 \, gP \, m^{-3}$
Excess sludge	Q_{ES}	=	$100 \, m^3 \, d^{-1}$
Solids retention time	θ_X	=	$10 \, d$
Activated sludge tank volume	V_{AS}	=	$2500 \, m^3$
Activated sludge concentration	TS_{AS}	=	$3500 \, gTSS \, m^{-3}$
Surplus sludge concentration	TS_{ES}	=	$8750 \, gTSS \, m^{-3}$
Phosphorus content			
of the activated sludge	i_P	=	$0.025 \, gP \, g^{-1}TSS$

All sampling devices, chemical analysis, etc. are carefully examined; the largest possible error remains with the flow measurements.
How large do you estimate the error in the reported flow rate to be?

P load in the influent:	$Q \cdot C_{P,in}$	=	$30{,}000 \, gP \, d^{-1}$
P load in the effluent:	$Q \cdot S_{P,out}$	=	$3{,}000 \, gP \, d^{-1}$
P load in the excess sludge	$Q_{ES} \cdot TS_{ES} \cdot i_P$	=	$21{,}875 \, gP \, d^{-1}$.

From influent = effluent one obtains a discrepancy
of $30{,}000 - 21{,}875 - 3000 = 5125 \, gP \, d^{-1}$.

With the choice of $Q_{in} = 8100 \, m^3 \, d^{-1}$ the discrepancy would disappear: this is a strong indication of a systematic measurement error. However an adaptation of Q_{ES} would also adjust the balance. Only a careful calibration of the two measuring systems can help.
Here it is ensured that the mass of the phosphorus in the system is many times smaller than the total influents or effluents. With a solids retention time of 10 days, this mass in comparison to the yearly load amounts is only approximately 3% ($365 \, d/\theta_X$) or $219 \, kgP$.
Since the concentration of excess sludge varies throughout the day, sampling for its analysis is critical and might be another source of uncertainty.

Chapter 12
Parameter Identification, Sensitivity and Error Propagation

Mathematical models used in the environmental engineering sciences usually require the identification of model parameters from experimentally obtained data. In this process experimenting is typically the most expensive part, and a rather complex and time-consuming task. Thus, we want to gain as much reliable information as possible from an experiment. For the planning of productive and yielding experiments we use the tools of sensitivity and identifiability analysis. We then use statistical methods to determine the most likely values of model parameters from the observed data. In addition we want to gain information on the uncertainty or our model predictions.

12.1 Parameter Identification

For the identification of parameters we use different experimental observations and try to obtain those values of the model parameters that lead to model predictions which correspond with the largest probability (maximum likelihood) to the true performance of the system. Assuming the random errors to be normally distributed, we obtain a standardized weighted sum of squared deviations (residuals, errors) between data and model prediction in the form of χ^2. Parameters are then estimated such that the computed value of χ^2 is minimized.

The statistical literature offers different options for objective functions and procedures to identify parameters from data, possibly even combined with previous knowledge (Bayes statistics). Here we only deal with the simplest options and assume that measurement errors are normally distributed and not subject to bias.

12.1.1 Basic Principles, Chi Square, χ^2

The identification of model parameters requires that we evaluate the differences of the model predictions and observed variables of the real system as a whole. Since not all measurements are of equal quality we may want to weight individual values, i. e., with their standard error. The smaller the weighted deviations are, the better the estimated values of the parameters will be. For normally distributed measurement errors, the minimization of the χ^2 (chi square) leads to the most probable set of parameters and thereby a good estimate of model behavior:

$$\chi^2 = \sum_{i=1}^{n} \left(\frac{y_{m,i} - y_i(p)}{\sigma_{m,i}} \right)^2 \tag{12.1}$$

χ^2 = chi square, sum of the squares of the weighted difference between the measured and the computed state variables [−]

$y_{m,i}$ = i^{th} measured value of a state variable in the real system, assumed to be a normally distributed random variable [y]

$\sigma_{m,i}$ = standard error of the measurement of $y_{m,i}$ [y]

$y_i(p)$ = result of model prediction which corresponds to the measured $y_{m,i}$ in kind, time, and space [y]

p = set of the model parameters [p]

n = number of available data points [−]

The χ^2 probability distribution (Table 12.1) is equivalent to the probability that the sum of the squares of n samples of the standardized normal distribution $N(0,1)$ obtains a certain value. If the individual measurement errors are normally distributed with $N(0,\sigma_m)$, then the measurement errors, standardized with the standard error σ_m, obey a standard normal distribution $N(0,1)$, and χ^2 obtained from Eq. (12.1) obeys a χ^2 distribution with $v = n - 1 - n_P$ degrees of freedom, where n_P corresponds to the number of parameters which were determined from the set of n data points.

Procedure in the parameter identification

• We start with an assumed set of parameters p (based on experience) and obtain for all measured state values $y_{m,i}$ the associated model prediction $y_i(p)$.
• We compute with Eq. (12.1) the associated, χ^2.
• Now we change the parameter values with a suitable search strategy to $p + \Delta p$ and again compute the model values $y_i(p + \Delta p)$.
• If the value of χ^2 decreased, then we improved the parameter set; if it increased, the probability that the model prediction corresponds to reality was reduced.
• This procedure is repeated until the minimum possible χ^2 is found (simulation software provides suitable, automated routines for this task).
• With the finally available parameter values we can compute the most probable system behavior which led to the measurements.

Table 12.1 Cumulative χ^2 distribution as a function of the degrees of freedom v $F(v,\chi^2)$. Reading example: if 20 numbers are drawn from a normal distribution $N(0,1)$, then the sum of the squares of these 20 numbers is in 95% of the cases smaller than 31.4 ($v = 20$ degrees of freedom, $\alpha = 0.95$)

v	α			
	0.80	0.90	0.95	0.99
1	1.6	2.7	3.8	6.6
2	3.2	4.6	6.0	9.2
3	4.6	6.3	7.8	11.3
4	6.0	7.8	9.5	13.3
5	7.3	9.2	11.1	15.1
6	8.6	10.6	12.6	16.8
7	9.8	12.0	14.1	18.5
8	11.0	13.4	15.5	20.1
9	12.2	14.7	16.9	21.7
10	13.4	16.0	18.3	23.2
12	15.8	18.6	21.0	26.2
14	18.2	21.1	23.7	29.1
15	19.3	22.3	25.0	30.6
16	20.5	23.5	26.3	32.0
18	22.8	26.0	28.9	34.8
20	25.0	28.4	31.4	37.6
25	30.7	34.4	37.7	44.3
30	36.3	40.3	43.8	50.9
40	47.3	51.8	55.8	63.7
50	58.2	63.2	67.5	76.2
60	69.0	74.4	79.1	88.4
70	79.7	85.5	90.5	100
80	90.4	96.6	102	112
90	101	108	113	124
100	112	119	124	136
150	164	173	180	193
200	217	226	234	249
500	526	541	553	576

With the minimum value of χ^2 we can now perform a test of whether the deviations between model prediction and measurements can be explained by independent and normally distributed measurement errors with the standard deviation σ_m. If that is not the case, in particular if the χ^2 becomes too large, the model may be subject to structural problems (see Sect. 12.4) or else we may have underestimated the standard error of the measurement procedures.

If we do not know the standard error of the measurements σ_m, then we may give weights to individual measurements based on our judgment of the reliability of individual results; reliable measurements receive large weights, whereas less reliable measurements receive smaller weights. The variable that is then minimized has the form:

$$X_{rms} = \sqrt{\frac{1}{n} \cdot \sum_{i=1}^{n} \left(g_i \cdot \left(y_{m,i} - y_i(p) \right) \right)^2} \tag{12.2}$$

X_{rms} = weighted *root mean square* of the deviation between measurement and model prediction [y]
g_i = weight of the ith measurement [–]

Frequently we do not have any reason to differentiate between individual measurements (or we might want to keep the procedure simple), in which case the choice of $g_i = 1$ is most common. With free choice of the weights g_i the resulting X_{rms} does not have further meaning. However:

$$\text{for } g_i = \frac{1}{\sigma_{m,i}} \text{ is valid } \chi^2 = n \cdot x_{rms}^2 . \tag{12.3}$$

For identically normally distributed and independent measurement errors of all measured values $y_{m,i}$ we can derive from both the χ^2 and x_{rms} an estimated value for the standard error of the measurements. It applies (without proof) for $g_i = 1$ that:

$$\sigma_{m,i}^2 = \frac{\chi^2}{n - n_P} = \frac{n}{n - n_P} \cdot x_{rms}^2 . \tag{12.4}$$

For $n \gg n_P$ one has $\sigma_{m,i} \approx x_{rms}$.

Example 12.1: Interpretation of χ^2

Case 1:
You have 20 data points available in order to estimate four parameters. The minimized χ^2 value is 18.7.
You want to test whether you can accept the hypothesis that the standardized measurement errors are really N(0,1) distributed, which would be a good indication for the fact that you have identified a model that is justified by the data.
The number of degrees of freedom is $v = 20 - 1 - 4 = 15$.
From Table 12.1 you obtain for $v = 15$ in 95% of the cases $\chi^2 < 25.0$. You can accept the hypothesis.
Case 2:
You are estimating the parameters of a model by minimizing χ^2. The resulting χ^2 is many times larger than the value that you would expect based on the number of degrees of freedom v under consideration with a confidence interval of 95%.
You must reject the hypothesis. The result is an indication that the model cannot be justified by the data; there may be some problems with the structure of the model or you have totally underestimated the standard error of the measured state variables. A more detailed analysis of the residuals (difference measurement –

model prediction) is necessary. If the residuals are not independent and normally distributed, this indicates that the structure of your model is not adequate. See also Sect. 12.4.1.

Example 12.2: Parameter identification in Berkeley Madonna

BM provides several options that allow the simultaneous and automatic identification of several parameters. The option *Curve Fit* is based on the minimization of x_{rms} based on the assumption that all data have equal weight $g_i = 1$. If several data sets are available, it is possible to choose an individual weight g_j for each dataset j. x_{rms} is available as a result and therefore allows the computation of χ^2. In addition the option Optimize allows the definition of any function to be minimized.

Example 12.3: Local minima of χ^2 may not yield optimal parameter values

It is possible that the minimization of χ^2 stops with a parameter combination that yields only a local but not a global minimum of χ^2. This means that the best parameter combination has been found inside (not at the edge) of a narrow range of parameter values. Jumping to significantly different parameter values may result in an even lower χ^2, but the optimization algorithm does not allow for such a jump.
An indication that local minima may exist is when largely different starting values of the parameters lead to largely different end results.

The strategies and the numeric methods for parameter identification are not discussed here. Generally they are computationally very demanding, because simulations with always new and improved parameters must be accomplished many times, in particular if many parameters are to be estimated.

12.1.2 Case Example: First-Order Reaction in a Batch Reactor

The following simple example describes the details of the procedure for parameter identification.

Task

You want to identify the degradation rate constant for a material that you know is degraded in a first-order reaction. The chemical analysis of the material is not very accurate, but you did not carry out an investigation in order to characterize the measurement errors. The experiment is performed in a batch reactor, with the concentration of the pollutant measured every 15 min over 5 h. The results are given in Fig. 12.1.

Fig. 12.1 Measured values of the pollutant concentration in a batch reactor

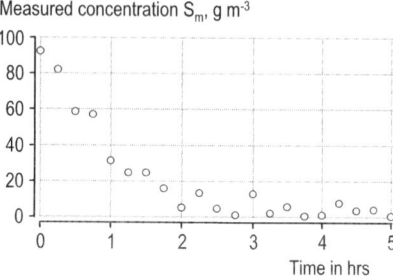

Model

The model for the degradation of the pollutant is simple. The balance equation for the batch reactor and its analytical solution have the form:

$$\frac{dS}{dt} = -k \cdot S \text{ solved analytically } S = S_0 \cdot \exp(-k \cdot t). \tag{12.5}$$

The model has two unknown parameters (the initial value S_0 and the rate constant k) and is based on the assumption that the degradation is first order.

Parameter identification

Since we have an analytical solution of the model equation available, we can use a spreadsheet. We do not know the standard error of the chemical analysis; we can, however, accept that this error is of equal size over the entire measuring range. Therefore we choose equal weights $g_i = 1$ for all measurements and minimize x_{rms} (Eq. (12.2)).

Table 12.2 shows a possible table for the computation of x_{rms}. The two parameters S_0 and k were first chosen, based on a subjective guess and then identified with the help of the optimization routine of the spreadsheet software such that x_{rms} becomes minimal.

Table 12.2 Table showing the computation of x_{rms}. Altogether $n = 21$ data points are given. The computed values $y_i(p)$ are obtained with the optimal parameter values of $S_0 = 97.4$ g m^{-3} and $k = 1.01$ h^{-1}

Time in h	Measurement $y_{m,i}$ in g m^{-3}	Simulation $y_i(p) = S_0 \cdot e^{-k \cdot t}$	$y_{m,i} - y_i(p)$ g m^{-3}	$(y_{m,i} - y_i(p))^2$ g^2 m^{-6}
0	91.7	97.4	−5.7	32.3
0.25	81.7	75.8	5.9	34.8
...	...	...	...	...
4.75	3.8	0.8	3.0	8.8
5.00	0.0	0.6	−0.6	0.4
			$\Sigma = 1.9$	$\Sigma = 478.2$
			Average $= 0.1$ g m^{-3}	$x_{rms} = 4.8$ g m^{-3}

Fig. 12.2 Comparison of the model prediction with the measured data

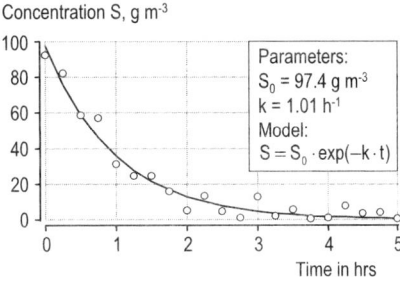

Results

As result of the parameter identification we obtain the two parameters of the model ($S_0 = 97.4$ g m^{-3}, $k = 1.01$ h^{-1}) and the minimized $x_{rms} = 4.8$ g m^{-3} as well as the average value of the measurement error $(y_i(p) - y_{m,i})_{avg} = 0.1$ g m^{-3}. With the assumption that the measurement errors are random and normally distributed, we can now compute the most probable course of the concentration with the identified parameters (Fig. 12.2). The average value of the measurement error lies within only 0.1 g m^{-3} of the expected value of 0 g m^{-3}. Thus, the obtained x_{rms} is a good estimation for the standard error of the analysis: $x_{rms} = 4.8$ g m^{-3}. This value must be corrected with the factor $\sqrt{n/(n-1-2)}$, because it was obtained from the data together with two parameters. Thus the standard error of the measurement of the pollutant concentration becomes $s_m = 4.8 \cdot (21/18)^{0.5} = 5.2$ g m^{-3}.

We cannot use χ^2 to test the quality of the model since we have obtained the standard error s_m of the analytical procedure from the data.

12.2 Introduction of an Extended Case Study

The further topics of this chapter are introduced on the basis of a rather straight-forward, simulated example: a batch reactor contains activated sludge which is aerated; the task is to determine the aeration rate and the oxygen consumption rate of the microorganisms.

Microorganisms in the activated sludge use oxygen, which in full-scale activated sludge systems is continuously delivered by aeration. The oxygen input of an aeration system depends on many parameters. Frequently it is necessary to determine the performance of the aeration and the rate of oxygen consumption of the microorganisms experimentally. Here a simple experiment is described which makes these two variables accessible.

Model

It is not the goal of this exercise to describe aeration in full detail. We characterize this process with the following simple rate equation:

$$r_{oxy} = k_1 a \cdot (S_{sat} - S) \tag{12.6}$$

r_{oxy} = oxygenation rate (introduction of oxygen) $[M_{O2} \, L^{-3} \, T^{-1}]$
$k_1 a$ = oxygen transfer coefficient $[T^{-1}]$
S_{sat} = saturation concentration of dissolved oxygen. This corresponds to the highest attainable concentration, if no oxygen is consumed $[M_{O2} \, L^{-3}]$
S = effectively existing concentration of dissolved oxygen, the state variable $[M_{O2} \, L^{-3}]$

In order to make our analysis simple, we choose a batch reactor as the experimental system and assume the oxygen consumption to be constant. The mass balance for oxygen has the form:

$$\frac{dS}{dt} = r_{oxy} + r_{O2} = k_1 a \cdot (S_{sat} - S) + r_{O2} \tag{12.7}$$

r_{O2} = consumption rate of oxygen (its value is negative, because O_2 is used) $[M_{O2} \, L^{-3} \, T^{-1}]$

Integrated, with the initial condition $S(t=0) = S_0$, this results in

$$\frac{S_{sat} + \dfrac{r_{O2}}{k_1 a} - S}{S_{sat} + \dfrac{r_{O2}}{k_1 a} - S_0} = \frac{S^* - S}{S^* - S_0} = \exp(-k_1 a \cdot t), \tag{12.8}$$

$$S^* = S_{sat} + \frac{r_{O2}}{k_1 a}. \tag{12.9}$$

Where S^* is the equilibrium concentration of the dissolved oxygen, which is reached asymptotically with constant r_{O2} and constant aeration $[M_{O2} \, L^{-3}]$. Solving for S results in

$$S = \underbrace{S^* \cdot (1 - \exp(-k_1 a \cdot t))}_{\text{Approach to equilibrium}} + \underbrace{S_0 \cdot \exp(-k_1 a \cdot t)}_{\text{'forgetting' the initial condition}} \quad \text{or} \tag{12.10}$$

$$S = \left(S_{sat} + \frac{r_{O2}}{k_1 a}\right) \cdot (1 - \exp(-k_1 a \cdot t)) + S_0 \cdot \exp(-k_1 a \cdot t). \tag{12.11}$$

If at time t_0 the aeration is turned off, $k_1 a = 0$ and the initial condition becomes $S = S(t_0)$. The solution is then

$$S = S(t_0) - r_{O2} \cdot (t - t_0). \tag{12.12}$$

Experiment

Figure 12.3 intoduces a possible experiment in which a time series of the oxygen concentrations can be measured. Here we work with simulated data, which has the advantage that we know the model and the correct parameter values. From experience we know that the dissolved oxygen electrode with the given temporal resolution is subject to a random, normally distributed measurement error with a standard deviation of $\sigma_m = 0.10 \, \text{gO}_2 \, \text{m}^{-3}$.

Figure 12.4 shows a time series of there measured (i.e., simulated) oxygen concentrations S_m with $n = 46$ available data points, distributed over 45 min. During the first 15 min the reactor is aerated; afterwards it is only mixed without aeration.

Task

A model is to be developed which is suitable to describe the behavior of the system in Fig. 12.3. The parameters of the model are to be identified and the uncer-

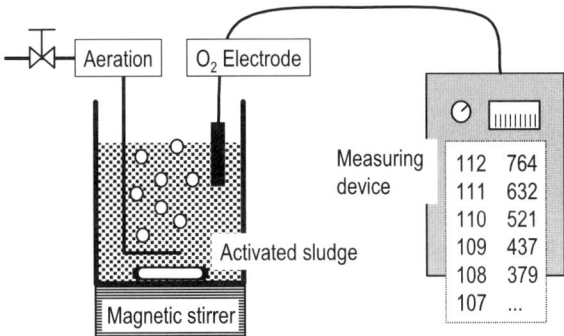

Fig. 12.3 Experimental equipment for the determination of the oxygen requirement

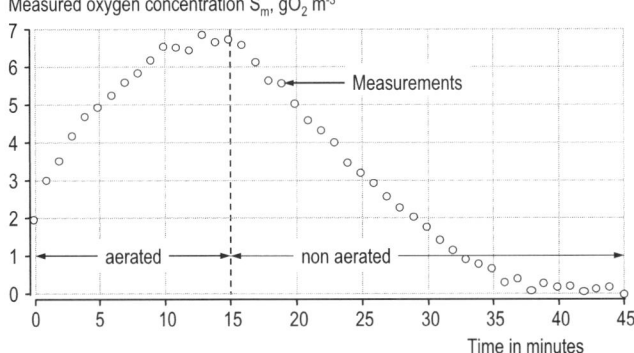

Fig. 12.4 Time series of the measured (actually simulated) oxygen concentrations

tainties of the model predictions are to be quantified and discussed. The kinetic parameters are of special interest, as they can be transferred to another system.

Based on this example we will develop and introduce the required systems analysis tools.

12.3 Sensitivity and Identifiability

In the case study of Sect. 12.2 we analyze for the time being only the first 15 min, when the aeration is still switched on. Here the identifiability of model parameters is limited. Local sensitivity functions indicate why this is the case.

12.3.1 Case Study

The model in Eq. (12.11) has a total of four parameters, which are to be determined from the data

$$S = \left(S_{sat} + \frac{r_{O2}}{k_1 a} \right) \cdot \left(1 - \exp\left(-k_1 a \cdot t \right) \right) + S_0 \cdot \exp\left(-k_1 a \cdot t \right) \tag{12.11}$$

S_{sat} = saturation concentration of the dissolved oxygen in the experimental system $[gO_2\, m^{-3}]$

r_{O2} = oxygen consumption of the microorganisms $[gO_2\, m^{-3}\, min^{-1}]$

$k_1 a$ = aeration coefficient $[min^{-1}]$

S_0 = initial concentration of for dissolved oxygen $[gO_2\, m^{-3}]$

In reality there is a fifth uncertain parameter: the time when the aeration is switched off. However, since we can measure the time much more accurately than e. g., the concentration of the dissolved oxygen, we accept the measured value as free of error and do not identify it from the data.

In order to keep the task simple we try to identify the four parameters from the measurements during the first 15 min of the experiment, when the aeration is switched on. Example 12.4 gives the code which is used for the identification of the parameters.

Example 12.4: Code for Berkeley Madonna for the implementation of the model equation (12.11)

The following code implements the model of the case study in BM. It uses the balance equation and not the analytic solution. The last line permits the aeration to be switched off after 15 min.

```
METHOD RK4        ; Integration with fourth-order Runge–Kutta
STARTTIME = 0     ; Beginning of the experiment
```

STOPTIME = 15   ; End of the experiment (with aeration) min
DT = 1   ; Time step min
Sm = #Data(time)   ; Read in the data
Ssat = 10   ; Estimated value for oxygen saturation $gO_2\,m^{-3}$
rO2 = −0.40   ; Estimated value for oxygen consumption $gO_2\,m^{-3}\,min^{-1}$
kla = 0.15   ; Estimated value for aeration coefficient min^{-1}
S0 = 2   ; Estimated value for initial oxygen concentration $gO_2\,m^{-3}$
INIT S = S0   ; Balance for oxygen concentration $gO_2\,m^{-3}$
d/dt(S) = on*kla*(Ssat − S) + rO2
limit S >= 0   ; Negative oxygen concentrations are not possible
on = if time <= 15 THEN 1 ELSE 0   ; Switch off the aeration after 15 min

Table 12.3 Identified parameters for different duration of the experiment and different initial values of the parameters (computed with BM). Models A and B were identified from the data of the first 15 min; model C is based on the data up to 30 min. S^* is computed with Eq. (12.9)

Parameter	Identification			Units
	Model A 0–15 min	Model B 0–15 min	Model C 0–30 min	
Number of data points n	16	16	31	–
O_2 saturation S_{sat}	2.91	11.10	9.06	$gO_2\,m^{-3}$
O_2 consumption r_{O2}	0.732	−0.665	−0.352	$gO_2\,m^{-3}\,min^{-1}$
Aeration k_la	0.171	0.171	0.179	min^{-1}
Initial concentration S_0	2.02	2.03	2.01	$gO_2\,m^{-3}$
ΔS_{rms}	0.113	0.113	0.114	$gO_2\,m^{-3}$
Equilibrium conc. S^*	7.19	7.21	7.09	$gO_2\,m^{-3}$

Table 12.3 summarizes the results of the identification for three cases A to C:

- In case A the permissible range for r_{O2} was not limited. The result is a physically impossible, positive value for r_{O2} and the saturation concentration is far from the expected value around $10\,gO_2\,m^{-3}$. The prediction of the model is compared with the data of the first 30 min in Fig. 12.5. The first 15 min of the experiment are well described, but obviously the identified parameters are not suitable for an extrapolation without aeration.
- In case B r_{O2} was limited to negative values. But, as shown in Fig. 12.5, these values are also not suitable for extrapolation. The initial concentration S_0 and the aeration coefficient k_la are nevertheless in both cases, A and B, identified with nearly identical values.
- In case C the data from the first 30 min were used for the identification. The identified model convinces over the entire range. Again S_0 and k_la are nearly identical to the values of the cases A and B.

The three cases show that the data of the first 15 min are obviously not sufficient for a unique identification of the parameters of the model. Only the use of a period without aeration leads to a parameter set that can describe the whole range.

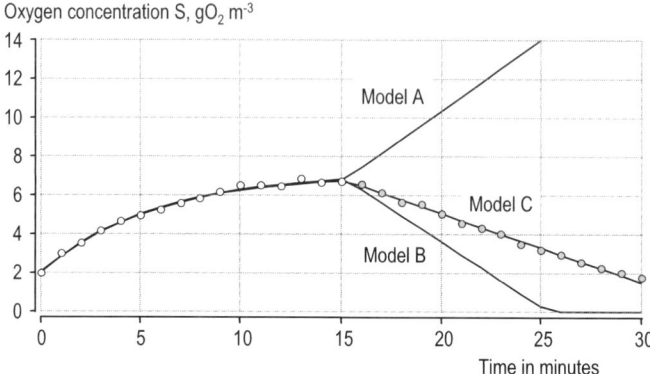

Fig. 12.5 Comparison of the simulation based on the models identified in the cases A, B and C of Table 12.3

The reason for the identification problems lies in the mathematical structure of the model. Equation (12.11) provides the analytical solution for the model for the first 15 min, when the aeration is still active:

$$S = \left(S_{sat} + \frac{r_{O2}}{k_1 a} \right) \cdot \left(1 - \exp(-k_1 a \cdot t) \right) + S_0 \cdot \exp(-k_1 a \cdot t) . \qquad (12.11)$$

Equation (12.9) combines the first term of Eq. (12.11) into the equilibrium concentration S^*:

$$S^* = S_{sat} + \frac{r_{O2}}{k_1 a} . \qquad (12.9)$$

The simplification of Eq. (12.11) in the form of Eq. (12.10) shows that this model has in reality only three parameters: S^*, $k_1 a$, and S_0,

$$S = S^* \cdot \left(1 - \exp(-k_1 a \cdot t) \right) + S_0 \cdot \exp(-k_1 a \cdot t) .$$

If these three parameters are known, then there are infinitely many combinations of S_{sat} and r_{O2} that result in the same concentration S^*. Table 12.3 shows that, for all three parameter sets, nearly identical equilibrium concentrations S^* result. A change of the saturation concentration S_{sat} can be fully compensated in this model by an adjustment of the oxygen consumption r_{O2}.

We don't always have an analytical solution of the model equation available and often the situation arises that the compensation of a change of a parameter can only approximately be compensated by the adjustment of another parameter.

Is there a method to help us to recognize problems with the parameter identification independently of the analytic solution of the model? The local sensitivity function discussed in the following section supports us in the treatment of this problem.

12.3.2 Local Sensitivity Functions

Sensitivity functions express how the prediction of a model reacts to a change of a model parameter. We use these functions (i) to determine whether a parameter is clearly identifiable, and (ii) to design experiments that permit us to determine model parameters efficiently.

Here only the local sensitivity, which indicates how the prediction of a model with fixed parameters reacts to small changes of the parameters, is introduced. This is in contrast to global sensitivity, which analyzes an entire parameter space.

Basic Principles

The *sensitivity* of a state variable y relative to a model parameter p expresses how strongly the prediction for the state variable changes (Δy), if the parameter is changed by a small amount Δp. From this description we derive that we can determine the sensitivity as the partial derivative of the state variable relative to the parameter:

$$\delta_{y,p} = \frac{\partial y}{\partial p} \approx \frac{\Delta y}{\Delta p} \tag{12.13}$$

y   = state variable, whose sensitivity is determined [y]
p   = parameter which affects the state variable y [p]
$\delta_{y,p}$ = local sensitivity function of the state variable y relative to the parameter p
        with the dimensions [y]/[p]

If we have an analytic solution for the model equation available, we can directly obtain the partial derivative in Eq. (12.13). Mostly we must, however, use numeric methods. Berkeley Madonna provides for all dependent state variables y_i as functions of all independent parameters p_j the appropriate sensitivity functions.

Since the units of $\delta_{y,p}$ differ from the units of y, an interpretation of this sensitivity function is sometimes difficult, and it is especially difficult to compare the sensitivity of different parameters. The absolute-relative sensitivity function, which has the unit of the state variable, is more favorable and is defined as follows:

$$\delta_{y,p}^{a,r} = p \cdot \frac{\partial y}{\partial p} . \tag{12.14}$$

It indicates in a linear approximation how much a state variable changes absolutely if a model parameter changes relatively by 100% (Fig. 12.6).

Numerically the sensitivity function is obtained by solving the model equations once for p and once for $p + \Delta p$. Then, the partial derivative can be approximated as

$$\delta_{y,p}^{a,r} \approx p \cdot \frac{\Delta y}{\Delta p} = p \cdot \frac{y(p + \Delta p) - y(p)}{\Delta p} . \tag{12.15}$$

Fig. 12.6 Interpretation of the absolute-relative sensitivity function (Reichert, 1995)

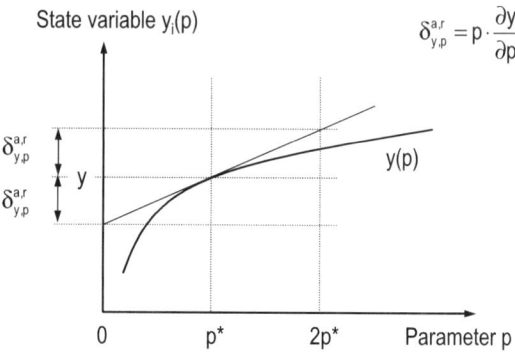

Δp is selected to be very small (e. g., 0.1% of p), so that the linearization does not introduce large errors. The computation of the sensitivity function for each additional parameter requires an additional simulation of the entire case, since for the partial derivative only one parameter can be changed at a time. However, the sensitivity of all state variables relative to a parameter can be computed from two simulation runs, since each simulation makes all state variables available at the same time.

Example 12.5: Analytic determination of the sensitivity function

If a material is degraded in a zero-order reaction in a batch reactor, then we obtain, with the initial condition $C(t=0)=C_0$, the concentration as follows:

$$C = C_0 - k \cdot t .$$

This model equation has two parameters, C_0 and k, and one state variable C.
What is the absolute-relative sensitivity of the state variable relative to the two parameters?

$$\delta^{a,r}_{C,C_0} = C_0 \cdot \frac{\partial C}{\partial C_0} = C_0 \cdot 1 = C_0 \quad \text{and} \quad \delta^{a,r}_{C,k} = k \cdot \frac{\partial C}{\partial k} = -k \cdot t .$$

A change of the initial value C_0 shifts the concentration curve of C upward proportionally to ΔC_0, whereas a change of k has, starting from the constant initial concentration C_0, an ever-larger effect with increasing time.
These two sensitivity functions correspond in our case study to the situation without aeration ($t > 15$ min, $C_0 = S(t=15$ min), $k=-r_{O2}$).

Example 12.6: Sensitivity functions in Berkeley Madonna

BM makes the partial derivative $\delta_{y,p} = \partial y / \partial p$ available as a sensitivity function. The absolute-relative sensitivity function $\delta^{a,r}_{y,p} = p \cdot \partial y / \partial p$ is, however, simpler to interpret. If we introduce a new, relative variable with the value of unity, we can compute the absolute-relative sensitivity as follows:

p = pabs * prel   ; We compute the value of parameter p
pabs = 8.72   ; We define the absolute value of p in units of p
prel = 1   ; We enter a relative variable, value of unity, dimensionless

$\delta_{y,p}^{a,r} = p \cdot \partial y / \partial p$ now arises as the partial derivative of $\partial y / \partial p_{rel}$ which corresponds to the sensitivity computed in BM. By the introduction of a fictitious relative variable for all parameters, we obtain all the absolute-relative sensitivity functions. The parameter p_{abs} must only be introduced separately if it is to be identified from data, otherwise it is sufficient to write $p = 8.72 * p_{rel}$.

Application to the Case Study

With the aid of Berkeley Madonna we obtain all sensitivity functions of the model; the result for the period of the first 15 min is given in Fig. 12.7. Beyond 1 min the sensitivity is largest for the saturation concentration S_{sat}, i.e., if any parameter is changed by a small amount, say 5%, S_{sat} will have the largest effect on the model output S.

From visual inspection of Fig. 12.7 it becomes clear that the sensitivity of S_{sat} and r_{O2} only differ by a proportionality factor but otherwise have exactly the same form, while the two other functions for k_1a and S_0 differ by more than a proportionality factor. From this observation we can derive that a change of S_{sat} can entirely be compensated by an appropriate adjustment of r_{O2}. Thus, these two parameters cannot be identified uniquely from the data at hand. Because the two sensitivities for k_1a and S_0 are not proportional to any other function, a unique identification of these values is possible. This is in line with the experience documented in Table 12.3. In the case study we have the analytical model equation (12.11) at hand. This permits us, with Eq. (12.14), to obtain an analytical form for the absolute-relative sensitivity functions for all four parameters:

$$\delta_{S,S0}^{a,r} = S_0 \cdot \frac{\partial S}{\partial S_0} = S_0 \cdot \exp(-k_1 a \cdot t), \tag{12.16}$$

$$\delta_{S,Ssat}^{a,r} = S_{sat} \cdot \frac{\partial S}{\partial S_{sat}} = S_{sat} \cdot \left(1 - \exp(-k_1 a \cdot t)\right), \tag{12.17}$$

$$\delta_{S,r_{O2}}^{a,r} = r_{O_2} \cdot \frac{\partial S}{\partial r_{O_2}} = \frac{r_{O_2}}{k_1 a} \cdot \left(1 - \exp(-k_1 a \cdot t)\right), \tag{12.18}$$

$$\delta_{S,k_1a}^{a,r} = k_1 a \cdot \frac{\partial S}{\partial k_1 a} = -\frac{r_{O_2}}{k_1 a} \cdot \left(1 - \exp(-k_1 a \cdot t)\right) + \left(S_{sat} + \frac{r_{O_2}}{k_1 a} - S_0\right) \cdot t \cdot \exp(-k_1 a \cdot t). \tag{12.19}$$

Absolute-relative Sensitivity δ_S of S, gO_2 m^{-3}

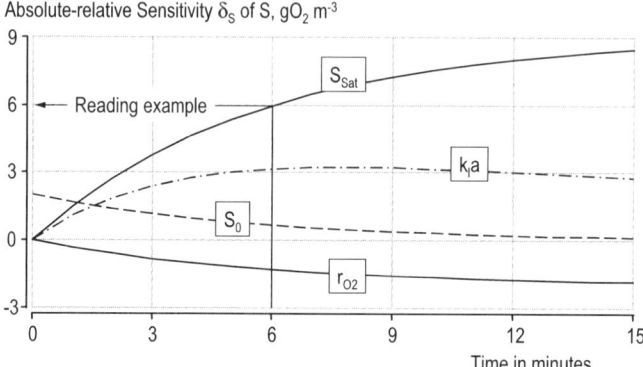

Fig. 12.7 Absolute-relative sensitivity functions for the four parameters in the case study for the first phase of the experiment, with aeration only. This figure is based on a numeric computation with BM for the parameter set C in Table 12.3. *Reading example:* if we increase the value of the oxygen saturation S_{sat} in the model by 5% and if all other parameters are kept constant, then after 6 min the oxygen concentration will increase by $0.05 \cdot 6 = 0.3$ gO_2 m^{-3}

Absolute-relative Sensitivity δ_S of S, gO_2 m^{-3}

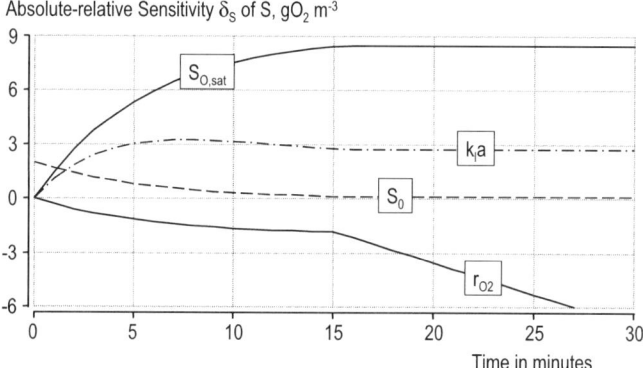

Fig. 12.8 Sensitivity functions similar to Fig. 12.7, however, for a longer time period (0–30 min) including a phase with the aeration turned off ($t > 15$ min)

Equations (12.17) and (12.18) show that these sensitivity functions differ only by a proportionality factor, while the structure of the two other functions is mathematically quite different, and therefore an identification of the two parameters is possible: a change of $k_l a$ or S_0 cannot be fully compensated by another parameter (or set of parameters).

If we now extend the time period up to 30 min, then the sensitivity functions show a completely different picture (Fig. 12.8). While S_{sat} does not affect the process any longer after the aeration is turned off, the influence of r_{O2} increases with time. Considering the entire range, the four sensitivity functions now show significantly differing behavior, which leads to the fact that all four parameters can be uniquely identified from the data over the first 30 min (actually even 20 min might be enough). This is also the experience documented in Table 12.3.

Summary: plotted local sensitivity functions are numerically easy to obtain. They help possible problems with the identifiability of parameters from experimental data to be recognized.

Unfortunately, linear (or even nonlinear) combinations of several different parameter values can also occasionally lead to identical solutions of the model. This may lead to problems with identifiability of parameters. Such problems are difficult to recognize visually and usually require a systematic procedure (see Brun et al., 2001).

Example 12.7: Linear combination of sensitivity functions

In a 10-km-long corrected river, a step has been built every 500 m to stabilize the sediment. These steps prevent the back mixing of water and materials. The river is characterized as follows:

Q = $4 \, m^3 \, s^{-1}$, constant effluent

h = $0.5 \, m$, average water depth

B = $10 \, m$, width of the river

In the river the following processes are of importance (the algae grow as a biofilm on the sediment):

Process	Dissolved oxygen S_O in $gO_2 \, m^{-3}$	Process rate ρ in $gO_2 \, m^{-3} \, d^{-1}$
Photosynthesis	+1	$K_P \cdot I / h$
Respiration	−1	K_R / h
Reaeration	+1	$K_B \cdot (S_{O,sat} - S_O) / h$

With

K_P = $0.1 \, gO_2 \, Wd^{-1}$, oxygen release per Watt day of light energy

I = $I_{max} \cdot (-\cos(2 \cdot \pi \cdot t))$, $I > 0$, available light energy in $W \, m^{-2}$

I_{max} = $1200 \, W \, m^{-2}$, maximum light intensity at midday

h = $0.5 \, m$, discharge depth in the river

K_R = $40 \, gO_2 \, m^{-2} \, d^{-1}$, respiration rate of the algae

K_B = $25 \, m \, d^{-1}$, reaeration constant of the river including steps

$S_{O,sat}$ = $10 \, gO_2 \, m^{-3}$, saturation concentration for oxygen

t = Time of the day, real time

Try to simulate with Berkeley Madonna the oxygen concentration in the diurnal variation at the end of the flow distance and as a length profile in the river at a certain time, and determine the sensitivity functions of the oxygen concentration relative to all model parameters.

Discuss the influence of the upper boundary condition of S_O on the discharge concentration of oxygen.

Discuss which parameters of the model you can determine from a continuously measured concentration curve of the dissolved oxygen in the effluent of the river. How can you improve the situation?

Hints:

Model the individual stretches between two steps as an array (cascade) of fully mixed reactors. Define a separate variable $S_{O,out}$, whose sensitivity you can compute.

Make for the time being a reasonable assumption for the upper boundary condition of S_O ($S_{O,in}$).

The sensitivity functions in Madonna have completely different orders of magnitude. Therefore you should determine the absolute-relative sensitivity functions.

Note that one sensitivity function can be formed as linear combination of two others.

This example has been suggested in similar form by Brun et al. (2001).

12.4 Model Structure

Differences between model predictions and observed data have various origins. They may be due to (i) an insufficient mathematical structure of the model, or (ii) incorrectly determined model parameters, and (iii) uncertainties of the model parameters as well as (iv) random and (v) systematic measurement errors. Here we discuss the problems with the mathematical structure of the model.

We select the model structure (mathematical dependences) based on the question to answer, our experience, and our knowledge of the processes that we assume are relevant for the behavior of the system to be analyzed. If the measured data are made consistent (by the elimination of outliers and systematic deviations, see Sect. 11.3) and the parameter identification is correctly performed, remaining deviations between the data and models are due to: (i) random measurement errors, (ii) uncertainty of the model parameters, (iii) inadequate structure of the model or (iv) experimental errors.

There is not a single correct model to describe the performance of a system. The suitable model structure depends on the question, the degree of detail, and the accuracy of the desired model prediction as well as on the available data (Fig. 12.9). Later, tests will be introduced that may point to an insufficient model structure. Depending upon the question to be answered, even models that are identified as

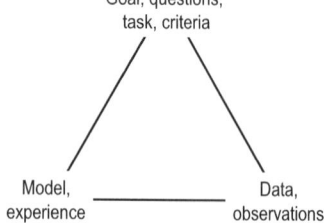

Fig. 12.9 The task, model structure, and available data depend on each other

insufficient by these tests might be able to meet our requirements; this is particularly the case if the measurements are clearly more accurate than our expectations regarding the accuracy of model predictions.

12.4.1 Structural Model Deviations

If we identify in the case study of Sect. 12.2 the four parameters of the model in Eq. (12.11) by using all available measured values (0–45 min) of the oxygen concentration, the parameter set D in Table 12.4 results, which clearly differs from the parameter set C in Table 12.4. Figure 12.10 shows that model D is not at all accurate for small oxygen concentrations; it cannot match the curvature (second derivative) of the trace of the data. The constant oxygen consumption will, after the stop of the aeration, always lead to a linear decrease of the oxygen concentration.

If we pursue the oxygen concentration S as its value strives towards 0, deviations between the model and data appear that cannot be eliminated with the simple model structure chosen here. If the behavior of the system in the range of small oxygen concentrations is of any interest, we must adapt the structure of the model. A frequently selected possibility extends the kinetics of oxygen consumption rate with a so-called Monod term:

$$r_{O2} = r_{O2,max} \cdot \frac{S}{K_{O2} + S} .$$

(12.20)

Table 12.4 Identified parameters for different models but identical data. Below the residuals are analyzed statistically (see text): Model D without Monod kinetics, model E with Monod kinetics, model F corresponds to the parameters that were used to simulate the measured values. Besides the parameter values the table also includes some statistical test parameters that will be discussed later

Parameter	Identification		Simulation	Units
	Model D 0–45 min	Model E 0–45 min	Model F 0–45 min	
Number of data points n	46	46	46	–
O_2 saturation S_{sat}	8.61	9.85	10.00	$gO_2\,m^{-3}$
O_2 consumption r_{O2}, $r_{O2,max}$	−0.329	−0.438	−0.450	$gO_2\,m^{-3}$
Aeration coefficient $k_l a$	0.194	0.152	0.150	min^{-1}
Initial concentration S_0	1.97	2.03	2.00	min^{-1}
K_O	–	0.996	1.00	$gO_2\,m^{-3}$
				$gO_2\,m^{-3}$
Measurement error ΔS_{rms}	0.176	0.097	0.100	$gO_2\,m^{-3}$
χ^2 (degrees of freedom)	142 (41 DF)	40.4 (40 DF)	46 (46 DF)	–
χ^2 test (95%)	negative	positive	–	$gO_2\,m^{-3}$
Equilibrium conc. S^*	6.91	–	–	–
Sign changes of residuals	11	25	25	–
Binominal distribution (95%)	negative	positive	positive	

Oxygen concentration S, gO_2 m^{-3}

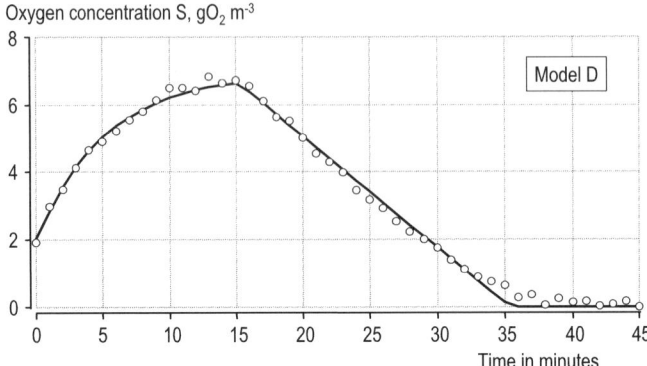

Fig. 12.10 Comparison of the measurements and the simulation for the parameter set D. The model prediction deviates systematically from the observations for small oxygen concentrations

Oxygen concentration S in g O_2 m^{-3}

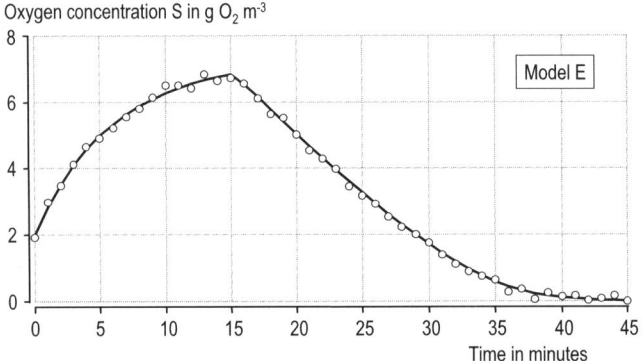

Fig. 12.11 Model E with improved model structure (Monod kinetics) follows the data clearly better than Model D in Fig. 12.10

Thus, the balance Eq. (12.7) becomes

$$\frac{dS}{dt} = r_{oxy} + r_{O2} = k_l a \cdot (S_{sat} - S) + r_{O2,max} \cdot \frac{S}{K_{O2} + S}. \qquad (12.21)$$

For Eq. (12.21) we do not have an explicit, analytic solution readily available. The model now has a total of five parameters: S_{sat}, $r_{O2,max}$, $k_l a$, S_0, and K_O which can all be identified reliably from the available data (parameter set E in Table 12.4; see also the local sensitivity functions in Fig. 12.14). The extended model is compared with the data in Fig. 12.11, a clear improvement over Fig. 12.10, especially as there are now no longer any systematic deviations at small concentrations. Structurally the model Eq. (12.21) is superior to the simpler model Eq. (12.7).

With the adjustment of the mathematical structure of the model, we succeeded in making a more reliable prediction. Here the adjustment is based on the experience with microbial processes and represents a refinement of the initially very simple model; the number of the parameters is increased by one. It is not primarily the larger number of parameters that leads to success, but the improved mathematical structure that was adapted to the problem and introduced nonzero second derivatives ($d^2S/dt^2 > 0$ for $t > 15$ min) once the aeration is stopped.

12.4.2 Simple Test Procedures

The statistical test procedures described here are very simple examples to exemplify the procedure. Statistical tests with improved power (to separate more sharply) and based on fewer, less restrictive assumptions are available. They would, however, require the introduction of extended statistical know-how for which there is insufficient space here.

Visual Test of the Residuals

The identification of the model parameters with the aid of the weighted sum of least squares is based on the assumption that the *residuals* (the difference between the model prediction and data) are normally distributed with an expected value of zero and a standard deviation equal to the standard error of the measurement process. In addition, the errors are assumed to be independent, i. e., not autocorrelated (see Sect. 14.10).

A simple test of whether the residuals are subject to unexpected anomalies is based on a diagram and the statistical characterization of the residuals. Figure 12.12 shows the residuals for the case study with the parameter sets D and E (Table 12.4). For model D, with an unsatisfactory structure, a noticeable deviation from random behavior results, while for model E the residuals provide the impression of a random distribution. The random measurement error of the oxygen concentration measurement process is subject to a standard error of $\sigma_m = 0.1$ gO_2 m^{-3}, which is clearly smaller than the standard deviation of the residuals in model D ($\sigma_{res} = 0.18$ gO_2 m^{-3}), while $\sigma_m = 0.1$ gO_2 m^{-3} explains the residuals of model E. The residuals from model D strongly indicate that structural problems exist.

χ^2 Test

If we know the standard error of the measurement process $\sigma_{m,i}$, we can test the hypothesis that the residuals ($y_{m,i} - y_i(p)$) correspond to a normal distribution with the associated standard error $\sigma_{m,i}$ with a χ^2 test (see Example 12.1). Usually this hypothesis is accepted if χ^2 is smaller than the 95% value of the cumulative χ^2 distribution. If χ^2 is larger than the 95% value, this is a strong indication that there are systematic deviations between the measured values and the model predictions that relate to structural problems of the model.

Residual, simulation – measurement, gO_2 m^{-3}

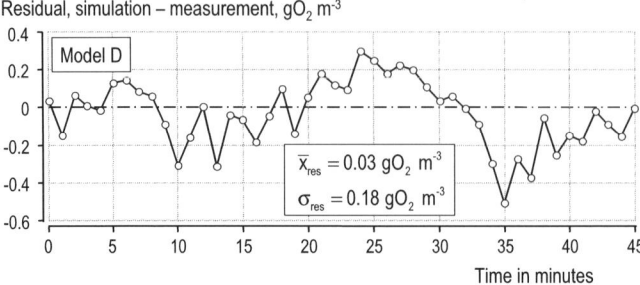

Residual, simulation – measurement, gO_2 m^{-3}

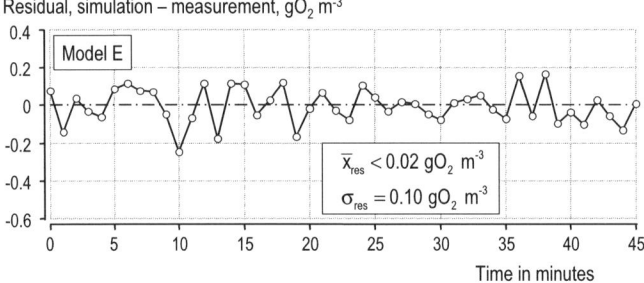

Fig. 12.12   Representation of the deviation between the model prediction and the measurements for the case study. *Above*: parameter set D, *below*: parameter set E of Table 12.4. See also the statistical characteristics of the residuals

For the parameter set D of the case study a χ^2 value of 142 results. With 46 data points and four identified parameters, $\nu = 46 - 1 - 4 = 41$ degrees of freedom remain. According to Table 12.1 the 95% limit is approximately 57. We must reject the hypothesis that the residuals are caused by random measurement errors with a standard error of $\sigma_m = 0.1$ gO_2 m^{-3}: The test suggests structural problems of the model. The result of the test is, however, strongly dependent upon the chosen value for σ_m.

For the model E with five parameters results the χ^2 value is 40.4. This is well below the limiting 95% value of 55.8 and corresponds with the expected value of 40 (= degrees of freedom). There is no indication of structural model deviations.

The disadvantage of this test is that we must know (or estimate) the standard error of the measurement process and thus we cannot estimate this value from the data.

Sign Test

The sign test is a very simple test, in which the number of sign changes of the residuals is counted (see Fig. 14.20). If the random measurement errors are symmetrically distributed (e. g., normally) with expected value 0, then we expect on average a sign change after each second residual. The number of sign changes is subject to a binomial distribution with a probability of $p = 0.5$ and $n - 1$ trials (n data points provide $n - 1$ possible sign changes).

The number of sign changes can be approximated for a sufficiently large number of data points ($n > 30$) by the following normal distribution (central limit theorem, Benjamin and Cornell, 1970):

$$f\left(n_{SC}\right) = N\left(\overline{n}_{SC}, \sigma_{SC}\right) = N\left(p \cdot (n-1), \sqrt{p \cdot (1-p) \cdot (n-1)}\right) = N\left(\frac{n-1}{2}, \sqrt{\frac{n-1}{4}}\right)$$

$$(12.22)$$

n_{SC} = number of sign changes $(n-1)$
p = probability of a sign change $= 0.5$
n = number of data points >30
σ^2 = variance

The 95% confidence interval of a normal distribution covers approximately two standard deviations (more exactly, 1.96) in both directions. Based on this we accept the hypothesis that the number of sign changes corresponds to a random distribution, if:

$$\frac{n-1}{2} - \sqrt{n-1} < n_{SC} < \frac{n-1}{2} + \sqrt{n-1} \, .$$

$$(12.23)$$

The parameter set D in the case study results in $n_{SC} = 11$ sign changes from $n = 46$ data points. The 95% confidence interval for random sign changes covers the range 16–29 sign changes. Thus, n_{SC} is outside of this interval and deviates significantly from the expected value of 22–23. This points to a nonrandom number of sign changes and thus to possible autocorrelation of the residuals or structural problems of the model.

If there is a good reason to presume that the structural problem can only lead to fewer than the expected number of sign changes, a one-sided test may be carried out, which enhances its power. In this case, instead of 1.96 we use 1.64 and Eq. (12.23) becomes:

$$n_{SC} > \frac{n-1}{2} - \sqrt{0.82 \cdot (n-1)} \, .$$

$$(12.24)$$

For model D there is no reason to assume that the structural deficiencies lead to additional oscillations of residuals; therefore we could apply Eq. (12.24)

Model E results in $n_{SC} = 25$, which is within the 95% confidence interval and does not suggest any structural problems of the model.

The advantage of this test is that it is only based on the assumption that the residuals are subject to a symmetrical distribution and does not require a standard error of the measurement process. This even allows the simultaneous analysis of different measured state variables. Its application is very simple. Unfortunately with a large number of data it reacts very sensitively to small degrees of autocorrelation in the residuals (see Fig. 14.20).

Example 12.8: Modern data loggers and online electrodes lead to autocorrelated data

Modern data loggers and continuously measuring electrodes permit data to be registered at very high temporal density. Many electrodes have response times of up to 1 min and more. If we register the measurements more frequently, the results will be autocorrelated. The statistical methods that we use to identify parameters and the remaining uncertainty of these parameters are based on the assumption of independent residuals. Autocorrelated residuals violate this assumption and lead to a gross underestimation of parameter uncertainties. See also Sect. 14.10.

Test for Autocorrelation

Autocorrelation describes the fact that elements of a time series may be correlated with previous elements (see Sect. 14.10). The simplest autocorrelation is of first order; in this case we determine the correlation coefficient between an element of the time series and its immediate forerunner and the slope of the linear regression is equivalent to the correlation coefficient. We then test whether we must reject the null hypothesis that no autocorrelation exists. Since structural problems typically lead to positive correlation, the test is one sided. For a number of data pairs $n > 6$, the following approximation of the t-test statistic is valid:

$$t = r \cdot \sqrt{\frac{n-2}{1-r^2}} \quad \text{with } t = \text{Student's t, } \nu = n-2 \text{ degrees of freedom.} \quad (12.25)$$

Figure 12.13 shows the first-order autocorrelation for models D and E. Clearly there is (statistically significant) autocorrelation in the residuals of model D. This means that, if the previous residual is large and positive, then there is a large chance that the actual residual is also large and positive. Autocorrelation identifies structure in the residuals and does not fulfill the assumptions we make when we identify the parameters and especially their uncertainty (see the next section) by minimizing χ^2. We may, however, still obtain the most likely parameter value.

For model E Fig. 12.13 does not show any significant autocorrelation in the residuals. Therefore we would accept the null hypothesis that there is no autocorrelation and therefore the residuals are independently distributed.

Example 12.9: Testing the significance of the autocorrelation of the residuals from model D

The correlation coefficient for first-order autocorrelation of the residuals from model D is $r = 0.64$ (Fig. 12.13). We want to test whether we must reject the null hypothesis that this correlation is not significant at a 95% level. We know that the correlation could only be positive, since a negative correlation would mean that the residuals must oscillate. Therefore the test is one sided.

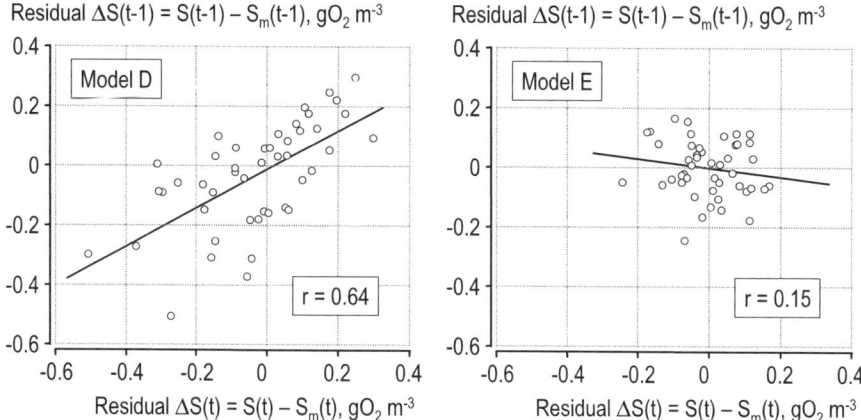

Fig. 12.13 Autocorrelation of the first order of the residuals $\Delta S = S - S_m$. *Left*: for model D with a significant correlation coefficient of 0.64. *Right*: for model E with a not significant correlation

There are 46 residuals, which yield 45 pairs to be correlated. The t-statistic according to Eq. (12.25) has $\nu = 45 - 2 = 43$ degrees of freedom and is $t = 5.46$.
A table for Student's t provides a critical value of $t_{crit}(\nu = 43, 0.95) = 1.68$. We must reject the null hypothesis since $t > t_{crit}$.
For model E with $r = 0.15$ we obtain $t = 0.99$. We therefore accept the null hypothesis.

12.5 Parameter Uncertainty

In the compilation of a model we make assumptions about the behavior of the system to be modeled. We use our knowledge about the active processes and derive a meaningful mathematical structure for the model. In the engineering sciences we adapt the model structure to the question; the simpler the model, the more economical is its application, but there is also a risk of not recognizing important aspects of the system. In the natural sciences we frequently want to learn about new processes; here the mathematical structure for the description is a priori still unknown and part of the research question. In both cases we must determine the associated parameters for the selected models. Thus we fit the model prediction to the experimental observations (see also Sect. 12.1). Since measurement errors always lead to remaining uncertainties about the identified model parameters, uncertainties remain about the model predictions even with a structurally optimized model.

If the residuals are normally distributed, minimization of χ^2 leads to parameter values that describe the most probable behavior of the system. In addition there exist less probable combinations of parameters which lead to less probable, but realistic predictions. Since the measurements are random variables, we cannot identify unique parameter values but we must accept that values of parameters are uncertain and must be characterized with an associated probability distribution. This section will deal with the problem of using the information contained in the residuals in order to estimate the distribution of the parameter values.

Error propagation and *parameter uncertainty* are closely related: uncertainty about the parameter values has consequences for model predictions. Measurement errors in the data lead to uncertainties of the parameter values. Here only the uncertainty of the identified parameters is discussed; error propagation is the topic of the next section.

12.5.1 Theoretical Background

The following theoretical derivation is the basis for the quantification of parameter uncertainty and linear error propagation. These methods are integrated into many system-analytic programs at different levels of complexity.

Symbols:

k = number of parameters of the model

n = number of measured values $y_{m,i}$

$y_{m,i}$ = measured value of a state variable of the model

σ_i = standard error of the measurement $y_{m,i}$

g_i = weight of the measurement y_{mi}, if σ_i is unknown. If possible the weights should be selected in the relationship of $1/\sigma_i$

$[\mathbf{y_m}]$ = vector of the n measured values

y_i = model prediction for the value of the measured state variable $y_{m,i}$

$[\mathbf{y}]$ = column of the model predictions in the places of $[\mathbf{y_m}]$

Δy_i = measurement error of $y_{m,i}$, $\Delta y_i = y_{mi} - y_i$

$[\mathbf{\Delta y}]$ = column of the n measurement errors $= [\mathbf{y_m}] - [\mathbf{y}]$

$s_{i,j}$ = covariance of the variables i and j

$[\mathbf{\Sigma}]$ = Variance covariance matrix

$[\mathbf{R}]$ = correlation matrix

For all matrices, the format is indicated as: number of rows · number of columns.

Assumption:

Individual random measurement errors $\Delta y_i = (y_{m,i} - y_i)$ are independent, i. e., they are not correlated, their covariance is zero. It follows that the variance covariance matrix is a diagonal matrix of the following form:

$$\Sigma_{n \cdot n}(\Delta \mathbf{y}) = \begin{bmatrix} \sigma_1^2 & 0 & 0 \\ 0 & \sigma_i^2 & \dots \\ 0 & \dots & \sigma_n^2 \end{bmatrix}_{n \cdot n} . \qquad (12.26)$$

The inverted covariance matrix Σ^{-1} has the form:

$$\Sigma_{n \cdot n}^{-1}(\Delta \mathbf{y}) = \begin{bmatrix} \dfrac{1}{\sigma_1^2} & 0 & 0 \\ 0 & \dfrac{1}{\sigma_i^2} & \dots \\ 0 & \dots & \dfrac{1}{\sigma_n^2} \end{bmatrix}_{n \cdot n} .$$

Example 12.10: Autocorrelated measurement errors

As indicated in Example 12.8, measuring errors of data collected with the aid of continuously registering electrodes may be subject to autocorrelation. This leads to the fact that Eq. (12.26) is not valid anymore. The theory of generalized least squares deals with these problems.

Autocorrelation goes in parallel with a reduction of the information content of the data (since some of the information depends on already known previous values). If we do not consider these autocorrelations, we will overestimate the information content of data and thus underestimate the resulting parameter uncertainty.

There is no space here to deal with this topic in detail. An easy pragmatic solution is to add a normally distributed random number with expected value $\mu = 0$ and standard deviation σ to the data, where σ is chosen large enough in order to obtain a nonsignificant autocorrelation.

General linear error propagation

The general, linear error propagation results from the first element of a Taylor expansion of the model $y(p)$ at $y_i(p)$; it has the form

$$\Sigma_{n \cdot n}(\Delta \mathbf{y}) = \left[\frac{\partial \mathbf{y}}{\partial \mathbf{p}^{\mathrm{T}}} \right]_{k \cdot n} \cdot \Sigma_{k \cdot k}(\mathbf{p}) \cdot \left[\frac{\partial \mathbf{y}}{\partial \mathbf{p}^{\mathrm{T}}} \right]_{n \cdot k}^{\mathrm{T}} \qquad (12.27)$$

$\Sigma_{k \cdot k}(\mathbf{p})$ = covariance matrix of the parameters. It is symmetrical and contains on the diagonal the variance of the parameters and in the other elements the respective covariance of the associated parameters.

The partial derivatives $\left.\dfrac{\partial y}{\partial p_j}\right|_{y_i}$ for all parameters p_j are taken at the location of the measured value $y_{m,i}$.

Example 12.11: Application of the general linear error propagation equation (12.27)

The application of Eq. (12.27) to a model with two dependent parameters, $y = f(p_1, p_2)$, has the following form:

$$\sigma_y^2 = \left(\frac{\partial y}{\partial p_1}\right)^2 \cdot \sigma_{p_1}^2 + \left(\frac{\partial y}{\partial p_2}\right)^2 \cdot \sigma_{p_2}^2 + 2 \cdot \frac{\partial y}{\partial p_1} \cdot \frac{\partial y}{\partial p_2} \cdot s_{p_1,p_2}$$

$$= \left(\frac{\partial y}{\partial p_1}\right)^2 \cdot \sigma_{p_1}^2 + \left(\frac{\partial y}{\partial p_2}\right)^2 \cdot \sigma_{p_2}^2 + 2 \cdot \frac{\partial y}{\partial p_1} \cdot \frac{\partial y}{\partial p_2} \cdot r_{p_1,p_2} \cdot \sigma_{p_1} \cdot \sigma_{p_2}$$

$r_{p1,p2}$ = correlation coefficient, $s_{p1,p2}$ = covariance of p_1 and p_2.
The uncertainty of the model prediction depends on: (i) the uncertainty of the individual parameters, (ii) the covariance of the two parameters, and (iii) the sensitivity of the model prediction on the two parameters. Depending on the signs of the sensitivities and the covariance, the uncertainty may increase or decrease with increasing covariance (see also Example 15.9).

The Special Case of the Gaussian Error Propagation

If the parameters are independent, $\Sigma_{k \cdot k}(\mathbf{p})$ is a diagonal matrix. Thus, for any position y_i, the result is:

$$\sigma_y^2 = \sum_{j=1}^{k} \sigma_{p_j}^2 \cdot \left(\frac{\partial y}{\partial p_j}\right)^2. \tag{12.28}$$

This corresponds to the Gaussian law of error propagation for independent parameters. If the parameters are determined from data, e. g., by minimizing χ^2, then they are typically not independent and Eq. (12.28) should not be used in this simple form because it may overestimate σ_y^2. In this case we should apply Eq. (12.27). If the parameters and their uncertainties are estimated based on experience, then we usually assume that these estimations are independent, because an expert cannot normally estimate the covariances. However, neglecting correlation in error propagation may lead to the underestimation of the uncertainty of model prediction (see Example 15.9).

Covariance of the parameters

For the determination of the uncertainty of the parameters, in particular their co-variance matrix, we solve Eq. (12.27) for $\Sigma_{k \cdot k}(\mathbf{p})$ by right multiplication of Eq. (12.27) with

$$
\left(\Sigma_{n \cdot n}^{-1}(\Delta \mathbf{y}) \cdot \left[\frac{\partial \mathbf{y}}{\partial \mathbf{p}^T} \right]_{k \cdot n} \right) \cdot \left[\left[\frac{\partial \mathbf{y}}{\partial \mathbf{p}^T} \right]_{n \cdot k}^T \Sigma_{n \cdot n}^{-1}(\mathbf{p}) \cdot \left[\frac{\partial \mathbf{y}}{\partial \mathbf{p}^T} \right]_{k \cdot n} \right]_{k \cdot k}^{-1} .
$$

The result is:

$$
\underbrace{\Sigma_{n \cdot n}(\Delta \mathbf{y}) \cdot \Sigma_{n \cdot n}^{-1}(\Delta \mathbf{y})}_{= I_{n \cdot n}} \cdot \left[\frac{\partial \mathbf{y}}{\partial \mathbf{p}^T} \right]_{k \cdot n} \cdot \left[\left[\frac{\partial \mathbf{y}}{\partial \mathbf{p}^T} \right]_{n \cdot k}^T \Sigma_{n \cdot n}^{-1}(\Delta \mathbf{y}) \cdot \left[\frac{\partial \mathbf{y}}{\partial \mathbf{p}^T} \right]_{k \cdot n} \right]_{k \cdot k}^{-1} =
$$

$$
\underbrace{\left[\frac{\partial \mathbf{y}}{\partial \mathbf{p}^T} \right]_{k \cdot n} \cdot \Sigma_{k \cdot k}(\mathbf{p}) \cdot \left[\left[\frac{\partial \mathbf{y}}{\partial \mathbf{p}^T} \right]_{n \cdot k}^T \cdot \Sigma_{n \cdot n}^{-1}(\Delta \mathbf{y}) \cdot \left[\frac{\partial \mathbf{y}}{\partial \mathbf{p}^T} \right]_{k \cdot n} \right] \cdot \left[\left[\frac{\partial \mathbf{y}}{\partial \mathbf{p}^T} \right]_{n \cdot k}^T \Sigma_{n \cdot n}^{-1}(\Delta \mathbf{y}) \cdot \left[\frac{\partial \mathbf{y}}{\partial \mathbf{p}^T} \right]_{k \cdot n} \right]_{k \cdot k}^{-1}}_{= I_{k \cdot k}}
$$

Thus:

$$
\Sigma_{k \cdot k}(\mathbf{p}) = \left[\left[\frac{\partial \mathbf{y}}{\partial \mathbf{p}^T} \right]_{n \cdot k}^T \cdot \Sigma_{n \cdot n}^{-1}(\Delta \mathbf{y}) \cdot \left[\frac{\partial \mathbf{y}}{\partial \mathbf{p}^T} \right]_{k \cdot n} \right]_{k \cdot k}^{-1} . \tag{12.29}
$$

If the standard errors of the measurements σ_i are not known, then the identification of the parameters frequently relies on weights g_i which are specified for each measured value $y_{m,i}$. The weights are to be specified as well as possible according to the following rule:

$$
\frac{g_1}{\sigma_1} \approx \frac{g_i}{\sigma_i} \approx \frac{g_n}{\sigma_n} \approx K . \tag{12.30}
$$

This requires that, with unknown σ_i, an appropriate (possibly subjective) estimate of the relative measurement errors is available.

If in place of the standard errors σ_i the weights g_i are used, we obtain instead of the inverse covariance matrix $\Sigma^{-1}(\Delta \mathbf{y})$ the following diagonal matrix:

$$
\mathbf{G}_{n \cdot n}(\mathbf{y}_{m,i}) = \begin{bmatrix} g_1^2 & 0 & 0 \\ 0 & g_i^2 & \cdots \\ 0 & \cdots & g_n^2 \end{bmatrix}_{n \cdot n} .
$$

The minimized sum of the weighted squared errors χ^2 then relates to the standard deviation of data from model prediction, which in the best case is equivalent to the measurement errors

$$\chi^2 = \sum_{i=1}^{n}\left(\frac{y_{m,i} - y_i}{\sigma_i}\right)^2 \quad \text{with expected value } \overline{\chi^2} = n \tag{12.31}$$

$$\chi^2 = \sum_{i=1}^{n}\left(g_i \cdot \left(y_{m,i} - y_i\right)\right)^2$$

with expected value $\overline{\chi^2} \approx \dfrac{n-k}{n} \cdot \sum_{i=1}^{n}\left(\dfrac{g_i}{\sigma_i}\right)^2 = (n-k) \cdot K^2$. $\tag{12.32}$

This provides the possibility to obtain the covariance matrix of the parameters as follows:

$$\Sigma_{k \cdot k}(\mathbf{p}) = \frac{\chi^2}{n-k} \cdot \left[\left[\frac{\partial \mathbf{y}}{\partial \mathbf{p}^T}\right]_{n \cdot k}^T \cdot \mathbf{G}_{n \cdot n}(\mathbf{y}) \cdot \left[\frac{\partial \mathbf{y}}{\partial \mathbf{p}^T}\right]_{k \cdot n}\right]_{k \cdot k}^{-1} . \tag{12.33}$$

The divisor (n–k) of χ^2 reduces the number of the degrees of freedom because the k parameters were determined from the data and therefore the adjustment of y_i to the data is on average better than in reality.

The covariance matrix Σ of the parameters has the following, symmetrical form:

$$\Sigma_{k \cdot k}(\mathbf{p}) = \begin{bmatrix} \sigma_1^2 & s_{1,i} & s_{1,k} \\ s_{i,1} & \sigma_i^2 & ... \\ s_{k,1} & ... & \sigma_k^2 \end{bmatrix}_{k \cdot k}$$

$s_{i,j} = s_{j,i} =$ covariance of parameters i and j

with $s_{x,y} = s_{y,x} = \dfrac{1}{n} \cdot \sum_{i=1}^{n}\left(\left(x_i - m_x\right) \cdot \left(y_i - m_y\right)\right)$

$$\tag{12.34}$$

Evaluation of $K = g_i/\sigma_i$ with Eq. (12.32) in the context of the adjustment of the data to the model provides an estimate of the variance σ_i^2 of the random error of the data $y_{m,i}$ relative to the model y_i, which includes various causes of deviations in addition to pure measurement errors. However, we lose the possibility to uncover structural problems of the model (systematic deviations) with the help of a χ^2 test, although we can still rely on sign changes and autocorrelation.

Correlation matrix

The elements of the covariance matrix have different dimensions $[S_{i,j}] = [p_i \cdot p_j]$, which makes their comparison and interpretation rather difficult. A dimensionless form is the correlation matrix $\mathbf{R}$, which replaces the covariance $s_{i,j}$ with the correlation coefficient $r_{i,j}$. The correlation matrix is also symmetrical and all diagonal elements are equal to unity (a parameter correlates perfectly with itself). The individual correlations $r_{i,j}$ are defined as:

$$r_{i,j} = \frac{s_{i,j}}{\sqrt{\sigma_i^2 \cdot \sigma_j^2}} = \frac{s_{j,i}}{\sigma_i \cdot \sigma_j} \approx \frac{s_{j,i}}{s_j \cdot s_i}. \tag{12.35}$$

Thus, the correlation matrix of the parameters has the form

$$\mathbf{R}_{k \cdot k}(\mathbf{p}) = \begin{bmatrix} 1 & r_{i,i} & r_{i,k} \\ r_{i,1} & 1 & \dots \\ r_{k,1} & \dots & 1 \end{bmatrix}_{k \cdot k}. \tag{12.36}$$

The parameters p_i are statistically independent if none of the correlations differs statistically significant from 0. If we identify the parameters from data, based on the minimization of the χ^2, then typically the resulting parameters are correlated. The larger the individual correlation is, the stronger the statistical dependence becomes. On the one hand parameters that are highly correlated can barely be uniquely identified; their values become quite uncertain. On the other hand uncertainties of model predictions as a consequence of parameter uncertainties are reduced if correlations are included in the uncertainty analysis.

Probability distribution of the parameters

This section provides hints for how to proceed but no extensive introduction.

With Eq. (12.26) we take the measured values $y_{m,i}$ as statistically independent random variables, which are typically normally distributed. With Eq. (12.29) or (12.33) we obtain the covariance and the most probable values of the parameters, but this does not include the actual shape of the probability distribution of the parameters, especially if the model is nonlinear.

A method that permits the shape of the distribution of the parameters to be determined is Markov chain Monte Carlo simulation (MCMC). In addition to data, this method can simultaneously consider *a priori* know-how (Bayesian statistics). We first estimate a provisional distribution of the parameters (the *prior distribution*) from experience. Then, this estimate is improved by choosing random parameter combinations in a Markov chain (random walk), evaluating the likelihood of the combination given the data, keeping the result depending on likelihood, and so compiling a multidimensional histogram of parameter combinations. With typically several thousand evaluations of parameter combinations, we obtain a *posterior distribution* of the parameter values. In the literature approaches that implement this strategy are called *MCMC* and the *Metropolis algorithm*.

Without this additional effort we must make assumptions about the shape of the distributions of the parameters, and thus a new source of error is introduced, e. g., in the Monte Carlo simulation that will be discussed later. If the most probable parameter values μ_{pi} and their covariance matrix $\Sigma_{pi,pj}$ are known from parameter identification, a multidimensional normal distribution with these characteristics is then frequently selected. It might be advantageous to use log-normal distributions in order to avoid negative parameter values.

12.5.2 Application to the Case Study

In order to obtain the uncertainty of the model parameters and their correlation matrix, the model Eq. (12.21) and the associated data (Fig. 12.4) were implemented in AQUASIM (Reichert, 1995). This program includes the routines required for this task.

Parameter uncertainty

Table 12.5 summarizes the expected values as well as the standard errors of the five model parameters of the case study. Model F provides the parameters that were used to simulate the data; these true parameter values differ by less than one standard error from the identified parameters. The largest relative uncertainty (coefficient of variation, cv) results for K_O, the saturation coefficient for oxygen in the Monod model. This is firstly due to the low sensitivity of this parameter and is secondly caused by the high correlation of this parameter with r_{O2} (see below).

Table 12.5 Estimated values and standard errors of the parameters for model E with Monod kinetics for 0–45 min (Table 12.4); obtained with AQUASIM (Reichert, 1995)

Parameter		Model E			Model F	Units
		Expected value μ_x	Standard error σ_x	cv	Simulation of data with:	
O_2 saturation conc.	S_{sat}	9.85	0.29	3%	10	$gO_2\,m^{-3}$
O_2 consumption rate	r_{O2}	−0.438	0.016	4%	−0.450	$gO_2\,m^{-3}\,min^{-1}$
Aeration coefficient	k_la	0.152	0.009	6%	0.150	min^{-1}
Initial concentration	S_0	2.03	0.078	4%	2.00	$gO_2\,m^{-3}$
Monod coefficient	K_O	0.996	0.144	14%	1.00	$gO_2\,m^{-3}$

Covariance matrix

Table 12.6 introduces the covariance matrix for model E. It has the disadvantage that the covariance values have different units; they can barely be compared to each other.

Table 12.6 Covariance matrix Σ_{kk} for the case study, model E. The units result from the product of the units of the parameters involved (matrix computed from the correlation matrix and the standard deviations of the parameters from the results of a simulation with AQUASIM, Reichert 1995)

Parameter	S_{sat}	r_{O2}	k_la	S_0	K_O
S_{sat}	0.084	−0.003	−0.0026	0.0113	0.022
r_{O2}	−0.003	0.00026	0.00008	−0.00014	−0.0022
k_la	−0.0026	0.00008	0.00009	−0.00044	−0.00056
S_0	0.0113	−0.00014	−0.00044	0.0061	0.00079
K_O	0.022	−0.0022	−0.00056	0.00079	0.021

Correlation matrix

Table 12.7 introduces the correlation matrix of the five parameters of model E. The high, negative correlation between the saturation concentration S_{sat} and the aeration coefficient k_la is remarkable ($r = -0.97$). A deviation in the identification of S_{sat} can be compensated to a very large degree by an adjustment of k_la. This also results from the sensitivities in Fig. 12.14: during the first 15 min, when the aeration is on, there is a small difference in the shape of the sensitivity. However, as soon as the aeration is off (15–45 min) these two parameters no longer influence the oxygen concentration; here their sensitivity is perfectly correlated. Since in the case study the kinetic parameters r_{O2} and K_O are of prime interest, the values identified for k_la and S_{sat} are sufficient.

The second remarkably high correlation exists between the oxygen consumption rate r_{O2} and the saturation constant K_O ($r = -0.96$). Here, too, an error in one parameter can nearly be fully compensated by the adjustment of the other. Figure 12.15 illustrates the sensitivity of these two parameters against each other; again a high correlation exists, which may also be deduced from Fig. 12.14. There is no significant correlation between the initial concentration S_0 and K_O; neither Fig. 12.15 nor the correlation matrix ($r = 0.07$) indicates a linear dependence.

Table 12.7 Correlation matrix **R** for Model E of the case study, 0–45 min. Obtained from AQUASIM (Reichert 1995)

Parameter	S_{sat}	r_{O2}	k_la	S_0	K_O
S_{sat}	1	−0.65	−0.97	0.50	0.53
r_{O2}	−0.65	1	0.52	−0.11	−0.96
k_la	−0.97	0.52	1	−0.60	−0.42
S_0	0.50	−0.11	−0.60	1	0.07
K_O	0.53	−0.96	−0.42	0.07	1

Absolute-relative Sensitivity δ_S of S, gO$_2$ m^{-3}

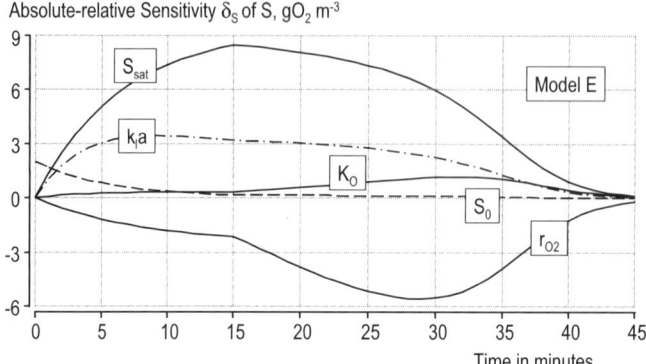

Fig. 12.14 The absolute relative sensitivities of the five parameters of the case study in model E point to a good identifiability of the parameters

Absolute-relative Sensitivity $\delta_{S,S_0}^{a,r}$ and $\delta_{S,r_{02}}^{a,r}$, gO$_2$ m^{-3}

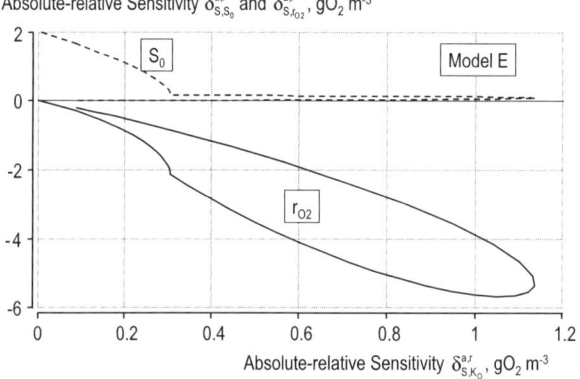

Fig. 12.15 Examples of the correlation of the absolute-relative sensitivity of K_O and r_{O2} as well as S_0

The experiment includes concentrations several times larger than K_O (Fig. 12.4). In this situation the Monod kinetics in Eq. (12.20) is dominated by $r_{O2,max}$ (see also the sensitivity during the first 20 min in Fig. 12.14):

$$r_{O2} = r_{O2,max} \cdot \frac{S}{K_O + S} \approx r_{O2,max} \cdot \frac{S}{S} \approx r_{O2,max} \quad \text{for} \quad S \gg K_O .$$

This leads to the fact that $r_{O2,max}$ (cv = 4%) can be identified much more accurately than K_O (cv = 14%). K_O influences the experiment only over a limited period of time and has only a modest influence on the results (small sensitivity).

12.6 Linear Error Propagation

In the application of models, parameter uncertainty leads to uncertainty in the model output (the computed state variables). Since in the environmental sciences the parameter uncertainties are often considerable, an analysis of the uncertainty of model predictions is of special importance.

We distinguish between linear error propagation, which results in simple cases in Gaussian error propagation, and nonlinear error propagation, which will rely on Monte Carlo simulation. Gaussian methods can only deal with small uncertainties, as long as linearization of the model is still adequate. Monte Carlo simulation can deal with large uncertainty but has high computational cost.

12.6.1 Basics

Equation (12.27) introduces in matrix notation the general relationship for linear error propagation; it is rarely used in this general form. The equation becomes simpler if we apply it to a single value of a state variable y_i:

$$\sigma_{y_i}^2 = \sum_{n=1}^{k} \left(\frac{\partial y_i}{\partial p_n} \cdot \sum_{j=1}^{k} \left(\frac{\partial y_i}{\partial p_j} \cdot s_{n,j} \right) \right). \tag{12.37}$$

In the special case where the parameters are statistically independent (the covariances are equal to 0), we can simplify Eq. (12.37) to the Gaussian error propagation Eq. (12.28):

$$y_i = f\left(p_1..p_k, y_1..y_m, t, x, y, z\right),$$

$$\sigma_{y_i}^2 = \sum_{j=1}^{k} \sigma_{p_j}^2 \cdot \left(\frac{\partial y_i}{\partial p_j} \right)^2. \tag{12.38}$$

y_i	=	dependent state variables
$\sigma_{y_i}^2$	=	variance of $y_i = s_{i,i}$
p_j	=	independent model parameters
$\sigma_{p_j}^2$	=	variance of p_j
$s_{i,j}$	=	covariance of the parameters p_i and p_j
k	=	number of parameters
t, x, y, z	=	system variables for time and space

Equation (12.38) is valid for all probability distributions of p_i, as long as f can be differentiated with respect to p_j (else the partial derivative is not available) and the parameters p_j do not correlate among each other (which frequently is only partially fulfilled).

Since the partial derivatives $\partial y_i/\partial p_j$ are available from the sensitivity analysis, the application of Eqs. (12.37) and (12.38) is rather simple. As a result we receive the most probable course of all the state variables $y_i = f(p_j)$ and an estimation of the standard deviation of other possible values around this most probable course. This standard deviation is a measure of the uncertainty of the model prediction as a consequence of the uncertainty of the model parameters. Gaussian error propagation is based on a local linearization of the model (Taylor expansion, tangent k-dimensional plane); it can therefore only be reliable in the close proximity of the model prediction. With large parameter uncertainties this kind of analysis is not reliable and should not be applied.

Equation (12.38) is valid for independent parameters. If the values are correlated (with nonzero covariance), the application of Eq. (12.27) is essential (see Example 12.11 for the case of two dependent parameters).

Example 12.12: Gaussian error propagation

The degradation of a material in a batch reactor with a first-order reaction leads to the following model equation:

$y(t) = C(t) = C_0 \cdot e^{-k \cdot t}$ with the two parameters: C_0, the initial concentration and k, the rate constant.

Both model parameters are uncertain; we estimate their average values and standard error as:

$\mu_{C0} = 100$ g m^{-3}, $\sigma_{C0} = 10$ g m^{-3} and $\mu_k = 2$ d^{-1}, $\sigma_k = 0.5$ d^{-1}. We do not expect a correlation of the two parameters, since our estimates are obtained independently.

What is the standard deviation of the model prediction as a consequence of the uncertainty in the parameters?

According to Eq. (12.38) we have

$$\sigma_y = \sqrt{\sum_{i=1}^{n}\left(\frac{\partial y}{\partial p_i}\cdot\sigma_{pi}\right)^2} = \sqrt{\left(e^{-k \cdot t}\cdot\sigma_{C0}\right)^2 + \left(-C_0 \cdot t \cdot e^{-k \cdot t}\cdot\sigma_k\right)^2} = e^{-k \cdot t}\cdot\left(100 + 2500 \cdot t^2\right)^{0.5} /$$

Figure 12.16 (left) shows the standard deviation of the model prediction y. On the right the result y is shown, based on the most probable values μ_{C0} and μ_k. In addition the standard deviation of the model prediction is added.

If we wanted to predict how an initial concentration of $C_0 = 100$ g m^{-3} is degraded over time, then we would have to set $\sigma_{C0} = 0$ and repeat the analysis.

Example 12.13: Application of Gaussian error propagation

For uncorrelated variables Eq. (12.38) may be applied to simple combinations as follows:

Addition and subtraction: $y = a \cdot x_1 \pm b \cdot x_2$ $\sigma_y^2 = a^2 \cdot \sigma_{x_1}^2 + b^2 \cdot \sigma_{x_2}^2$

Multiplication and division: $y = x_1 \cdot x_2 / x_3$ $\dfrac{\sigma_y^2}{y^2} = \dfrac{\sigma_{x_1}^2}{x_1^2} + \dfrac{\sigma_{x_2}^2}{x_2^2} + \dfrac{\sigma_{x_3}^2}{x_3^2}$

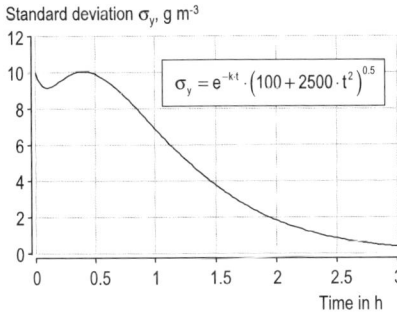
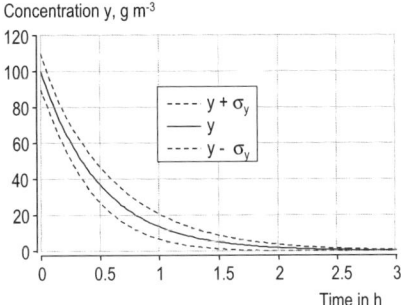

Fig. 12.16 *Left*: standard error of the model prediction based on Gaussian error propagation. *Right*: error bands around the expected time course of the decay of the concentration

Application:

We want to evaluate the 95% confidence interval of the concentration of the COD in an industrial wastewater. We must first dilute the wastewater in order to be able to analyze it.

The dilution amounts to a factor $f_D = 10$ with a standard deviation of $\sigma_D = 0.2$.

The result of the analysis amounts to $C_{COD} = 260 \, g \, m^{-3}$ with a standard error of $\sigma_{COD} = 10 \, g \, m^{-3}$.

The expected value of the wastewater concentration is:

$$C_{WW} = C_{COD} \, f_D = 2600 \, gCOD \, m^{-3}$$

The standard deviation amounts to (multiplication):

$$\sigma_{CWW}^2 = \left(\frac{\sigma_D^2}{f_D^2} + \frac{\sigma_{COD}^2}{C_{COD}^2} \right) \cdot C_{WW}^2 = 12'704 \text{ and } \sigma_{CWW} = 113 \, g \, COD \, m^{-3}.$$

The 95% confidence interval is: $C_{WW} = 2600 \pm 1.96 \cdot 113 = 2600 \pm 220 \, gCOD \, m^{-3}$. 21% of the variance of the uncertainty originates from dilution, and 79% from the chemical analysis.

Disadvantages of linear error propagation

Linear error propagation based on the general Eq. (12.37) is applicable with different distributions of the parameters (thus also with non-normally distributed parameters), as long as the partial derivatives (local sensitivity functions) exist. Its application becomes problematic, however, if the standard deviations of the parameters are so large that the nonlinearity of the model which is frequently present begins to take effect.

If the parameter uncertainty is not normally distributed (which typically is true for nonlinear models), then we only obtain the standard deviation of the model uncertainty; the distribution of the uncertainty remains unknown.

12.6.2 Application to the Case Study

For the model Eq. (12.21) of the case study we do not have an analytical solution available, therefore the partial derivatives (local sensitivity) are not available in analytical form either. We must use the error propagation based on the numerically determined sensitivities.

The procedure

The following procedure is successful:

- We identify the parameters of the model.
- We compute or estimate the associated covariance matrix. If the result can reasonably be considered to be a diagonal matrix, then the parameters are statistically independent and Eq. (12.38) applies, otherwise we use Eq. (12.37).
- We compute the local sensitivity for all parameters and the state variables of interest.
- We evaluate the associated linear error propagation equation. A possible code is introduced in Example 12.14.

Example 12.14: Code for BM for the computation of the error propagation after Eq. (12.37)

Procedure:

1. With BM we write a file that contains the sensitivities. This file must be slightly edited, e. g., in EXCEL to introduce a first line with the indices of the parameters: 0,1,2,3,...k
2. The covariance matrix is brought into the same form.
3. The two matrices are read into BM as data

The code for the computations then has the following form:

```
STARTTIME = 0              ; Beginning of simulation
STOPTIME = 45              ; End of simulation
DT = 1                     ; Time step 1 min, according to input
; as data were read in with a base of one per minute:
; #Sen(46,5) effective format (47·6) Sensitivity, partial derivative ∂S/∂p versus
                                      time
; #Cov(5,5) effective format (6·6)   Covariance matrix, Table 12.6
k = 5                      ; Number of parameters
EC[1..k,1..k] = #Sen(time,i)*#Sen(time,j)*#Cov(i,j)
                           ; Computation of the error contributions
sigma = sqrt(arraysum(EC[*]))  ; Standard error, summation of error contribu-
                                 tions
```

Result

Figure 12.17 shows the uncertainty of the model predictions (confidence interval) based on the identified parameters. The consideration of the covariance of the parameters leads to a large reduction of the standard deviation of possible simulation results. This is particularly large in this case study, because two of the ten correlations are very large and only two are negligibly small.

Figure 12.18 shows that the probable range of the model predictions is clearly larger than the range of measurements if the covariance of the parameters is neglected. The "data" were simulated with a standard error of $\sigma_m = 0.1$ gO$_2$ m^{-3}; the standard error of the simulation results becomes on average clearly larger (Fig. 12.17). If we would have to estimate parameter uncertainty as experts, we would approximately reach this unfavorable result, because covariance can hardly be estimated in a new context – and each experiment is a new context for experts.

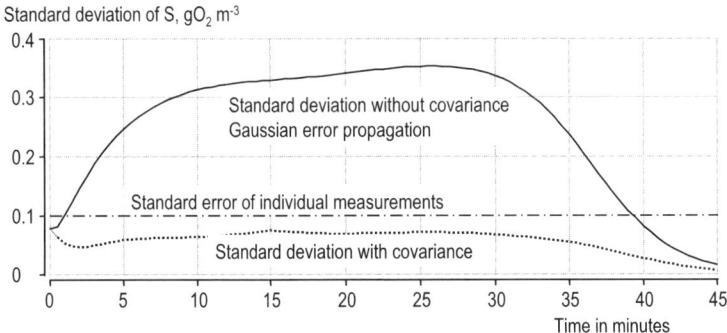

Fig. 12.17 Comparison of the standard deviation of the model predictions with and without consideration of the covariance of the parameters

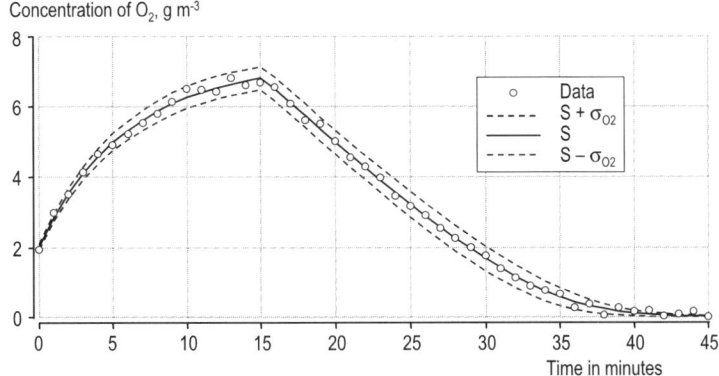

Fig. 12.18 Uncertainty (confidence interval, one standard deviation) of the oxygen concentration obtained with uncorrelated parameters and linear error propagation

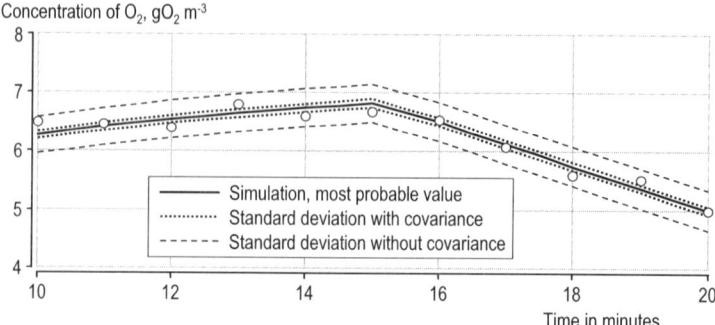

Concentration of O_2, $gO_2\ m^{-3}$

Fig. 12.19 Comparison of the confidence interval with and without consideration of the covariance of the parameters (extract from Fig. 12.18)

The decrease of the standard deviation with consideration of the correlation of the parameters leads to a value that is smaller than the standard error of the measurement (Fig. 12.17). This results from the number of measurements; the more measured values that are available, the smaller the uncertainty of the model prediction becomes. Figure 12.19 shows that with consideration of the covariance the remaining range of the possible model predictions (the confidence interval) is clearly covered by the range of the measured values.

12.7 Nonlinear Error Propagation

In the environmental sciences models are frequently nonlinear and the parameter uncertainties are relatively large. If we must estimate the uncertainties, we often select non-normal distributions that reflect our experience (i. e., evenly distributed ranges or other meaningful distributions). Monte Carlo simulation can deal with all these situations.

12.7.1 Monte Carlo Simulation

The name Monte Carlo simulation (MC simulation) is given to a method that permits distributions of the uncertainty of a model prediction as a consequence of the uncertainty of the model parameters to be obtained. On the basis of their statistical distribution, random combinations of model parameters are selected and used to make model predictions. Subsequently the results of several hundred full simulations are analyzed.

MC simulation allows the propagation of even large errors through highly nonlinear models. However, this is only possible at great computational costs; the model may have to be fully evaluated for hundreds of parameter combinations. In addition the procedure requires that the simulation software supports MC simulation, which is not the case for many technical simulation programs.

The random choice of the model parameters within their statistical distribution becomes challenging if the parameters are not statistically independent but correlated. If such correlations are known and quantified but neglected, this leads exactly as in the linear error propagation to (mostly conservative) overestimation of the prediction errors (see Fig. 12.19).

Monte Carlo simulation has the advantage that the error propagation is analyzed with the full model and not with a linearized form. In the environmental engineering sciences, the estimation errors of the model parameters are frequently quite large, and thus error prediction with a linearized model is often unreliable.

Example 12.15: Computation of π with stochastic simulation

A simple way of obtaining the value of π with the help of stochastic simulation is based on the following strategy:
Random points in the unit square of x and y from (0,0) to (1,1) are selected. Subsequently, it is determined whether these points lie within the unit circle with center at (0,0) and radius = 1. Because the probability is proportional to the area, the number π can be derived from the result. With an increasing number of simulations the computed variable becomes ever more accurate.
Code for Berkeley Madonna:

```
STARTTIME=1
STOPTIME=1E+7                 ; 10^7 trials
DT=1                          ; Counter
DTout=1E+5                    ; Output interval
r2=random(0,1)^2+random(0,1)^2 ; Radius^2 of a random point
INIT Success=0               ; Counter of successes
NEXT Success=if r2<=1 THEN Success+1 ELSE Success
NPI=4*Success/Time           ; Estimate of π
```

The resulting successes obey a Poisson distribution. This allows us to obtain the expected value and its standard deviation:

$$\mu=\frac{n_S}{n}=\frac{\text{Successes}}{\text{Trials}} \quad \text{and} \quad \sigma_\mu=\sqrt{\frac{\mu^2\cdot(1-\mu)+(1-\mu)^2\cdot\mu}{n}}=\sqrt{\frac{\mu-\mu^2}{n}}.$$

The expected value for π is $\mu_\pi=4\cdot\mu$ and its standard error is $S_\pi=4\cdot\sigma_\mu$.
In order to obtain an estimate of π with a standard error of less than 0.001, we must choose about $n=3\cdot10^6$ trials (choosing STOPTIME accordingly).
With a more sophisticated choice of the location of the trials we could approach good accuracy faster but at the cost of more programming. This will be discussed in Sect. 12.7.2 under the topic of sampling methods.

Example of a simple Monte Carlo simulation

Similar to Example 12.12, the uncertainty of modeling the degradation of a material in a first-order reaction in a batch reactor is to be determined. The analytical form of the model equation is:

$$C(t) = C_0 \cdot e^{-k \cdot t} .$$
(12.39)

The model has two parameters: the initial concentration C_0 and the rate constant k. The uncertainty will be obtained with the aid of a MC simulation. The two parameters are uncertain; we estimate mean value and standard deviation as $\mu_{C0} = 100$ g m^{-3}, $\sigma_{C0} = 10$ g m^{-3} and $\mu_k = 2$ d^{-1} and $\sigma_k = 0.5$ d^{-1}. We assume they are normally distributed and statistically independent.

Table 12.8 Code for BM for the execution of a Monte Carlo simulation for the computation of the uncertainty in the model prediction after Eq. (12.39)

```
1 STARTTIME = 0              ; Beginning of the simulation
2 STOPTIME = 2               ; End of the simulation, h
3 DT = 0.05                  ; Time step of the model evaluation, h
4 init k = normal(2,0.5)     next k = k    ; Stochastic choice of k, h⁻¹
5 init C0 = normal(100,10)   next C0 = C0  ; Stochastic choice of C₀, g m⁻³
6 C = C0*exp(-k*time)        ; Model equation
```

Table 12.8 gives the code for Berkeley Madonna for the computation for the model Eq. (12.39) with a random choice of the two model parameters. On line 4 a value for k is determined at the beginning of each simulation from a normally distributed random number and kept constant throughout this simulation. In each new call of the program another value is determined. C_0 is determined similarly. The MC simulation is processed in a special routine (batch run), in which the program is called many times. The individual results can be evaluated afterwards either as individual computations (Fig. 12.20) or based on statistical summaries (Fig. 12.21).

Because Eq. (12.39) is a nonlinear function, the MC simulation results in a skewed distribution of individual simulations around the simulation with the most probable parameters. Figure 12.21 clearly shows that the minimum and maximum values are not symmetrically distributed around the expected value. Only at the beginning with $t \approx 0$, where the error range is dominated by the normally distributed initial concentration C_0, is the distribution still symmetrical.

Table 12.9 compares the results from the linear error propagation and the MC simulation. The relatively large uncertainty, especially of the reaction rate constant k, leads to significant differences between the two approaches. The average result of a MC simulation is not necessarily equal to the most probable result of the simulation (the modal value) because the resulting distribution of the results is asymmetrical. Generally, only a comprehensive analysis of the distribution of the

Fig. 12.20 Principle sketch for a Monte Carlo simulation. Here only ten of several hundred realizations are shown

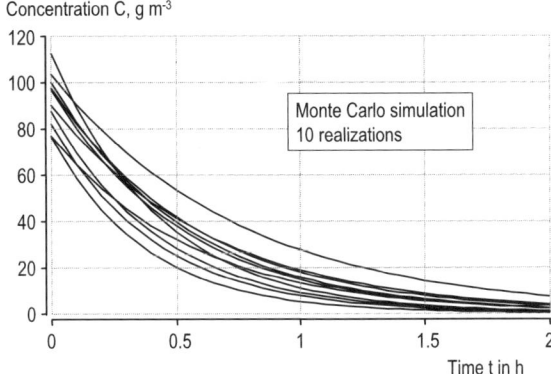

Fig. 12.21 Statistical evaluation of 1000 realizations of Eq. (12.39)

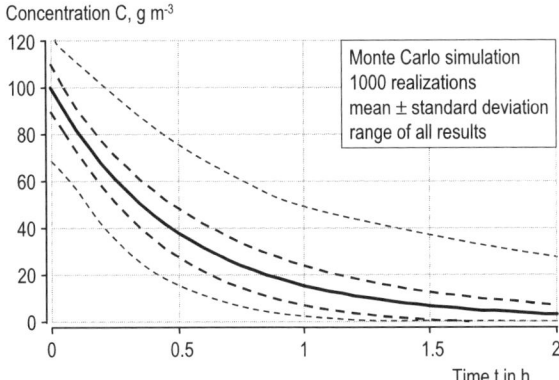

Table 12.9 Comparison of the results from Gaussian error propagation and an MC simulation (based on 1000 realizations) with normally distributed, independent parameters $C_0 \sim N(100,10)$ and $k \sim N(2,0.5)$ in Eq. (12.39); simulation with BM

Time	Gaussian error propagation			Monte Carlo simulation				
h	μ_G	σ_G	cv%	μ_{MC}	σ_{MC}	cv%	Min	Max
0	100	10	10	100	10	10	66	130
0.5	36.8	9.9	27	37.8	10	26	16.7	74
1.0	13.5	6.9	51	15.2	7.9	52	3.2	50
1.5	5.0	3.8	76	6.5	5.3	82	0.61	34
2.0	1.8	1.8	100	2.9	3.4	117	0.11	23

individual results can lead to the most probable value. MC simulation has the advantage that we can estimate the entire distribution of the uncertain results if the distributions of the parameter uncertainty are known. Linear error propagation only provides the moments of these distributions.

Example 12.16: Accident with toxic materials

Not all fish react equally sensitively to toxic materials. Assume that fish react as follows to exposition (the integral of the concentration over time) to a pollutant: the tolerated exposition E_T up to an unacceptable damage of an individual is normally distributed: $E_T \sim N(\mu_{ET},\sigma_{ET}) = N(750 \text{ g min m}^{-3}, 50 \text{ g min m}^{-3})$.

After an accident the concentration of the toxic material in a river has the following time course: $C = C_0 \cdot e^{-k \cdot t}$.

Since the observation of the accident is unreliable, the pollutant load can only be indicated inaccurately. The following estimated values are available:

C_0 = evenly distributed in the range $5 < C_0 < 15 \text{ g m}^{-3}$

k = normally distributed with expected value $\mu_k = 0.02 \text{ min}^{-1}$, $\sigma_k = 0.002 \text{ min}^{-1}$

What is the probability that an individual fish becomes affected by the pollutant? (approximately 7.3%)

This is a typical situation where uncertainty has to be evaluated. On the one hand the load (exposition) is uncertain and on the other hand the carrying capacity (tolerable load) of an individual fish is uncertain. We look for the probability that, for an individual, the load exceeds the capacity.

The code for BM is as follows:

```
METHOD RK4                              ; Integration with fourth-order Runge–Kutta
STARTTIME = 0                           ; Beginning of the simulation
STOPTIME = 1000                         ; End of simulation, min
DT = 1                                  ; Time step, min
init Et = normal(750,50) next Et = Et   ; Random choice of tolerable exposure,
                                          g min m⁻³

init C0 = random(5,15)
next C0 = C0                            ; Random choice of initial pollutant concen-
                                          tration g m³
init k = normal(0.02,0.002) next k = k ; Recession constant for pollutant concentra-
                                          tion, min⁻¹
C = C0*exp(−k*time)                     ; Time course of the pollutant concentration
init E = 0  d/dt(E) = C                 ; Exposure of the individual
Damage = if E > Et THEN 1 else 0        ; Check whether damage occurs
```

This code is repeated, say, 1000 times (Batch Run) and the expected value of damage is recorded.

See also Example 15.5

12.7.2 Sampling Methods

In a Monte Carlo simulation we must choose the parameters from different distributions and possibly consider the correlation of the parameter values in many random combinations; this process is called sampling. The more efficiently the selected parameters permit reliable distributions of the model results to be developed, the smaller the cost of computation becomes. There are sampling strategies that keep the cost of computation small.

For complex models the cost of computation frequently limits the possibilities for careful error analysis because it is not possible to accomplish the computations several hundred times.

Random Choice of Parameters

The simplest method to choose a random set of parameters relies on the statistical distributions of the parameters (marginal distributions) and selects each individual parameter randomly and fully independently of the others. Simulation programs make some functions available that provide random numbers that satisfy certain distributions.

Example 12.17: Random numbers in Berkeley Madonna

The program BM makes the following random numbers available:
random(a,b), random(a,b,seed): evenly distributed random number within the range form a to b. A fixed value entered as seed always results in the same series of random numbers. This is convenient to test a code. Be aware that seed is interpreted as a rounded integer.
normal(μ, σ), normal(μ, σ,seed): normally distributed variable with average μ and standard deviation σ
Transformation makes variables that obey other distributions available, such as the log-normal distribution: *exp(normal(μ, σ))* or an exponential distribution: *–log n(random(0,1))/λ*
For discrete distributions the *binomial(p,n,seed)* and *poisson(p,seed)* distributions are available.
A randomly generated integer in the range from n_1 to n_2 is obtained from $INT(random(n_1, n_2 + 1))$.

Figure 12.22 shows how a random value of a parameter that obeys any distribution can be generated from the cumulative distribution and an evenly distributed random number in the range from 0 to 1. This procedure can be realized analytically for many distributions; alternatively it may be possible to interpolate between well-defined nodes that are entered as data into the program. If only a histogram of the distribution of a parameter is known (as might be the case in a bootstrap procedure), then a cumulative distribution that is the basis for the determination of an accordingly distributed random number can be obtained. For discrete events, either a cumulative stair function develops or alternatively discrete values may be stored in an array (vector), and the index of a random element can then be obtained from a randomly generated integer.

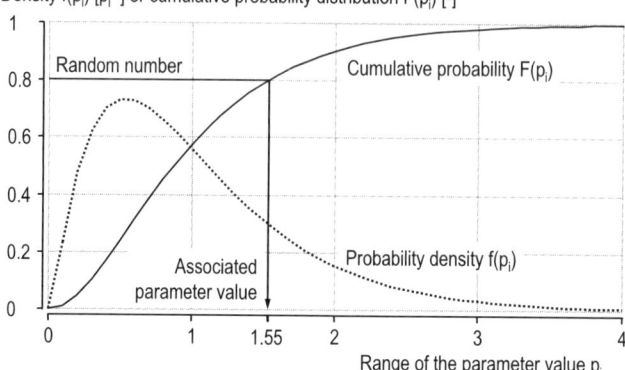

Fig. 12.22 Determination of random number p_i from an evenly distributed random number in the range of 0 to 1 and the cumulative density $F(p_i)$. *Reading example*: an evenly distributed random number x within the range 0 to 1 has the value 0.8. The cumulative probability distribution of the transformed random number has the value $F(p_i) = 0.8$. The transformed random number p_i has the value 1.55

The disadvantage of a purely random choice of parameters is that correlations between parameters are not considered. The procedure assumes that all covariances disappear ($s_{pi,pj} = 0$). If covariances are known, this has serious consequences (see below).

Two Correlated Parameters

Obtaining corresponding random parameter sets is not trivial if the covariance matrix of the parameters Σ is known and the covariances $S_{pi,pj} \neq 0$. For two normally distributed parameters this situation is shown in Fig. 12.23. The marginal distributions of the two parameters x and y each point to a large standard deviation ($\sigma_x = 33$, $\sigma_y = 10$); in contrast the distribution of the residuals around the regression line is much smaller: $\sigma_{Res} = 5.3$. The covariance or the correlation is not zero.

For two normally distributed parameters the following equations are valid:

$$y_i = \mu_y + \frac{r_{x,y} \cdot \sigma_y}{\sigma_x} \cdot (x_i - \mu_x) + \varepsilon(0, \sigma_{Res}), \tag{12.40}$$

$$\sigma_{Res}^2 = \frac{n}{n-2} \cdot (1 - r_{x,y}^2) \cdot \sigma_y^2, \tag{12.41}$$

where y_i is a normally distributed random variable that is scattered around the regression line. ε characterizes the residual (the deviation from the regression line) and is normally distributed. n counts the number of available data pairs. In Example 12.18, Eqs. (12.40) and (12.41) are used to perform an MC simulation with correlated parameters.

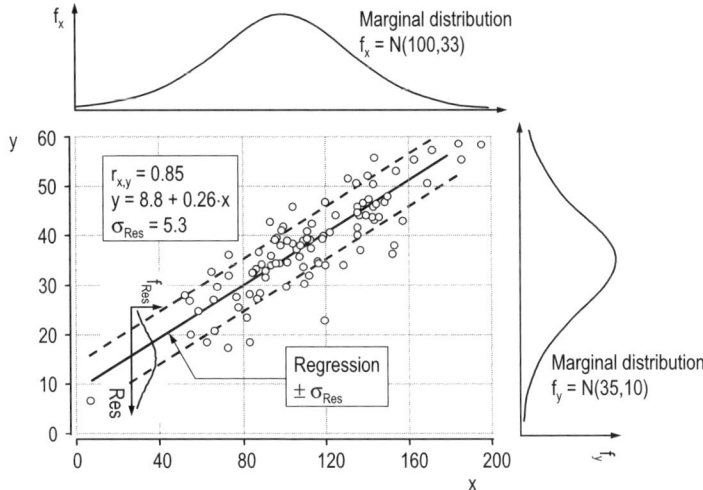

Fig. 12.23 Two series of simulated correlated data, each with their marginal distribution and the distribution of the residuals around the regression line (see Table 12.10)

Table 12.10 Definition of the symbols and parameter values of the correlated data in Fig. 12.23

Parameter	Distribution	Mean	Variance	Correlation matrix **R**	
X	normal	$\mu_x = 35$	$\sigma_x^2 = 10^2$	1	$r_{y,x} = 0.85$
Y	normal	$\mu_y = 100$	$\sigma_y^2 = 33^2$	$r_{x,y} = 0.85$	1

Example 12.18: MC simulation with two correlated parameters

The simple example of an MC simulation in Table 12.8 is to be supplemented by a correlation between the parameters. Altogether $n = 10$ experiments have been conducted and the most probable parameter values have been identified. The analysis indicated a high correlation between the two parameters k and C_0 of $r_{k,Co} = 0.95$. *What is the range in which further results are to be expected?*

```
STARTTIME = 0              ; Beginning of the simulation
STOPTIME = 2               ; End of the simulation, hrs
DT = 0.05                  ; Time step of the model evaluation, hrs
n = 10                     ; Number of observations
mueC0 = 100  sigC0 = 33    ; Marginal distribution of C0, normal
muek = 2  sigk = 0.5       ; Marginal distribution of k, normal
corr = 0.95                ; Correlation of C0 and k
sigRes = sqrt(n/(n-2)*(1-corr^2)*sigk^2)
                           ; Standard deviation of the residuals, Eq. (12.41)
init C0 = normal(mueC0,sigC0)
next C0 = C0               ; stochastic variation of C0, g m^-3
```

Fig. 12.24 MC simulation with correlated parameters. Compare Fig. 12.21 for the same situation without correlation

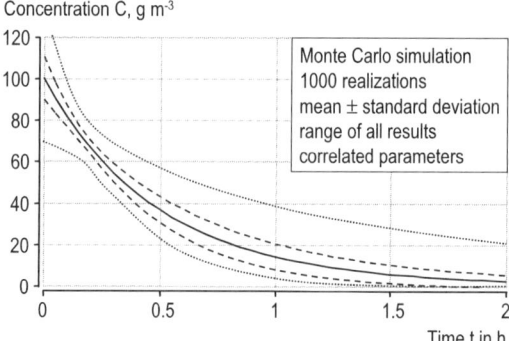

Init k = muek + corr*sigk/sigC0*(C0 – mueC0) + normal(0,sigRes)
next k = k   ; Eq. (12.40)
C = C0*exp(–k*time)   ; Model equation

The results are shown in Fig. 12.24. In comparison with Fig. 12.21 the range covered is narrowed.

Several Correlated Parameters

If the MC simulation is to be made with several correlated parameters, possibly even with different types of marginal distributions, obtaining appropriate sets of parameters becomes fastidious; the details cannot be covered here. The simulation package UNCSIM (Uncertainty simulation, Reichert, 2004) provides the program *randsamp.exe,* which generates such parameters and makes them available in a text file, which can then be edited and adapted to the needs of simulation programs (for the BM code see Example 12.22). A prerequisite is of course that the correlation matrix and the distributions of the parameters are provided.

Number of MC Simulations

For large models the expenditure and costs of the many computations of a MC simulation may become inhibiting. Reducing costs and desired accuracy are therefore competing requirements. Morgan and Henrion (1990) give estimated values for the necessary number of simulation runs based on the approximation to the binomial distribution by a normal distribution. The relationship gives the 95% confidence interval for a selected percentile of the results:

$$m \ge p \cdot (1-p) \cdot \left(\frac{2}{\Delta p} \right)^2 \quad \text{and} \quad m \ge \frac{9}{p \cdot (1-p)} \tag{12.42}$$

m = required number of simulation runs
p = percentile of the results to be obtained [−], fraction of the results, that is
     smaller than the selected value p
2 = stands for the 95% confidence interval (more exactly 1.96)
Δp = the desired percentile p will, in 95% of the MC simulations, lie within the
     range of the $p - \Delta p$ to $p + \Delta p$ percentiles

The approximation of the binomial distribution by a normal distribution is reliable only if $m \cdot p \cdot (1-p) > 9$ (Stahel, 2002). If this inequality is not satisfied, m should be increased. An interesting aspect of Eq. (12.42) is that the number of simulation runs necessary m is independent of the number of variable parameters.

Example 12.19: Computation of the number of necessary simulation runs

We want to ensure with a confidence of 95% that the estimated 80th percentile of the model prediction is within the range of the true 75th to 85th percentile.
$p = 0.8$, $\Delta p = 0.05$ $m = 256$ $m \cdot p \cdot (1-p) = 41$
We want to ensure with a confidence of 95% that the estimated median (50th percentile) lies within the range of the $49^{th} - 51^{st}$ percentile.
$p = 0.5$, $\Delta p = 0.01$ $m = 10'000$ $m \cdot p \cdot (1-p) = 2500$
We want to ensure with a confidence of 95% that the estimated 99th percentile is larger than the 98th percentile.
$p = 0.99$, $\Delta p = 0.01$ $m = 396$ $m \cdot p \cdot (1-p) = 3.92 < 9$, m must be increased to 900

Reduction of the Variance of the Predicted State Variables

The results of a MC simulation are the statistically distributed, simulated state variables, which are considered to be random variables and are evaluated depending upon the question asked. These resulting statistical distributions are themselves uncertain; thus the expected values of the target variables are subject to scatter when the entire MC simulation process is repeated. This uncertainty can be reduced, e. g., by increasing the number of simulation runs m in an MC simulation. Optimized sampling strategies succeed in reducing this variance efficiently at reduced computational cost.

The desired result of an MC simulation is the statistical distribution of the possible model outcomes. The classical, simplest MC simulation is based on purely random sampling of the parameters. With a large number of simulation runs, the distributions of the results become stable, i. e., the variance of the target variables is small.

Example 12.20: Variance of the target variables

A model supplies as a result the necessary volume V of a reactor. From $n = 100$ MC simulations the individual values amount to an average of $V_m = 5000 \, m^3$ with a standard deviation of $\sigma_V = 1000 \, m^3$. *How large is the 95% confidence interval of the mean value?*

The standard deviation of the average value is $\sigma_{V_m} = \dfrac{\sigma_V}{\sqrt{n}} = 100 \text{ m}^3$. The 95% confidence interval covers, for normally distributed results, the range $5000 \pm 196 \text{ m}^3$. If we are to reduce the confidence interval to $\pm 50 \text{ m}^3$, we have to conduct approximately $n = 1600$ simulations.

Latin Hypercube Sampling

The goal of the sampling process is not primarily the randomness of the parameter combinations per se but rather the even probable distribution of the parameter sets over the entire parameter space.

UNCSIM by Reichert (2004) makes several efficient sampling strategies available. In the literature, *Latin hypercube sampling* (*LHS*) is the most frequently mentioned approach (but not necessarily the most efficient).

The principle of LHS is explained for two parameters in Fig. 12.25. For each parameter n equally probable sections are chosen. A sample is composed by choosing for each parameter at random a section that has not yet been selected. The parameter value is then determined according to Fig. 12.22 for either the center of the section or for a random position within the section.

For k parameters the k-dimensional *hypercube* is divided in appropriate elements rather than the unit square as given in Fig. 12.25. If the number of parameters is not very large and the model is approximately linear, then a very efficient approximation of the final distributions of the target values results from this kind of sampling.

Example 12.21: Designing a measurement campaign

The routine control of a wastewater treatment plant yields the following information, which we want to use to estimate the sludge loading B_{TS} of the activated sludge plant:

Parameter	Unit	Distribution	Mean μ	Stand. dev. σ	Correlation matrix Q	BOD$_5$	TS$_{AS}$
Q	m^3 d^{-1}	Normal	5000	1000	1	−0.75	−0.60
BOD$_5$	g m^{-3}	Normal	250	50	−0.75	1	0
TS$_{AT}$	g m^{-3}	Normal	3000	150	−0.60	0	1

Q = flow rate, BOD_5 = flow-weighted daily mean concentration, TS_{AT} = total solids in aeration tank
The volume of the aeration tank is $V_{AT} = 1000 \text{ m}^3$.

The sludge loading rate is defined as $B_{TS} = \dfrac{Q \cdot BSB_5}{V_{AT} \cdot TS_{AT}}$ (the F/M ratio).

Fig. 12.25 Schematic diagram to explain *Latin Hypercube sampling*: evenly distributed ranges are chosen without replacement

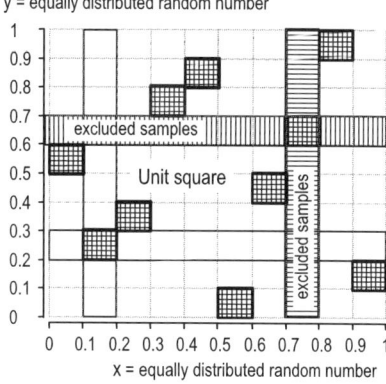

We are planning a measuring campaign that is expected to allow us to estimate the mean sludge loading rate B_{TS} with an error of less than 5% at the 95% reliability level. Since all measurement procedures have to be recalibrated, we are interested in the following question:

How many daily samples are required in order to obtain the indicated reliability and precision of the result?

Procedure:

With the aid of an MC simulation based on existing measurements, we obtain the standard deviation of the computed sludge loading rate. Since all measuring devices are recalibrated, this is only an estimate; we must obtain new reliable measurements, but the old values are sufficient to plan the new campaign.

With the aid of *randsamp.exe* (UNCSIM, Reichert, 2004), we obtain 1000 pseudo-random sets of parameters that consider correlation and use LHS to approach the distribution as well as possible.

In a table (i. e., in EXCEL) we obtain 1000 values of possible sludge loading rates, which we sort in increasing order and analyze statistically.

Figure 12.26 shows the cumulative distribution of the simulated B_{TS} values, which nearly follows a normal distribution. The results are summarized in the following table:

	Symbol	Value	Units
Mean from MC simulation	$B_{TS,m}$	0.44	$gBOD_5 \, g^{-1}TS \, d^{-1}$
Standard deviation from MC simulation	s_{BTS}	0.080	$gBOD_5 \, g^{-1}TS \, d^{-1}$
Allowable 95% confidence interval of mean	$\pm 0.05 \, \mu$	0.022	$gBOD_5 \, g^{-1}TS \, d^{-1}$
Allowable standard deviation of mean[1]	σ_m	0.011	$gBOD_5 \, g^{-1}TS \, d^{-1}$
Number of required measurements $= (s_{BTS}/\sigma_m)^2$	n	53	–

[1] Confidence interval/1.96

From this analysis it appears that a weekly sample taken over 1 year and distributed randomly over weekdays is required to obtain the information needed. Before

Fig. 12.26 Simulated distribution of the sludge loading rate B_{TS} based on routine data (1000 MC samples with correlated parameter values)

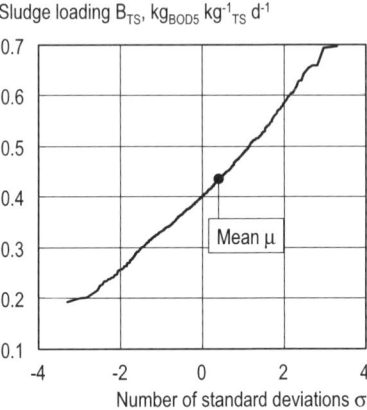

we could accept a shorter campaign with more intensive sampling (say a daily sample over two months), we would have to test the old data and analyze for seasonal and weekly trends.

12.7.3 Application to the Case Study

MC simulation provides us with the possibility to obtain the distribution of the model predictions as a consequence of parameter uncertainty. This permits even for nonlinear systems (such as the case study) the confidence intervals of the simulation to be obtained.

Task

How large is the 95% confidence interval of the simulation of the experiment of the case study described in Sect. 12.2?

Code for the MC simulation

In Example 12.22 a code and the procedure for the MC simulation of the case study is shown. The code is based on the assumption that the uncertainty of the parameters is normally distributed. For uncorrelated parameters we can use the random-number generator of Berkeley Madonna. For more than two correlated parameters the program *randsamp.exe* (UNCSIM, Reichert, 2004) allows 1000 correlated, multidimensional normally distributed parameter sets to be obtained. Since the computing time for the model is very small, we do not use a special sampling procedure (e. g., LHS).

Example 12.22: Code for the Monte Carlo simulation of the case study with BM

The case study is to be implemented in Berkeley Madonna in such a way that a Monte Carlo simulation (a) with uncorrelated and (b) with correlated parameters can be accomplished.

(a) Uncorrelated parameters (tested)

```
METHOD RK4
STARTTIME = 0.0
STOPTIME = 45
DT = 1
```

{the following random values of the parameters are normally distributed with indication of μ and σ. At the beginning of the computation a random parameter value is specified with INIT p_i = normal(μ_{pi},σ_{pi}), which is kept throughout the single simulation with NEXT p_i, the parameter values correspond to model E in Table 12.4 covariance matrix (Table 12.6) or correlation matrix (Table 12.7) are not considered}

```
init Ssat = normal(9.85,0.29)         next Ssat = Ssat          ; Stochastic S_sat
init rO2max = normal(−0.438,0.016)    next rO2max = rO2max      ; Stochastic r_O2,max
init kla = normal(0.152,0.0093)       next kla = kla            ; Stochastic k_la
init S0 = normal(2.03,0.078)          next S0 = S0              ; Stochastic S_0
init KO = normal(0.996,0.144)         next KO = KO              ; Stochastic K_O
rO2 = rO2max*S/(KO + S)                                         ; Reaction kinetics
init S = S0                                                     ; Initial value
d/dt(S) = if time <= 15 then kla*(Ssat − S) + rO2 else rO2      ; Balance equation
```

{The MC simulation is processed with the help of a batch of run in which no more parameters are changed, since these are already varied with the random choice of the initial values.}

(b) Correlated parameters

A file with 1000 correlated parameter combinations is obtained from *randsamp.exe* from UNCSIM (Reichert, 2004) and edited in EXCEL. In BM this file is read in as data with the name #P.

```
Run = 1                ; Counter of the simulation run which will increase from 1 to
                         1000
Ssat = #P(run,1)       ; S_sat is the first parameter in the list
rO2max = #P(run,2)     ; r_O2,max is the second parameter
kla = #P(run,3)
S0 = #P(run,4)
KO = #P(run,5)
```

{The MC simulation can now be accomplished by controlling the variable run from 1 to 1000 in a batch run}

Number of necessary simulation runs

If we want to determine the borders of the 95% confidence interval with a maximum 1% deviation (94–96%), we must after Eq. (12.42) choose m=975 simulation runs in order to obtain the distribution of the results with the desired accuracy.

Distribution of the results and confidence interval

Figure 12.27 shows the traces of 25 realizations of MC simulations. The results of the MC simulation are shown against a probability scale in Fig. 12.28. After 20 min the results are normally distributed, which is a consequence of the nearly linear model for the period of $0 < t < 20$ min. After 45 min the hyperbolic relationship of the Monod kinetics has a dominant influence, which results (especially for the uncorrelated parameters) in a skewed distribution and an increased average relative to the median value. For independent parameters, the resulting distribution is closely approximated by an exponential distribution.

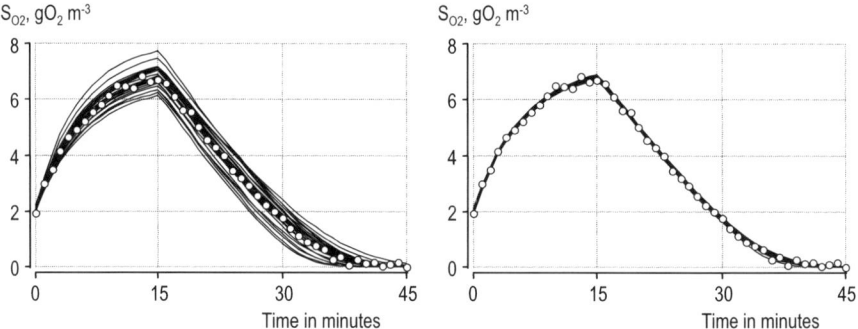

Fig. 12.27 Comparison of 25 MC realizations of the simulation with uncorrelated, normally distributed parameters (*left*) and correlated, normally distributed parameters (*right*); see also Fig. 12.19 and Table 12.7

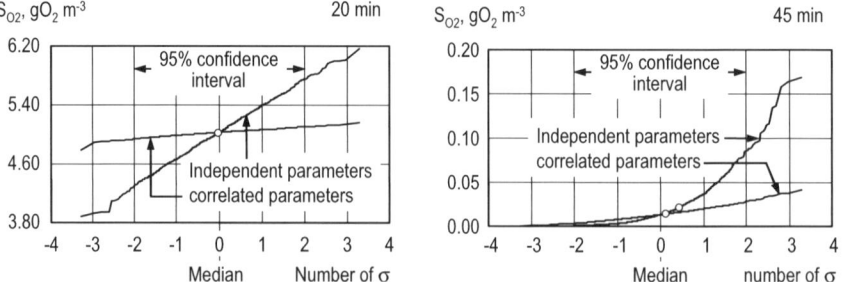

Fig. 12.28 Comparison of the distribution of the results of a MC simulation of the case study with independent and correlated parameters for m=1000 realizations. *Left*: t=20 min, *right*: t=45 min (probability net)

Table 12.11 Statistical characterization of the distribution of the results of the MC simulation after 20 min and after 45 min

Time	20 min		45 min	
Sampling of parameters	Uncorrelated	Correlated	Uncorrelated	Correlated
Average value, mean	5.03	5.03	0.020	0.014
Standard deviation	0.36	0.041	0.022	0.0062
Median (50% value)	5.02	5.03	0.013	0.013
Confidence interval 2.5%	4.32	4.94	0.00053	0.0035
Confidence interval 97.5%	5.71	5.10	0.077	0.026

Table 12.11 summarizes the statistical characteristics of the distributions in Fig. 12.28. After 20 min, the average and median values agree very well, the distribution is close to normal, and the empirical confidence interval is symmetrical. By considering the correlation of the parameters, the width of the confidence interval is reduced from 1.39 to only $0.16\,gO_2\,m^{-3}$. After 45 min the difference between the median and the mean for uncorrelated parameters exceeds 50%. The confidence interval becomes narrower (in absolute terms) but strongly asymmetric for both parameter sets.

12.8 Correlated Parameter Values: A Word of Caution

In the extensive case study presented above, consideration of the correlation between estimated parameter values resulted in a decreasing uncertainty range for model predictions (see Figs. 12.19 or 12.27). This is however not always the case.

Equation (12.27) written for two parameters for a specific value of a single state variable has the form (see Example 12.11):

$$\sigma_y^2 = \left(\frac{\partial y}{\partial p_1}\cdot\sigma_{p_1}\right)^2 + \left(\frac{\partial y}{\partial p_2}\cdot\sigma_{p_2}\right)^2 + 2\cdot r_{p_1,p_2}\cdot\left(\frac{\partial y}{\partial p_1}\cdot\sigma_{p_1}\right)\cdot\left(\frac{\partial y}{\partial p_2}\cdot\sigma_{p_2}\right). \quad (12.43)$$

Whether σ_y increases or decreases as a consequence of the correlation term in Eq. (12.43) depends on the combination of signs of the correlation coefficient r and the sensitivities (partial derivatives) relative to the parameters. Only a case-by-case analysis can provide further insight into this question.

Example 12.23: Uncertainty in nitrification performance of an activated sludge system

Nitrification in an activated sludge system includes the oxidation of ammonium to nitrite and the further oxidation of nitrite to nitrate by two different populations of autotrophic bacteria. Thus the production of nitrate depends on the reaction rates of the two reactions in series. The sensitivities (partial derivatives) of the nitrate concentration relative to both rate constants are positive. In addition increasing temperature will have a positive effect on both reactions, i.e., they become faster; the two rate constants are positively correlated when the water temperature changes. Thus if temperature variation is not carefully considered in a design procedure, positive sensitivities of nitrate concentration coincide with a positive correlation of the uncertainties of the parameters. According to Eq. (12.43), this increases the uncertainty in the resulting nitrate concentration.

In this example it will be difficult to obtain the uncertainties of the two rate constants (σ_{p1}, σ_{p2}) and their correlation ($r_{p1,p2}$) since they are not primarily dependent on experimental error but rather on the temperature variation, which is not sufficiently characterized.

12.9 Summary of Model Identification

Model identification is the procedure by which we use experience with a real system (measurements, observations) to identify the structure and parameters of a mathematical model such that the model can describe the behavior of the real system in a broad field with as small deviations as possible.

Figure 12.29 compares our information about the real system with the information that we gain from a model and, if necessary, improve by parameter identification or adaptation of model structure. It is remarkable that our knowledge of the reality is distorted by measurement errors both of the input and the output quantities. In the model computations the input deviates from the real loads and our predictions of the output differs again from our observations because of measurement errors. Only stochastic models or statistical analysis can deal systematically with these deviations.

Modeling is the art and science of abstracting and simplifying wherever possible without losing relevant information. The real systems are affected by disturbances which we cannot or do not want to measure but which influence their be-

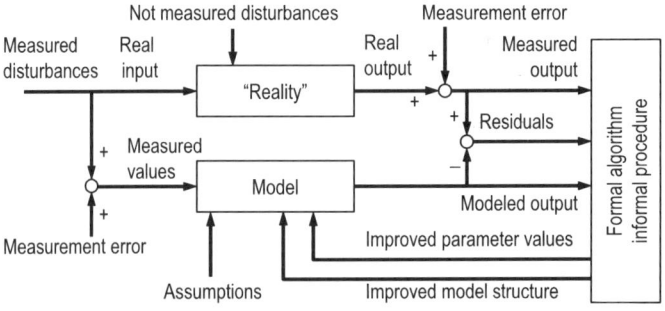

Fig. 12.29 Information flow in model identification (Beck, 1983, adapted)

havior nevertheless. Because no direct information about these effects is at our disposal, we must replace missing observations with modeling assumptions, whereby we introduce additional uncertainties.

The process of model identification is shown in Fig. 12.30. We start from the observation of the real system and compile a first model, for which for the time being only estimated parameters are available. In a first loop we identify the most probable parameter values. Subsequently, we test whether the model structure can cope with our observations without systematic deviations. If a suitable structure and its parameters are identified, the model is calibrated. We then examine the uncertainties of the predictions of the model. If these fulfill our requirements, we have successfully developed a validated model; otherwise we must obtain add-itional data.

The problem we work on, the questions we ask, the model structure we de-velop, the parameter values we obtain, the accuracy we achieve, and the available data all have tight reciprocal effects on each other (see Fig. 12.9).

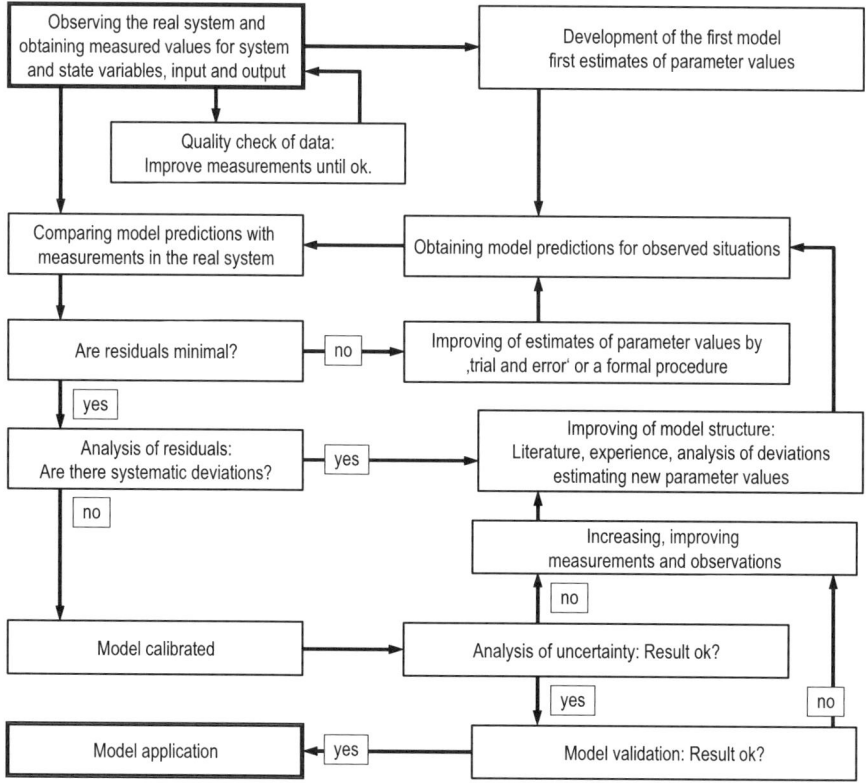

Fig. 12.30 The procedure in model identification

Chapter 13
Process Control Engineering

In the operation of technical systems we have various degrees of freedom, which we use in order to come closer to our operating goals. In process technology we try to optimize the performance of a procedure; to do this we develop strategies or guidelines according to which we use these degrees of freedom; subsequently, we apply these strategies to the operation of the system, by either automation or manual interference.

Process control engineering has the task of converting our operating goals and strategies into practical application. To do this, we use mathematical algorithms (strategies), technical elements (actuators, control members), and sensors (measuring systems). A basic understanding of process control helps us:

- *to apply simple operating strategies successfully to the manual control of systems;*
- *to search for possible ways of optimizing operation, e.g., with the help of simulation;*
- *to recognize possibilities and obstacles in drafting control strategies for more advanced control systems;*
- *to communicate successfully with control engineers.*

In addition, process control technology provides a general understanding of problems in the operation of plants.

Here only a short introduction to the simplest principles of process control technology is given.

13.1 Examples of Operating Strategies

The following examples show different aspects of control engineering on the basis of descriptive examples.

13.1.1 Adjusting the Water Temperature of a Shower

If we want to adjust the water temperature of a shower in a hotel, we have two degrees of freedom in operating the unknown mixing tap: the flow rate Q and the position of the temperature lever S. We start with the highest possible flow Q_{max} and choose the highest possible temperature T_{max}. As soon as we feel an increase of the temperature, we reduce T and possibly also Q. Since the showering system is unknown to us, the first correction will not be sufficient. A second one follows, possibly before the first one has fully affected the temperature (the lag in the controlled system is relatively long, i. e., it takes some seconds between the change of the position, the filling of the hose, and the reaction of the measurement system – our feeling of the temperature on the skin). After several corrections, we succeed in obtaining the desired temperature. The controlled variable settled down after several oscillations. Our strategy worked satisfactorily depending upon our patience, or we got scalded under too hot a shower.

At home, we have experience with our own shower; we are experts. We successfully reach the desired temperature with only a single correction. We can even consider a preheated, just used shower without any problems.

The specific characteristics of a plant (time constants, degrees of freedom, state variables, measured values, etc.) determine the reaction of the plant to our operating strategies. Expert knowledge (control engineering) improves these strategies and helps to achieve our goals faster and better.

In control engineering, we learn to use our knowledge systematically for the improvement of our strategies and their transfer into application.

13.1.2 Operation of an Activated Sludge System

In the operation of an activated sludge system, the operator tries to keep the activated sludge concentration X in the activated sludge tank as constant as possible. By daily adjustments of the flow of excess sludge, she tries to keep the observed concentration X_{obs} within the range of a target value X_{Target}, which again is obtained from experience, or prescribed by the responsible engineer. If the effective concentration X_{obs} becomes too small, then the treatment suffers; if it is too large, there is a danger that activated sludge is lost from the secondary clarifier as a consequence of high flow during rain events. Figure 13.1 shows the situation of the treatment plant operator:

- Daily, she measures the effective concentration of the activated sludge X_{obs}. The measured value is available with a delay of approximately 2 h after sampling.
- She then compares the observed value X_{obs} with the desired value X_{Target} and decides according to her experience (represented in the lower diagram) on an adjustment of the excess sludge removal.

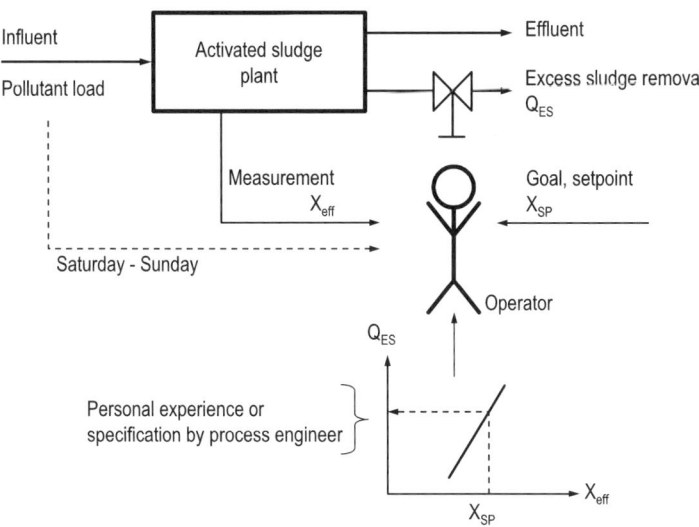

Fig. 13.1 Schematic representation of the control loop in which the operator keeps the concentration of the activated sludge in an activated sludge system at a constant value

• It may be that the operator even considers the day of the week. She may know that during the weekends fewer pollutants are discharged and therefore less sludge is produced. On Friday she will thus reduce the excess sludge withdrawal accordingly.

Since the concentration of the activated sludge reacts very slowly to small changes in the withdrawal rate (the characteristic time constant being the solids retention time), a measuring frequency of once per day is sufficient to reach the target concentration with good accuracy. In addition the long delay between sampling and the availability of the result is negligible compared to the solids retention time, which is typically on the order of 10 days.

13.1.3 Summary

The characteristic properties of the two process control loops (shower, activated sludge system) are:

• The system that we want to control has a behavior for which we have a model concept (i. e., we approximately understand its behavior). There are characteristic times within which the system reacts in an expected way to our interferences.
• There is an observation or a measurement which provides us with information about the state of the controlled system and thus provides us with feedback about the success of our interference.
• We have a strategy according to which we decide on our interferences.

- A lack of understanding of the system can lead to oscillations in the behavior of the controlled system.
- Expert know-how improves the performance of the control system.

13.2 Control Path and Control Loop

Feedforward control works forward, in the direction of flow; we speak of a control path. Feedback control works against the direction of flow (feedback control), in which case we speak of a control loop.

In Fig. 13.2 crosslinking of a system with its environment is schematically represented. Three kinds of links or effects exist:

- Disturbances (z) or inputs are influences that affect the system from the outside and we cannot exert any short-term influence on these variables. In a wastewater treatment plant, these are, e. g., the influent flow rate, the pollutant concentrations, and the temperature. We also call them the loads.
- Control variables (y) are influences that we can manipulate (adjust, control, change). In operation, they represent degrees of freedom. Examples for treatment plants are the chemical dosage, the aeration rate, and the excess sludge removal.
- State variables or output variables (x) stand for the output, the product of the process, the obtained performance which is measurable under the conditions in the system. In a treatment plant, e. g., these include the existing oxygen concentration and the remaining pollution in the effluent.

All three groups of variables can be functions of time and elements of vectors that include many variables. If x, y, and z are temporally invariant, we speak of the steady state, a situation which is not of relevance for process control as discussed here.

We select the control variables (y) according to certain strategies. We differentiate between feedforward and feedback strategies. We speak of:

- *An open-loop* or *feedforward control or control path*, if $y = f(z, t)$
- *A control loop* or *feedback control*, if $y = f(x, t)$
- A control loop with *disturbance compensation*, if $y = f(z, x, t)$

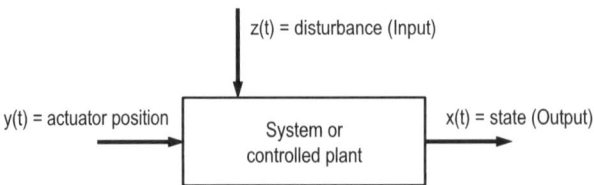

z(t) = disturbance (Input)

y(t) = actuator position | System or controlled plant | x(t) = state (Output)

Fig. 13.2 Schematic representation of a system with disturbances and control and state variables

Figure 13.3 shows a scheme of a *control path*. The control considers the variable disturbances, but has, however, no information available about the success of the control strategy or about its effect on the states x. Obviously, we need well-established experience or a very reliable model of the behavior of the system, in order to specify a feedforward control strategy. Examples of feedforward control are:

- Shooting by artillery: the first projectile must be fired correctly, based entirely on models of flight path, knowledge of the characteristics of the projectile, existing temperature, wind velocities and directions, etc. in order to find its way to the target. Any interference after firing is impossible (only the second projectile can be corrected into target by observation – this is then a feedback, control mechanism).
- In the production of drinking water, chlorine may be added to spring water for disinfection up to a constant concentration. The flow of spring water is measured, and the dosing equipment is controlled accordingly. The concentration of the dosed chlorine solution must be known, but the necessary model for the computation of the required dosage (mixing calculation) is so simple that an online control of the success of the dosing does not seem to be necessary.

Efficient strategies for feedforward control are relatively easy to define, if a reliable model for the behavior of the system under variable load is available. Since there is no feedback, there are no induced instabilities, if meaningful control strategies are implemented.

Figure 13.4 schematically shows a feedback *control loop*. Here the state variable, the system output or controlled variable x is measured and compared with a setpoint w (reference variable). Using the offset $e = w - x$, the automatic controller determines the new position y of the actuator (controlled member) according to the defined control strategies (algorithms). Thus, there exists the permanent possibility to survey the success of the control, i.e., to observe its effect on the controlled variable x, and to improve the control.

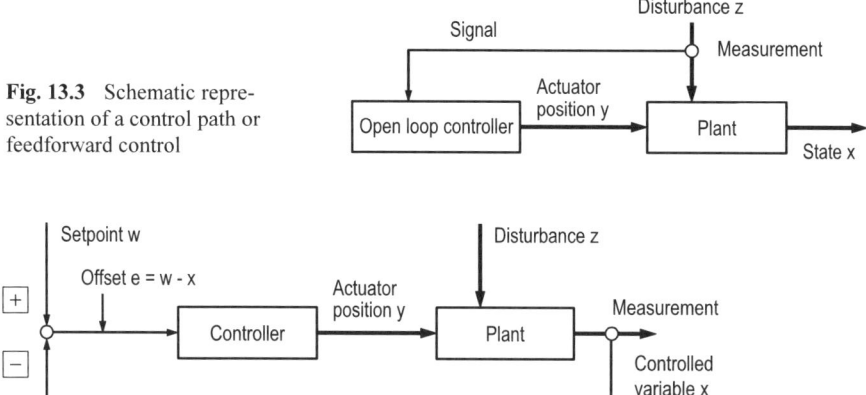

Fig. 13.3 Schematic representation of a control path or feedforward control

Fig. 13.4 Schematic representation of a feedback controller or a control loop

Automatic control loops are based on feedback, which may lead to unstable behavior of the system, i. e., that it begins to up-swing itself. Examples of control loops are the two examples in Sect. 13.1 (the shower and excess sludge removal).

Example 13.1: Control of the oxygen concentration in an aeration tank

In activated sludge tanks, in the course of diurnal variation, different amounts of oxygen are used. In order to save energy, the aeration in the tank is adapted to the oxygen consumption. The concentration of the dissolved oxygen is constantly measured with an electrode and the air flow is increased as soon as the concentration drops, or reduced if the concentration is too high. Air flow is changed either by adjusting the capacity of the blowers or by positioning a slider gate (valve). Since the success of this control is continuously observed by following the changing concentration of the dissolved oxygen, and since this signal is used to decide on the control, a closed loop results, we speak of a *feedback controller*.

Example 13.2: Feedforward control of phosphate precipitation

Simultaneous precipitation is the name of a procedure for the precipitation of phosphates in wastewater treatment. Iron salts are dosed into the influent to an activated sludge system, which results in the precipitation of phosphates simultaneously with biological treatment. Frequently, the iron salt is dosed proportionally to the phosphate load. This requires measurement of the influent flow rate as well as the phosphate concentration contained in the wastewater. This information is used to control the amount of iron salt dosed. *We speak of a control path* or *feedforward control*, because the success of the precipitation (the effluent concentration) is not used to influence the dosing, thus no closed loop exists.

Similarly temporal control of the iron dosage is also used: during the day with increased and at night with reduced dosage. Here, too, no continuous examination of the success of phosphate precipitation is intended.

Example 13.3: pH control

In an industrial operation, wastewater with a highly variable pH value is produced (range 3–11). The wastewater may, however, only be discharged into the public sewer with a pH value in the range 6–9. This rule is to be fulfilled by dosing acid or base into the waste.

What information is necessary in order to set up a feedback control?

What information is necessary in order to make use of feedforward control?

Using your chemical know-how, what would you select to fulfill the requirement: feedforward or feedback control?

Example 13.4: Maintaining the water pressure in a water distribution network

If in a water distribution network no possibility exists for building a reservoir at sufficient altitude, frequently the capacity of the main pumping station is continuously adapted (by changing its rotation speed) in such a way that the pressure in the net is held within narrow limits despite variable consumption.
Is this a feedforward or a feedback control? What is the variable disturbance of the system, the controlled variable, the actuator?

Example 13.5: A skidding car, an unstable control loop

As a driver of a vehicle you may know the situation that you want to evade an obstacle. In fear you turn the steering wheel too strongly (position), you see the consequence (state), and correct again too strongly (feedback); the vehicle becomes unstable and begins to skid.

13.3 Step Response of a Subsystem

The step response characterizes the effect of a fast change of a variable disturbance (input z, position y, etc.) on the output x of a subsystem.

If we analyze an automatic control loop in more detail, the picture in Fig. 13.5 develops. The measuring system converts the signal of the sensor into an electric signal, which the automatic controller can process. Depending upon the kind of the measurement this can introduce delays or inaccuracies (further variable disturbances) into the control loop: if we measure, e. g., the concentration of the dissolved oxygen with the help of an electrode, then it takes 15–60 s until the electrode emits a reliable, fully equilibrated electrical signal; if chemical analysis are necessary (e. g., the determination of the phosphate concentration), then delays of over 15 min may develop. A change of the correcting variable y takes effect only after the delay required by the actuator to actually reach the new required position, i. e., until a slide-gate position is adapted or a motor speed has really changed.

Thus, in an automatic control loop several individual subsystems (so-called transfer elements) work in series, all with their own characteristic behavior and

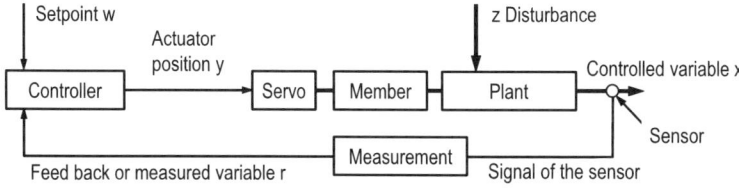

Fig. 13.5 Break down of an automatic control loop into subsystems

generated delays between input and output. The behavior of these individual elements is frequently dependent on their mechanical, physical, and reaction-kinetic boundary conditions. In an automatic control loop, it is primarily the controller itself (the element that derives the correcting value y based on an offset e) which offers degrees of freedom in the choice of its characteristics. A goal of control engineering is to use these degrees of freedom in such a way that the entire system will behave as well as possible in the way we want.

In the hydraulic characterization of reactors with the aid of residence time distributions, we determined the cumulative residence time distribution $F(\tau)$ as an answer for the effluent concentration (state variable x) to a step change of the inlet concentration (variable disturbance z). We may speak of a step response of the reactor that indicates the temporal connection between a disturbance in the influent and an answer in the effluent. This step response supports us in the analysis and understanding of the hydraulic behavior of reactors. Similarly, we can characterize the time dependence of the subsystems identified above with the aid of a step response:

- The connection between the input and output of a transfer element (reactor, measuring system, control member, controller, etc.) can be characterized by its *step response* and its *transient response* (Fig. 13.6).

Here we discuss only transfer elements, whose step response approaches a steady state, i. e., it reaches a fixed value after some time. There are transfer elements that do not reach a steady state such as reservoirs: a jump in the influent leads to a faster filling of the reservoir, which reaches, however, no steady state (here

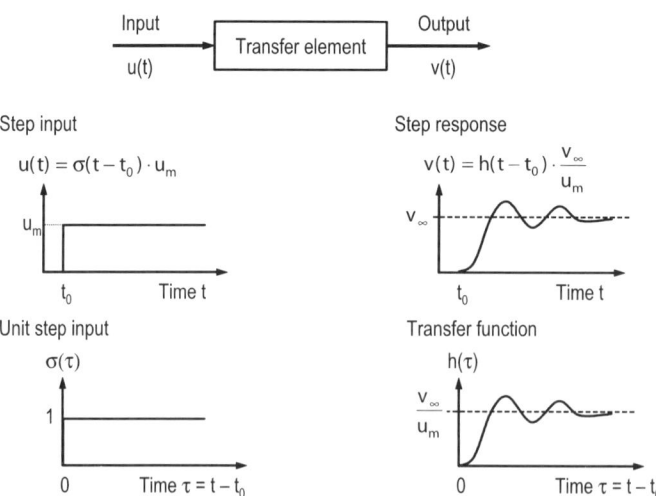

Fig. 13.6 *Above*: Input of a jump u(t) of amount u_m and the output of a step response v(t) with the equilibrium value v_∞ (after a long time) for a non integrating transfer element. *Below*: standardized unit jump $\sigma(\tau)$ (dimensionless) and transient response $h(\tau)$ with the dimensions of $[h] = [v]/[u]$

we could not determine a residence time distribution, because the system is not in a hydraulic steady state and the response would be time dependent).

Figure 13.6 shows an example of a step response that initially oscillates but then approaches a steady state. This behavior is possible in systems with feedback. If we expose a transfer element to a unit step $\sigma(\tau)$ (a dimensionless step of size unity) the step response is called the *transient response h(τ)*. In this case:

$$h(\tau) = v(t - t_0) / u_m \text{ whereby the units of } [h] = [v] / [u]. \tag{13.1}$$

For very small jumps, we can linearize the behavior of the transfer element. A linear relationship then applies between the variable size u_m of the jump and the associated step response v_∞, i. e., the transient response $h(\tau)$ is independent of the (small) size of the jump. One then has

$$v(t - t_0) = u_m \cdot h(\tau). \tag{13.2}$$

Example 13.6: Cumulative distribution of the residence time as transient response

The cumulative distribution of the hydraulic residence time $F(\tau)$ corresponds to the step response of a system to a step change of a nonreactive material in the influent $u(t)$ (tracer, disturbance) which is then made dimensionless. Thus, we characterize the hydraulic behavior of a reactor (a transfer element) or, better, the behavior of an inert tracer subject to the hydraulic transport processes in the reactor. The output signal $v(t)$ corresponds to the concentration of the tracer in the effluent of the reactor. Here u and v have the same units, and after a long time v reaches the value of u_m or $v_\infty = u_m$. The maximum value of h becomes $v_\infty/u_m = 1$ (dimensionless). The transient response $h(\tau)$ becomes identical to the cumulative distribution of the residence time of the water $F(\tau)$ (see Sect. 7.3.2).

The step response of a nonoscillatory transfer element is shown in Fig. 13.7. Contrary to reactor hydraulics, where we interpret residence time distributions as probability density functions, step responses in control technology are characterized by their dead time T_t, equivalent dead time T_u, and transitory period T_g:

- the time interval between the step of the input $u(t)$ and the beginning of the response in the output $v(t)$ is characterized as the *dead time* T_t.
- the *equivalent dead time* T_u and the *transitory period* T_g characterize the adaptation of the system to the new situation. They are empirically defined as illustrated in Fig. 13.7 and require that we identify the inflection point W of the step response.

The transition from the dead time T_t to the equivalent dead time T_u cannot normally be identified very accurately. The two times are frequently added up to calculate the dead time or delay time.

Example 13.7: Delay time and transition period in ideal reactors without reaction

The transient response for an inert material (a tracer) corresponds to the cumulative hydraulic residence time distribution of $F(\tau)$ (Example 13.6). Thus, if we have

Fig. 13.7 Sketch for the definition of the dead time T_t, the
equivalent dead time T_u and the
transitory period T_g (also build-
up time) of a transfer element.
W represents the inflection point
of a nonoscillating transient
response $h(\tau)$. The dead time T_t
designates the time period between the step of the input and
the beginning of the change of
the output

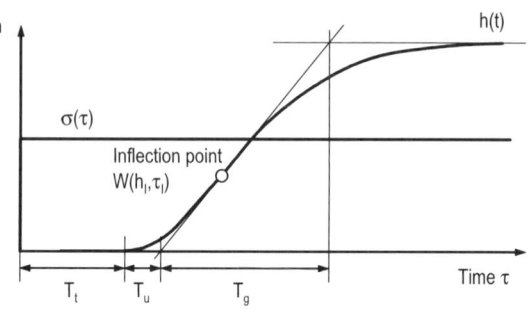

an analytic function available for $F(\tau)$ we can derive analytical equations for T_u
and T_g, see Sect. 7.4.2.

For a cascade of stirred tank reactors with n equal reactors Eqs. (7.27) and (7.28)
provide the location and gradient of the tangent in the inflection point:

$$\tau_W = \tau(f_{max}) = \theta_h \cdot \frac{n-1}{n} ,$$

$$\left.\frac{dh}{dt}\right|_{\tau_W} = f_{max} = \frac{n}{\theta_h} \cdot \frac{(n-1)^{n-1}}{(n-1)!} \cdot \exp(1-n) ,$$

which results in $T_g = \left(f_{max}\right)^{-1} = \frac{\theta_h}{n} \cdot \frac{(n-1)!}{(n-1)^{n-1}} \cdot \exp(n-1)$.

There is also an analytic solution for T_u. However, it has the form of a sum over
many expressions and it is advantageous to obtain the result numerically, e. g.,
with BM.

Example 13.8: Residence time distribution and transient response

For a cascade of six equal stirred tank reactors in series, the residence time distribution is obtained. The mean hydraulic residence time of the cascade amounts to 6 h.
*What are the dead time, equivalent dead time, and the transition time of this
reactor?*
To answer this question you should use your know-how about the residence time
distributions.
*What are T_t, T_u and T_g for a stirred tank reactor, a plug-flow reactor, and a plug-
flow reactor with dispersion?*

Example 13.9: Step response of a shower to a change of the temperature setpoint

*How does a shower react to a fast change of the temperature setpoint in the mix-
ing tap?*
First, there will be some dead time (transport delay) during which the contents of
the showering hose is replaced with new water. Then, there will be an adaptation

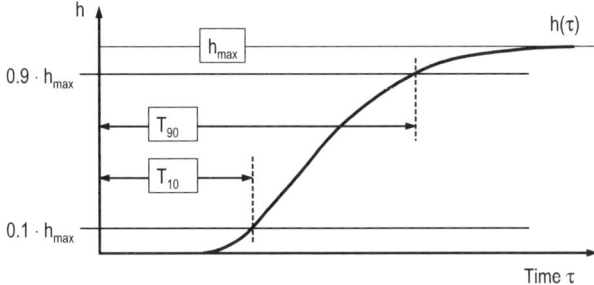

Fig. 13.8 Definition of T_{10} and T_{90} rather than dead time, equivalent dead time, and transitory period

period until all lines are warmed up and the new temperature is constant (in equilibrium). We characterize this adaptation time with an equivalent dead time and a transitory period.

Example 13.10: T_{10} and T_{90} as alternative characteristic times of a transient response

Today a transient response is increasingly characterized with the times T_{10} and T_{90}. T_{10} indicates how much time elapses after the step until the step response stably reaches at least 10% of the maximum step response. T_{90} refers to 90% of the maximum answer (see Fig. 13.8)

Determination of Transfer Functions

Transient responses can be determined experimentally if a system is disturbed with a step function, as we have already discussed in the context of the direct determination of the cumulative hydraulic residence time distribution of a reactor system. However, since in automatic control signals with different units are used, $h(\tau)$ may not be dimensionless, unlike $F(\tau)$.

If experiments can endanger the plant (e.g., a nuclear power plant) or, if the system has not been built yet, such experiments cannot be carried out. In such cases the system must be modeled in detail and the appropriate experiments will be accomplished as simulations only.

Example 13.11: Transient response of a control member or a whole plant

If the transient response of a pump that is controlled by variable revolutions (speed) is to be determined, then the control signal of the measured flow (e. g., a current or a voltage) can be raised artificially from a low to a high level by disconnecting the line and introducing an electronic device. The measured response of the flow through the pump then allows the transient response of this subsystem to be derived.

The transient response of a whole plant can be obtained by disturbing the entire plant and then following the transient response, e. g., in the effluent. An example is the induction of a step change of the ammonium concentration in the influent to a nitrifying wastewater treatment plant by constantly dosing a concentrated ammonium solution. The response of the ammonium concentration in the effluent allows the transient response curve of the entire plant between the location of ammonium addition and the point of measurement in the effluent to be derived.

13.4 Step Response of a Controlled System

In order to understand the characteristics of a controlled system, it frequently makes sense to obtain the step response of the controlled variable x for a step change of the correcting variable y. We differentiate between transfer elements without delay and transfer elements with delay of first and higher order.

Here only controlled systems are discussed for which a fixed disturbance z or y asymptotically results in a constant controlled variable x (systems that can reach a steady state).

13.4.1 Controlled Systems Without Delay

In transfer elements without delay the controlled variable x reacts immediately to a change of the correcting variable y. While controlled systems (the systems and reactors that interest us) rarely work without delay, many measuring systems are close to being free of delay (e. g., conductivity electrodes, flow measurements, and optical sensors all react nearly instantaneously to changes). Figure 13.9 shows the transient response for a delay-free, ideal transfer element. *We speak of a transfer element of zero order.*

While we can very easily realize transfer elements without delay in a simulation program, real transfer elements always react with a delay. Whether a delay must be simulated depends primarily on the time constants (the temporal behavior) of the real system. Delay-free controlled systems do not have a tendency to become instable (to swing), whereas strongly retarding transfer elements tend to induce oscillations and might even become instable.

Example 13.12: Transfer elements with small delay

If we adjust the opening of a water tap (y), then the flow rate (x) reacts nearly immediately to the change. The transfer element is nearly of zero order.
If we want to adjust a certain water flow at the tap, then this procedure can start to swing: we open the tap too strongly, correct too strongly, must back correct, etc. Here the entire system is of concern, the control loop it no longer free of delay. Our perception of the undesired result (too much flow) and our answer are slow in

Fig. 13.9 The zero-order controlled system does not have any delay; the step response reflects this

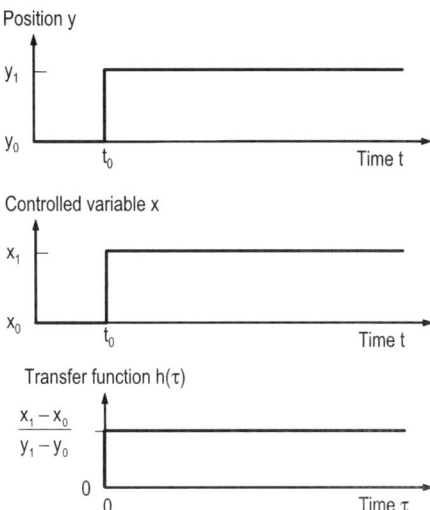

comparison to the answer of the flow rate to a change of the position of the tap. The measuring system and the controller are thus not delay free, and the system tends to swing.

We can only stabilize the water flow if we carefully observe the flow over time and derive from this process a suitable position of the tap. As humans we learn this quickly, whereas technical automated systems use time-dependent controllers which we will introduce later (Sect. 13.6.2).

13.4.2 Controlled Systems with Delay

Automatic control theory was primarily developed to deal with electric and electronic circuits. Here condensers represent storage elements, which lead to delays. In urban water management we frequently deal with water-filled reactors that are storage elements for individual materials (accumulation in material balances).

If the controlled system contains any reservoirs which must be filled by the controlled variable x, then this variable reacts only with a delay to a change of the environment (the position y or load z). The temporal connection between input and output or x and y determines the order of the delay (Fig. 13.10): an individual stirred tank reactor (or condenser) represents a delay of first order, cascades of n equal stirred tank reactors represent for inert materials delays of nth order. The higher the order, the more difficult the design of a controller becomes. Figure 13.11 shows step responses of different order. The dead time T_t has an effect that is comparable to a high-order delay: plug-flow reactors and dead times (Fig. 13.7) make control more difficult.

Fig. 13.10 Step responses of different T members

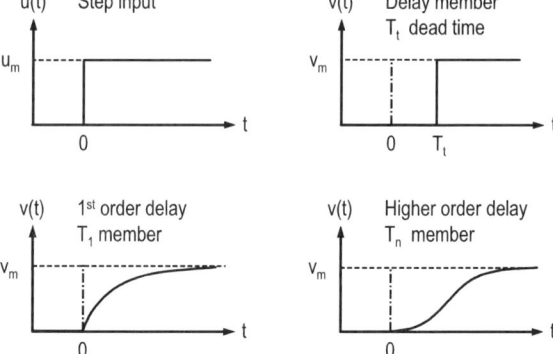

Fig. 13.11 Step responses with increasing order (in analogy to the cumulative residence time distribution of a cascade of stirred tank reactors)

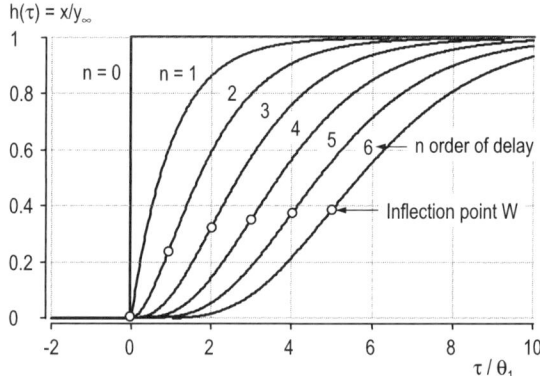

The step response of the temperature of the water in the shower (Sect. 13.1.1) contains both a dead time T_t (the residence time of the water in the hose) and a response of higher order (the slow adjustment of the temperature, because the plant must adapt, be heated, to the new temperature).

In Table 13.1 the characteristic times for the step responses of an inert tracer are indicated for a cascade of stirred tank reactors. Table 13.2 shows, based on experience, that the control of a loop becomes ever more difficult as the order of the delay increases (increasing dead time T_t).

Example 13.13: Controllability of a system

An activated sludge tank consists of a cascade of three equal stirred tank reactors which have, including the return sludge, a hydraulic residence time of 1 h each. In the third basin, the ammonium concentration is measured with the help of an online analyzer. If the ammonium concentration in the effluent (third basin) is low, the aeration in the first basin is to be stopped to obtain some denitrification. The ammonium concentration is available with a delay of 15 min (dead time, reaction time), and the aeration is to be restarted with at least 5 min delay after a rise of the ammonium concentration is observed in the effluent (which reduces overheating of the motors).

Table 13.1 Transition time T_g, equivalent dead time T_u, inflection point T_W, T_{10}, and T_{90} of a cascade of n equal stirred tank reactors with an average delay of θ_{tot} (hydraulic residence time), see Figs. 13.7, 13.8 and 13.11) (Mann et al., 2000, extended). $n = \infty$ corresponds to a plug-flow reactor or real dead time T_t

n	T_g/θ_{tot}	T_u/θ_{tot}	T_g/T_u	T_W/θ_{tot}	T_{10}/θ_{tot}	T_{90}/θ_{tot}	T_{90}/T_{10}	Stability of control
1	1.00	0.00	∞	0.00	0.11	2.30	21.90	very good
2	1.36	0.14	9.65	0.50	0.27	1.94	7.32	good
3	1.23	0.27	4.58	0.67	0.37	1.77	4.83	fair
4	1.12	0.36	3.13	0.75	0.44	1.67	3.83	
5	1.02	0.42	2.44	0.80	0.49	1.60	3.29	poor
6	0.95	0.47	2.03	0.83	0.53	1.55	2.94	
7	0.89	0.51	1.75	0.86	0.56	1.50	2.70	
8	0.84	0.54	1.56	0.88	0.58	1.47	2.53	
9	0.80	0.56	1.41	0.89	0.60	1.44	2.39	
10	0.76	0.59	1.29	0.90	0.62	1.42	2.28	very poor
∞	0.00	1.00	0.00	1.00	1.00	1.00	1.00	

Table 13.2 Controllability of transfer elements, empirical values (Mann et al., 2000)

$T_g/(T_u + T_t)$	<1.2	1.2–2.5	2.5–5	5–10	>10
Controllability	very poor	poor	fair	good	very good

Is this a difficult task to fulfill?

The controlled system is approximately third order, therefore from Table 13.1 we obtain

$\theta_{tot} = 3$ h, $n = 3$, $T_g/\theta_{tot} = 1.23$, $T_u/\theta_{tot} = 0.27$

$T_g = 3.69$ h, $T_u = 0.81$ h, the total dead time amounts to $T_t = 0.33$ h (15 + 5 min).

$T_g/(T_u + T_t) = 3.24$. According to Table 13.2, the controllability of this system is only fair. It will not be possible to maintain the effluent concentration very close to the setpoint. A shortening of the dead times would be of advantage.

Example 13.14: Modeling of a delay

In the following BM code the input signal V_{in} (here generated with an autoregressive model, AR(1)) is retarded by a dead time T_t and a cascade of sixth order.

```
n=6                       ; Order of the delay >= 1
THtot=0.5                 ; average delay
Tt=0.25                   ; Dead time
TH=(THtot-Tt)/n           ; Delay for one order
init Vin=2.5              ; Input signal: the example is an AR(1) model
next Vin=Vin*0.8+random(0,1)
init V[1..n]=Vin          ; Delay of the signal in nth order
d/dt(V[1..n])=if i=1 then (Vin-V[i])/TH else (V[i-1]-V[i])/TH
Vout=delay(V[n],Tt)       ; Addition of the dead time, retarded signal
```

Input and delayed output signal

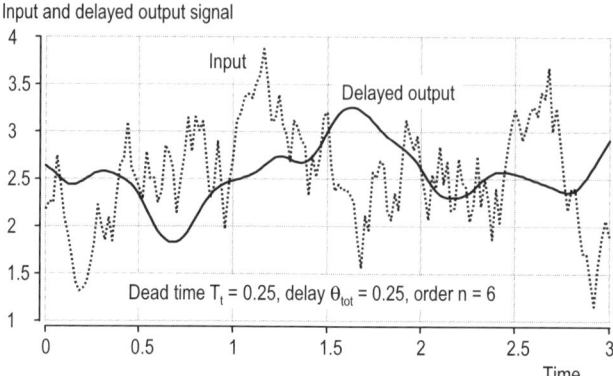

Fig. 13.12 Example of a retarded signal, dead time $T_t = 0.25$, $\theta_{tot} = 0.25$, order of the delay $n = 6$

Figure 13.12 compares the input and the attenuated, retarded output signal. In this case $T_g/(T_u + T_t) = 0.65$. According to Table 13.2 this results in a situation that is very difficult to control. The trend of nearly opposite directions between the signal (entrance) and retarded exit supports this analysis; there is hardly any synchronous information from the influent left in the effluent.

13.4.3 Controlled Systems with Dead Time

Measuring systems, the response times of control members, and flow distances with the character of plug-flow reactors lead to longer dead times and make the control of systems more difficult. In water technology, start up of motors, chemical analyses, and long transportation lines are of particular importance.

Example 13.15: The measuring system of BOD_5

If we want to use BOD_5 as a measured variable x for the control of a wastewater treatment plant, the measuring system leads to a time delay (dead time) between a change of the controlled variable x and the answer of the measured variable r of at least 5 days (until the results are available). Obviously a plant cannot be controlled (with neither feedforward nor feedback) with the control variable BOD_5. The diurnal variation requires a much shorter measuring cycle.

Example 13.16: Dead times of electric motors

Electric motors cause high current peaks when starting up and are thereby warmed up. Both effects are undesired. In order to control these phenomena, the switching frequency of these motors is limited, e.g., to only one startup in any 10 min period. This leads to additional dead time.

With modern power electronics, the frequency of alternating current can be controlled. This permits the speed of the motors to be adapted to the momentary requirements. Thus, full stop–start up cycles are abolished, and extra dead times are no longer necessary.

13.5 Characteristic Curves of a Controlled System

The characteristic curve of a controlled system shows graphically the dependence of the constant controlled variable x (output) on the constant correcting variable y (input). It refers to the steady state and can be represented for different constant disturbances z as a characteristic diagram.

If we compute or observe the relationship between load z, position y, and controlled variable x for the steady state of a system, then a picture develops, as presented in Fig. 13.13: we obtain a characteristic line for a certain disturbance (load) and altogether a characteristic diagram of the controlled system.

In Fig. 13.13, we can define an operating point of the automatic control loop with x_0, y_0, and z_0. If we only analyze the close surroundings of this operating point, we can linearize the characteristic behavior and thereby simplify the mathematical analysis. The control range indicates the range of the controlled variable x that can be measured reliably and within which range the correcting variable y can physically be changed (from the minimum to maximum possible position).

Example 13.17: Computation of a characteristic diagram

You are to compute the characteristic diagram for the following system:
An anaerobic digester is fed with $50–100 \, m^3 \, d^{-1}$ raw sludge from the primary clarification of a treatment plant. Its temperature varies during the year in the range 10–20°C. A heating system with an output of $1 \cdot 10^6–3 \cdot 10^6 \, kcal \, d^{-1}$ is available. The heating power is controlled as a function of the temperature in the digester. The desired value is in the range 30–37°C. Heat of 1000 kcal can warm $1 \, m^3$ of anaerobic sludge by 1°C. You may neglect possible heat losses as a consequence of radiation.

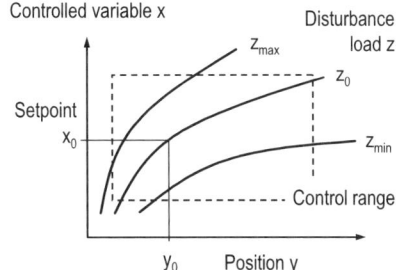

Fig. 13.13 Characteristic diagram of a controlled system: in order to reach the desired steady state with the load z_0 at the setpoint x_0, the position y_0 must be selected

The characteristics have two parameters: reduce them into a single load variable. Draw the characteristic diagram for the maximum and the minimum load. Do you have a suggestion for a better signal to be used for this control? How would you design an open-loop control path?

Example 13.18: Characteristic diagram for an aeration system

Draw the characteristic diagram for the aeration system of an activated sludge tank. The air input is to be controlled in such a way that, despite variable oxygen requirements, a constant oxygen concentration is maintained.

13.6 The Standard Automatic Controller

Automatic controllers are built into control loops to convert our conceptional control strategies for plants into application. Thereby automatic controllers should not lead to new problems (instability) and they should be able to follow the setpoints reliably and with small deviations.

Controllers have the task of constantly comparing the controlled variable with the setpoint, which may be time dependent and which is given from the outside. If an offset (a deviation between the setpoint and controlled variable) develops, the controller must supply a correcting variable $y(t)$ that is suitable to eliminate or at least decrease this deviation. In the standard automatic control loop (Fig. 13.14) the controller supplies a correcting variable $y(t)$ due to the development of the offset or error signal $e(t) = w(t) - x(t)$.

We differentiate between continuous and discontinuous automatic controllers. In a *continuous controller* the correcting variable $y(t)$ can take any value within a reasonable range, whereas a *discontinuous controller* permits only certain fixed positions (e. g., on/off or minimum/medium/maximum).

Digital signals are not continuous. This short introduction to automatic control cannot deal with the specific problems of digital automatic controllers. As a good first approximation, digital controllers (e. g., those implemented in process control systems) are similar to the analog controllers discussed here.

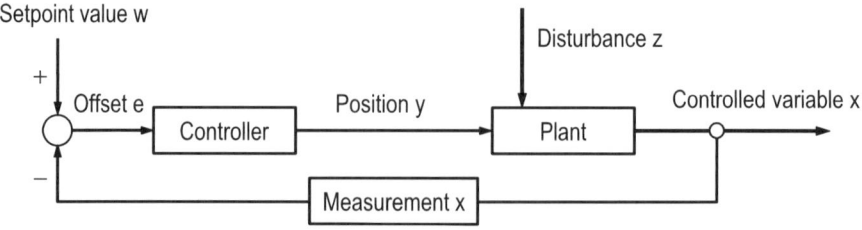

Fig. 13.14 The standard automatic control loop

13.6.1 The Two-Position Controller (A Discontinuous Controller)

The two-position controller is a discontinuous automatic controller which has only two positions, ON or OFF (or HIGH and LOW). In order to limit the number of switching processes, they are usually operated with a differential gap Δx that leads to a characteristic hysteresis curve.

A characteristic curve of a two-position controller is shown in Fig. 13.15. If the controlled variable x exceeds the setpoint x_S by half the differential gap Δx, then the correcting variable y is shifted up (or down), whereas if the controlled variable falls more than $\Delta x/2$ below the setpoint the controller switches to low (or high). Figure 13.16 shows the course of the position y(t) and the controlled variable x(t). Due to delays in the measuring or other subsystems some overshooting of the differential gap may occur. The same may happen if the new position is not sufficient to correct the controlled variable back towards the setpoint.

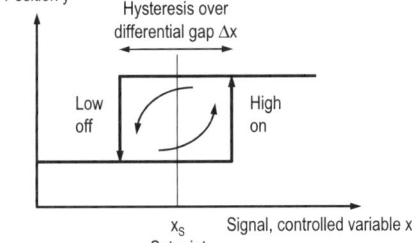

Fig. 13.15 Characteristic curve of a two-position controller with differential gap Δx

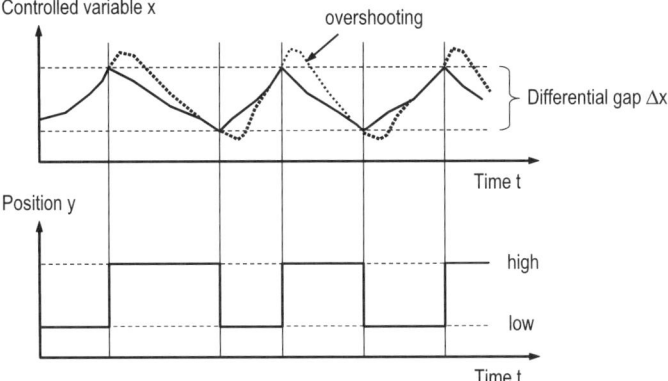

Fig. 13.16 Time course of the controlled variable x and the position y of a two-position controller. The controlled variable can overshoot due to dead times and delays (*dotted lines*)

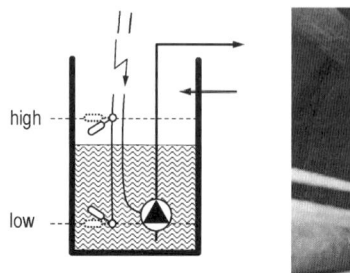

Fig. 13.17 *Left*: pump with on/off control by upper and lower floats. *Right*: example of a pump well with two pumps

---

Example 13.19: Two-position controller in a sewage pumping station

A typical application of a two-position controller is the operation of a pump. In a pumped well the water level is measured (controlled variable x). If the water level exceeds an upper limit, then the pump is switched on (position y). If the water level falls below a lower limit, then the pump is switched off. The volume of the pump well (switching gap Δx) is designed in such a way that the switching frequency does not become too large (see Fig. 13.17).

Example 13.20: Design of a two-position controller

Due to diurnal variations the activated sludge in an activated sludge tank uses between 150 and 350 gO_2 m^{-3} d^{-1} (r_{O2}). Sketch a two-position controller that maintains the oxygen concentration S_{O2} in a completely mixed activated sludge tank between 1.5 and 2.5 g_{O2} m^{-3}.
The oxygen input is controlled by the $k_l a$ value of the aeration equipment and is:
$r_{Aer} = k_l a \cdot (S_{O2,sat} - S_{O2})$ with $S_{O2,sat} = 10$ gO_2 m^{-3}.
How large is the minimal and the maximal required $k_l a$ value?
What is the signal of the controlled variable? When shall the controller switch to what position?
What is the switching interval if the oxygen requirement amounts to $r_{O2} = 200$ gO_2 m^{-3} d^{-1} (you may neglect the flow through the reactor)? How can you extend the switching interval?
Integration of the mass balance equation leads to an exponential function. As a simplification you can fix the oxygen input with the value at the setpoint (average oxygen concentration).

Example 13.21: Implementation of a two-position controller in BM

The following code implements a two-position controller with retarded signal in BM:

```
{Delayed two-position controller, parameters (tested)}
x = S                    ; Replace the controlled variable, here S
```

kla = y   ; Replace the resulting position, here kla
n = 1   ; Order of the delay, must be >0
T = 0.002   ; Delay time
Tt = 0   ; Additional dead time
y_min = 10   ; Minimum position
y_max = 100   ; Maximum position
x_min = 1.5   ; Lower setpoint
x_max = 2.5   ; Upper setpoint

{Two-position controller}
init Sig[1..n] = 0   ; Delay of the controlled variable
d/dt(Sig[1..n]) = if i = 1 then (delay(x,Tt) − Sig[1])*n/T else (Sig[i − 1] − Sig[i])*n/T
Signal = Sig[n]   ; Delayed controlled variable
init y = y_min
next y = if Signal > x_max then y_min else if Signal < x_min then y_max else y
                                 ; Control equation, possibly must exchange y_max and
                                 y_min

13.6.2 Continuous Automatic Controllers

*With continuous automatic controllers, the position y can take any value within
the permissible range; this frequently allows the controlled variable x to be
bought quite close to the setpoint. The standard automatic controller is the so-
called PID controller, which combines the instantaneous condition (P member)
with the past history (I member), and the possible future development (D member)
into one single control rule.*

If either the disturbance variable z (the load) or the setpoint w is changed, this has
an effect on the controlled variable x and the error signal e. The automatic controller must now adjust the position y in such a way that the error signal disappears.
The continuous standard automatic controller is the PID controller, which is subject to the following functions:

$$\Delta y = y - y_0 = \underbrace{K_P \cdot e}_{P-member} + \underbrace{K_I \cdot \int e \cdot dt}_{I-member} + \underbrace{K_D \cdot \frac{de}{dt}}_{D-member} \qquad (13.3)$$

Δy = deviation of the position from the expected value y_0
y = new position
y_0 = expected value for the position y, if e = 0
e = error signal (e = w − x)
x = controlled variable
w = setpoint = reference value
K_P = transfer coefficient (gain) for the proportional member
K_I = transfer coefficient for the integral member
K_D = transfer coefficient for the differential member

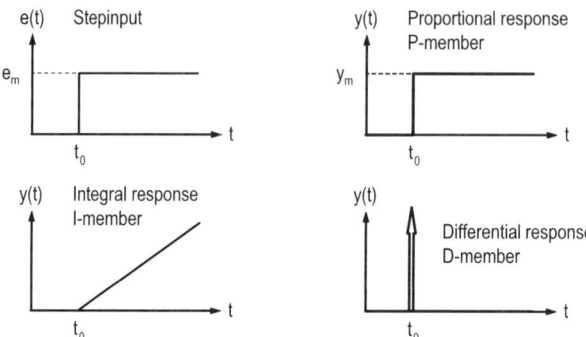

Fig. 13.18   Step responses of the three members of the PID controller

Often not the entire PID controller is used, but depending upon its task one or two members are implemented, e. g., a P controller with $K_P \neq 0$ and K_I, $K_D = 0$ or a PI controller with $K_D = 0$.

Figure 13.18 shows the step responses of the three members of a PID controller:

- The P member reacts immediately and proportionally to a change of the offset e by adjusting the position and then keeping it.
- The I member reacts by increasing the adjustment of the position as long as the offset e remains.
- The D member gives a unique, large impulse (Dirac pulse) to the position whenever the offset e changes. This member alone is not suitable as an automatic controller by itself; in combination with a P and/or an I member it may, however, improve the characteristics of the controller.

Proportional Controller

Proportional action controllers (P controller) are simple to implement, particularly in simulation programs. They have the disadvantage that a permanent offset e remains and that they tend to become unstable if the controlled systems contain delays of higher order or dead times.

Proportional controllers use only the P member of the standard PID controller Eq. (13.3), thus, the following applies:

$$y - y_0 = K_P \cdot (w - x) = K_P \cdot e . \tag{13.4}$$

Figure 13.19 shows the characteristic curve of a P controller. The position y is usually limited for physical reasons to a certain range (y_{min} to y_{max}). The term *gain* applies to the gradient K_P of the control characteristic.

Fig. 13.19 Characteristic curve of a P controller. Because the offset e is defined as $c = w - y$, the transfer coefficient K_P has a minus sign, if the controlled variable x is used instead of e as a coordinate

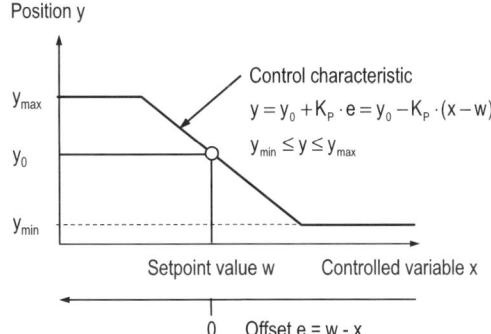

Permanent offset. The correcting variable y must be adapted to the load z. The proportional controller allows for different action (different positions y) only if the offset e changes (Fig. 13.19). Thus a change of the load leads to a permanent change of the offset. The proportional action controller cannot maintain a zero offset e with variable load; it serves its purpose only if a deviation of the controlled variable from the setpoint is permissible. Even in the steady state there will be a permanent offset that becomes ever smaller with an increasing gain or transfer coefficient K_P. However, a large gain enhances the danger of instabilities.

Instability of a P controller. The larger the gain K_P the larger the adjustment Δy of the position y will be when the offset e changes. A large change of y, however, leads to a large and rapid change of the controlled variable x, which will again affect the automatic controller (positive feedback). With delayed signals, the control loop may begin to swing and then become unstable. If we reduce K_P, the permanent offset becomes larger, but the automatic controller has the tendency to stabilize. In the design of an automatic P controller one can, e.g., in an experiment or in a simulation, try to find the critical gain $K_{P,crit}$ that begins to make the system unstable. In operation one would then adjust to $K_P = 0.5 \cdot K_{P,crit}$ to avoid instability.

Figure 13.20 shows a simple experiment with an automatic P controller. The goal is to dose a chemical to the influent of a reactor composed of a cascade of five stirred tank reactors such that the effluent contains a residual concentration $S_{C,5}$ close to the setpoint of $w = 1$ g m^{-3}. The experiment is disturbed by a variable influent in the sequence 20, 40, and 16 m^3 h^{-1}. The results of the simulation show:

- Between 0 and 4 h the plant swings slowly towards a steady state. A permanent offset remains with approximately $e = -0.4$ g m^{-3} of the chemicals in the effluent;
- Between 4 and 7 h with high influent, the plant rapidly reaches a new steady state with a small permanent offset of $e = -0.1$ g m^{-3}. For this situation, the automatic controller appears to be suitable;
- Starting from 7 h with a small influent and accordingly a large residence time in the system, the automatic control loop begins to swing with increasing amplitude and finally becomes unstable.

Fig. 13.20 *Above*: a simple plant with a P automatic controller: a cascade with five equal stirred tank reactors. Chemicals are dosed to the first reactor at a high concentration so that the setpoint of w = 1 g m^{-3} is reached in the last reactor. *Below*: the results of a simulation with variable influent (V = 5 × 1 m^3, Q = 16–40 m3 h^{-1}, Q$_C$ = 0.05 + 0.05·(w − S$_{C,5}$), S$_C$ = 1000 g m^3)

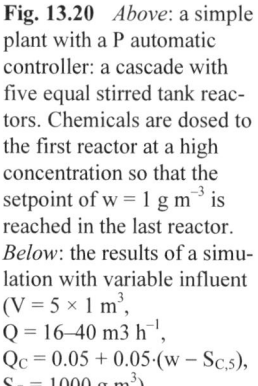
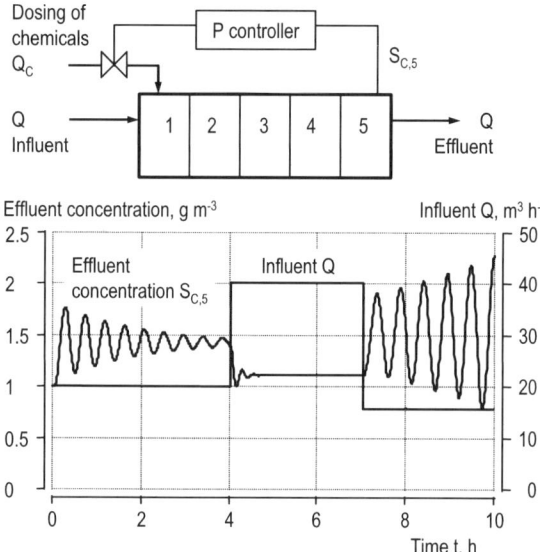

In this control loop the gain (transfer coefficient) K$_P$ must be reduced in order to avoid instabilities. This will, however, result in a larger permanent offset of the controlled variable. Alternatively an improved automatic controller (e. g., a PI controller) can be used.

Example 13.22: Implementation of a simple P automatic controller in BM

The following code implements a P automatic controller without delay

```
KP=0.05              ; Transfer coefficient, gain
y_0=0.05             ; Resting position, expected value of the position
y_min=0              ; Minimum value of the position
y_max=0.2            ; Maximum value of the position
W=1                  ; Setpoint
x=S[n]               ; Add signal in place of S[n]
QC=y                 ; Add position of controlled member in place of QC
y=y_0+KP*(w−x)       ; Control equation, w−x=e=error signal
limit y<=y_max       ; Keep the range of possible positions
limit y>=y_min
```

Example 13.23: Design of a proportional controller

The oxygen input OE into a batch reactor is to be controlled such that the oxygen concentration amounts to approximately $5 \, gO_2 \, m^{-3}$. The value of OE can vary between 100 and $300 \, gO_2 \, m^{-3} \, d^{-1}$, and the oxygen concentration is not to differ more than $1 \, gO_2 \, m^{-3}$ (permanent offset) from the setpoint.

It is good practice to deduce the parameters of the controller based on a characteristic control diagram:

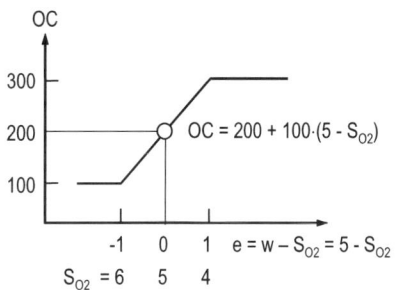

OE_{min} and OE_{max} are given, as is the range of the controlled variable (S_{O2}). The gradient of the straight line (the transfer coefficient K_P, gain) results from the selected extreme points of the control characteristic.

This control characteristic does not provide any information on the stability of the system. This aspect depends on the properties of the system, especially the delays within the entire automatic control loop.

Example 13.24: Permanent offset

In a batch reactor, microorganisms consume $r_{O2} = 150 - 250\,gO_2\ m^{-3}\ d^{-1}$. The oxygen input OE $[gO_2\ m^{-3}\ d^{-1}]$ compensates the oxygen consumption. It is controlled by an automatic proportional controller, based on the measured oxygen concentration S_{O2} with the following control characteristic (see also Example 13.23):
$OE = 200 + K_P \cdot (w - S_{O2})$ $w = 5\,gO_2\ m^{-3}$.
How large do you have to select K_P, so that the lasting offset remains smaller than $1.0\,gO_2\ m^{-3}$? What sign does the lasting offset have (positive or negative)? How large is the measured oxygen concentration if $r_{O2} = 250\,gO_2\ m^{-3}\ d^{-1}$? How large is the oxygen requirement r_{O2} if a stationary oxygen concentration of $4.8\,gO_2\ m^{-3}$ is measured?

Example 13.25: Delayed automatic P controller

Figure 13.21 shows a mechanical proportional action controller. Influent and effluent are determined by the position y and the water level (controlled variable) x. Assume that $Q_{in} = K_{in} \cdot y$ and $Q_{out} = K_{out} \cdot x$.

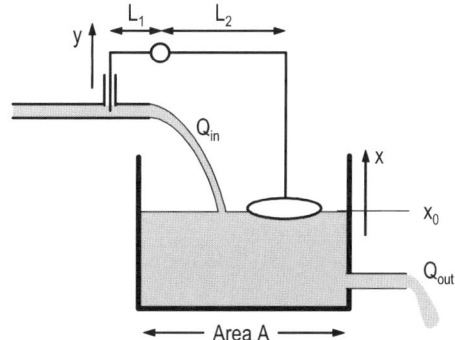

Fig. 13.21   Example of a mechanical automatic P controller

What order does the delay (storage of water) in this automatic control loop have? Write the equation for the control characteristic of the mechanical controller. How does the controlled variable (the water level x) change with time (this may be solved analytically)? What is the permanent offset of the controlled variable if the setpoint is $w = x_0$?

Integral Controller

Integral controllers use the I member of the standard automatic controller; normally this is realized in combination with a P member as a so-called PI controller, in which case Eq. (13.3) reduces to

$$y - y_0 = K_P \cdot (w - x) + K_I \cdot \int (w - x) \cdot dt = K_P \cdot e + K_I \cdot \int e \cdot dt . \qquad (13.5)$$

Figure 13.22 shows the step response of an I and a PI controller. The I controller adapts the adjustment of the position Δy as long as an offset e remains. Since the correction initially starts with 0, this automatic controller reacts only slowly, but it can eliminate the offset completely. The PI controller includes a P member, which accelerates the adjustment because now the correction does not begin at zero anymore, but includes according to the characteristics of the P member an initial step change. This results in an acceleration of the reaction time of the order of the so-called *reset time* $T_I = K_P / K_I$.

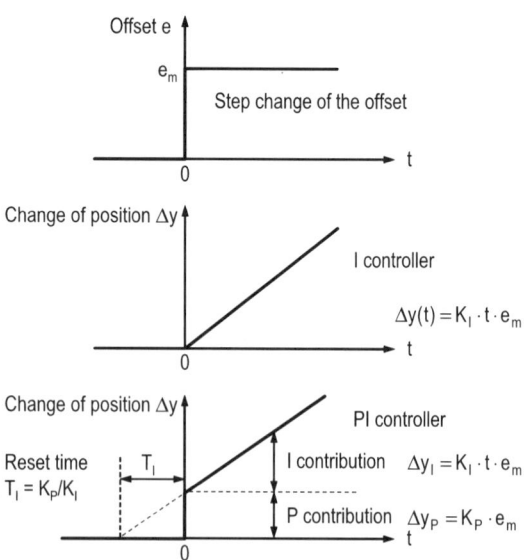

Fig. 13.22 Step response of an I and a PI controller. Definition of the reset time T_I

Example 13.26: An integrating, self adjusting system

In the system shown below, a constant quantity of water Q_{in} flows into a reservoir. The effluent quantity Q_{out} is independent of Q_{in}, so only the opening and the pressure h determine the effluent.

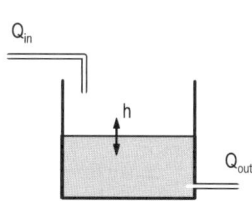

The water level will adapt until the difference between the influent and effluent disappears. The volume of the reservoir integrates over the offset $(Q_{in} - Q_{out})$ and changes the position of the water level, until the offset disappears. Thereby the surface of the reservoir plays the role of the (inverse) transfer coefficient K_I: the larger the surface, the slower the reaction becomes.

By partially filling the system at the beginning of the experiment (initial condition), a faster adjustment of the effluent to the control value results (equilibrium). That corresponds to a P member in the control procedure and can be described with a reset time T_I.

Case Example of a PI Controller (Simulated)

In a batch reactor, microorganisms (activated sludge) use oxygen according to Fig. 13.23a, after 0.5 d an additional pollutant is added, which causes an immediate increase of the oxygen consumption. The oxygen supply (here proportional to the value of $k_I a$) is (i) kept constant, (ii) controlled with a P controller, and (iii) controlled with a PI controller each with a setpoint of $w = 2 \, gO_2 \, m^{-3}$.

The resulting oxygen concentration is shown in Fig. 13.23b. With constant aeration results a large variation of the concentration S_{O2}. The P controller can reduce the offset, but a permanent residual error remains which depends on the momentary oxygen consumption. The PI controller reaches the desired value w without permanent offset after some short oscillations.

The aeration intensity (here $k_I a$) is shown in Fig. 13.23c. The constant value was chosen at $k_I a = 40 \, d^{-1}$ (not shown). With the P controller results a linear relationship between the offset e and the position $k_I a$. For the PI controller results a complicated control process which converges in each case on a point with $e = 0$. The control characteristics of the three experiments are shown in Fig. 13.23d. The PI controller uses a whole range of control and finds in the end a position with zero offset.

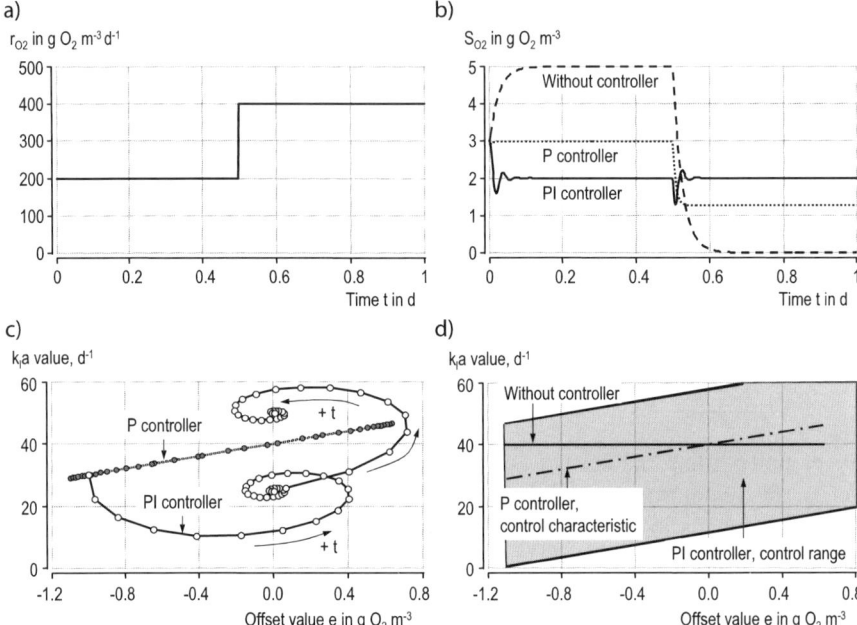

Fig. 13.23 (a) time course of oxygen consumption, (b) time course of the oxygen concentration with and without P or PI controller, (c) course of the k_la value versus offset over time for the P and the PI controller (points have a distance of 0.002 d), (d) control characteristic: a constant k_la value without controller, a straight sloped line for the P controller, and the control range for the PI controller

Differential Controller

Differential controllers make use of the D member of the standard automatic controller. They are not implemented alone, but supplement the characteristics of the P and I automatic controllers. They lead to a fast reaction to a developing offset, because the derivative includes changes as they start to appear. The D member tends to look into the future.

The control equation of a D automatic controller has the form

$$y - y_0 = K_D \cdot \frac{de}{dt} = -K_D \cdot \frac{dx}{dt} . \tag{13.6}$$

Differential automatic controllers alone make little sense. They are independent of the setpoint and consider only changes in the offset but not the amount of the offset. In the steady state, they would thus permanently permit any size of remaining offset, since $dx/dt = 0$.

Fig. 13.24 Reaction of a P and a PD controller to a linear change of the offset (contrary to Fig. 13.22, this is not a step response)

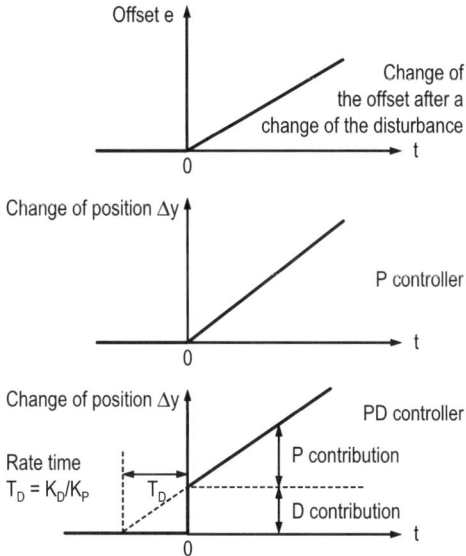

Example 13.27: Driving and controlling a car

An example of a D controller is our behavior if we react in shock to the skidding of our car. We observe a fast change of the driving direction and immediately react with counter steering without waiting for information on a new state of the car, but exclusively due to the information that our car begins to turn. We will most probably not hit the exact correct amount of control – an offset is bound to remain. The danger that the system becomes unstable is very large, in particular because we react in shock with a large transfer coefficient K_D.

Figure 13.24 shows the linear change of the offset with time that results as a consequence of an increase of the load z. With a P controller, the position y would also increase linearly with time (center graph). The PD controller includes a substantial reaction immediately after an offset starts to develop, thus the PD controller is accelerated relative to the P controller by the *rate time* T_D (or *derivative time*). PD controllers adjust faster to disturbances than P controllers; they cannot, however, prevent the permanent offset, which would require an I member.

Example 13.28: Rate time T_D and transfer coefficients

Derive a relationship between the rate time T_D and the two transfer coefficients K_D and K_P. The starting point is the control Eq. (13.3).

13.6.3 Comparison of the Standard Controllers

Figure 13.25 compares qualitatively the step responses of the different standard controllers. The I controller is slow: the controlled variable overshoots strongly and an offset does not remain. The P controller reacts faster, but cannot compensate for the entire offset. The combination in the PI controller can react fast and eliminate the offset. The PID controller reacts fastest; it is the most expensive controller and places the highest requirements on the adjustment and startup (during a startup procedure of a system and an automatic controller, the system frequently reacts extraordinarily dynamical, which causes additional problems). The D member can only approximately be realized; it will always contain some delays.

13.6.4 Implementation of a PID Controller in Berkeley Madonna

The code in Table 13.3 implements a standard PID controller in BM and includes a dead time as well as a delay of higher order. In the model of the system, the controlled variable (here S) must be defined as a time-dependent state variable and the correcting variable (here $k_l a$) as a simple variable (parameter). After the definition of the parameters of the controller, the computation of the delay of the controlled variable follows, with a dead time of T_t and a delay of nth order with a total time T.

The variable controller_on $= 0$ permits the controller to be taken out of operation; the controlled variable will then have the value y_0. Minimizing χ^2 with the help of the optimization routine permits estimates of suitable transfer coefficients to be estimated for the controller (K_P, K_I, and K_D). They may have to be adjusted.

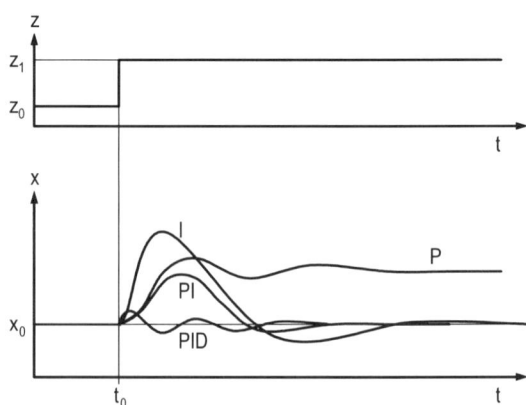

Fig. 13.25 Step response for different automatic controllers after a jump of the variable disturbance

Table 13.3   Code for the implementation of a PID controller in Berkeley Madonna (tested)

```
{PID controller, parameters}
x = S                      ; Substitute time-dependent controlled variable in place of S
kla = y                    ; Substitute controlled member in place of kla
n = 8                      ; Order of the delay, must be >0
T = 0.003                  ; Delay time
Tt = 0.004                 ; Additional dead time
y0 = 30                    ; Expected value of the position y, resting position
ymin = 5                   ; Minimum value of the position y
ymax = 50                  ; Maximum value of the position y
w = 2                      ; Setpoint for x
KP = 25                    ; Gain for the P controller
KI = 2500                  ; Gain for the I controller
KD = 0.0                   ; Gain for the D controller
Controller_on = 1          ; Switching the controller on/off, 1 = on, 0 = off
{PID controller, delay and control equation}
init Signal[1..n] = w      ; Delay for controlled variable y, consideration of the dead
                             time Tt
Signal'[1..n] = if i = 1   then (delay(x,Tt) – Signal[1])*n/T
                           else (Signal[i – 1] – Signal[i])*n/T
e = w – Signal[n]          ; Offset, delayed
init Int_e = 0             ; Integral for I controller, only, if controller is active
d/dt(Int_e) = if (y < ymax and y > ymin) then e else 0
y = y0 + Controller_on*(KP*e + KI*Int_e – KD*Signal'[n])     ; Control equation
limit y >= ymin            ; Keeping lower limit of the position
limit y <= ymax            ; Keeping upper limit of the position
{PID controller, optimization of gains}
init chi2 = 0              ; χ² can be minimized for the estimation of the optimal gains
d/dt(chi2) = e^2
```

13.6.5 Disturbance Variable Compensation

In the case study on the activated sludge plant in Sect. 13.1.2, the operator may, apart from the control variable of the activated sludge concentration X_{eff}, also consider the load of the plant (e.g., the pollutant load over the weekend, the weather, etc.) in the sense of a feedforward control that is used to choose an appropriate setpoint. The operator may adapt her empirical value for the gain of her proportional control equation for the computation of the excess sludge removal at weekends, or reduce the setpoint w of the activated sludge concentration with expected rain in order to prevent overloading of the secondary clarifiers (Fig. 13.26). In this case we speak of *disturbance variable compensation,* which combines feedforward with feedback control in the form of $y = f(x,z,t)$.

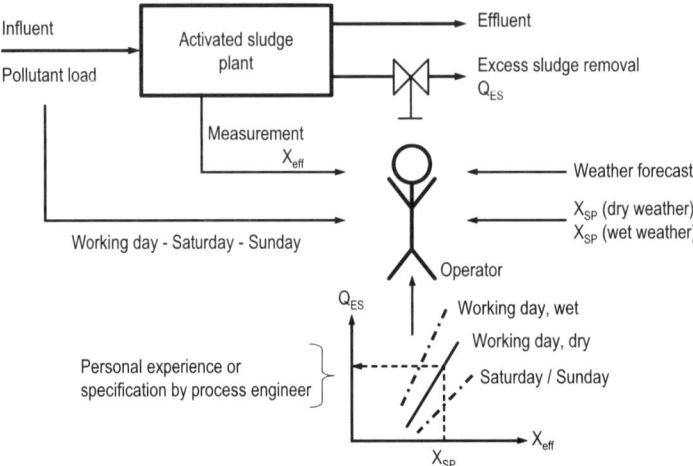

Fig. 13.26 Example of a disturbance variable compensation: during wet weather the setpoint is adapted and for weekends the control characteristic is adapted (see also Fig. 13.1)

13.6.6 Optimal Adjustment of a PID Controller

Optimal adjustment of automatic controllers makes the best use of the investment in a measuring system, an automatic controller, and control members. The adjustment is optimal if the automatic control loop does not become unstable in any possible operating situation and the offset is permanently kept as small as possible with small delay.

In the literature several procedures describe how the parameters of a PID controller are to be selected. A prominent example is that of Ziegler and Nichols (1942), who suggest the following two procedures.

(i) Experiments or simulations with an automatic control loop

We operate the automatic controller first as a pure proportional controller with a small K_P value ($K_I = K_D = 0$), which does not lead to any instabilities in the control loop. Now we gradually increase K_P, until the control loop reacts with a continuous oscillation with constant amplitude. The control loop is now at its so-called stability limit. We determine the associated gain $K_{P,}$ and the period of the oscillation T_{crit}. With these two values, suitable parameters for the automatic controller can be obtained from Table 13.4.

Table 13.4 Suggested choice of the parameters of a PID controller based on experiments with the P controller at the stability limit of the control loop. From a slightly oscillating P controller, the critical gain $K_{P,crit}$ and the associated period of the resulting oscillation T_{crit} are determined (Ziegler and Nichols, 1942)

Type of controller	K_P	K_I	K_D
P	$0.5 \cdot K_{P,crit}$	–	–
PI	$0.45 \cdot K_{P,crit}$	$\dfrac{K_P}{0.83 \cdot T_{crit}}$	–
PID	$0.6 \cdot K_{P,crit}$	$\dfrac{K_P}{0.50 \cdot T_{crit}}$	$0.125 \cdot K_P \cdot T_{crit}$

Table 13.5 Suggested choice of parameters of a PID controller based on a step response curve which is used to obtain the proportionality coefficient $K_S = \Delta x / \Delta y$, the total delay time T_u (incl. dead time) as well as the transition period T_g (Ziegler and Nichols, 1942)

Type of controller	K_P	K_I	K_D
P	$\dfrac{T_g}{K_S \cdot T_u}$	–	–
PI	$0.9 \cdot \dfrac{T_g}{K_S \cdot T_u}$	$\dfrac{K_P}{3.3 \cdot T_u}$	–
PID	$1.2 \cdot \dfrac{T_g}{K_S \cdot T_u}$	$\dfrac{K_P}{2.0 \cdot T_u}$	$0.5 \cdot K_P \cdot T_u$

(ii) Computation based on a transient step response curve

Starting from steady-state operation a step response is measured or simulated. For this a small change of the position Δy is induced and the response of the controlled variable Δx is recorded. Now we obtain the proportionality constant $K_S = \Delta x_\infty / \Delta y = v_\infty / u_m$ for this step response according to Fig. 13.6 and the total delay time (including the dead time) as well as the transitory period T_g according to Fig. 13.7. With these parameters the suitable parameters of the automatic controller can be obtained from Table 13.5.

Example 13.29: Control of the return sludge concentration in an activated sludge system

The concentration of the return sludge of an activated sludge system is to be controlled to a constant value with the help of a P controller. The simple system represented in Fig. 13.27 is chosen as a model of the plant. The solids concentration X_{AT} in the completely mixed activated sludge tank is accepted as constant; the (ideal) secondary clarifier has no volume and completely separates and concentrates all solids without time delay in the return sludge.

Fig. 13.27 Definitions in a simple activated
sludge system with an ideal secondary
clarifier

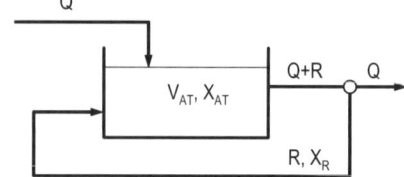

Questions:
Which variable in Fig. 13.27 is:

- the controlled variable x
- the offset e
- the position of the controlled member y
- the setpoint w
- the disturbance z

What is the form of the transient response of the secondary clarifier for a step
change of the variable disturbance z or for a step change of the position of the
control member y?
What is the form of the control characteristic of the P controller?
How large is the permanent offset at steady state?
How large can you choose the gain K_P and still expect the loop to be stable?
How would you solve this problem with a feedforward controller?
In reality this task would be much more difficult to solve. The activated sludge
tank would not be completely mixed and the secondary clarifier would delay the
response of the return sludge concentration to a change of the hydraulic load: what
would become more difficult in reality?

13.7 Case Study: Control of Oxygenation
in an Activated Sludge Plant

*The following simulated example shows the effect of different standard automatic
controllers in a simple system. The emphasis of the example is on the adjustment
(choice of parameters) for the automatic controller.*

13.7.1 Task

In a completely mixed activated sludge tank (Fig. 13.28) the oxygen concentration
is continuously measured. The blower output is to be controlled in such a way that
the setpoint for the oxygen concentration $w = 2 \, gO_2 \, m^{-3}$ is maintained with small
offset. The control member $k_l a$ is proportional to the measured air flow rate. The
oxygen transfer is (see Sect. 9.4.4):

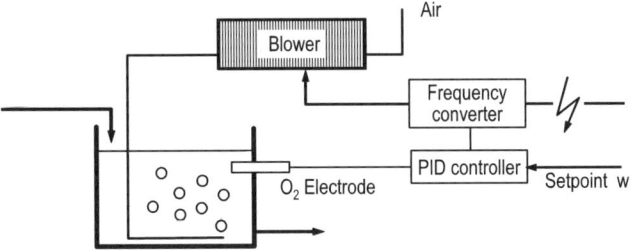

Fig. 13.28 Schematic representation of the controlled system. By an oxygen electrode the dissolved oxygen is measured; a PID controller affects the electronic system for the control of the output of the blowers. The oxygen input is proportional to the air flow

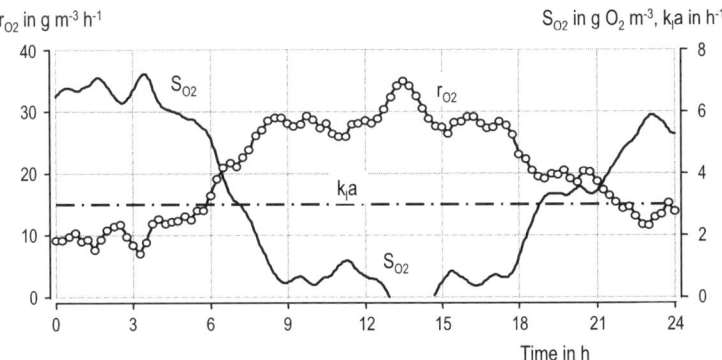

Fig. 13.29 Diurnal variation of the measured oxygen consumption rate r_{O2} and the computed course of the oxygen concentration S_{O2} if the aeration were kept constant with $k_la = 3\,h^{-1}$

$$oe = k_la \cdot (S_{sat} - S_{O2}) \qquad (13.7)$$

oe = oxygen input per reactor volume $[M_{O2}\,L^{-3}\,T^{-1}]$
k_la = oxygen transfer coefficient $[T^{-1}]$
S_{sat} = saturation concentration of dissolved oxygen. Corresponds to the highest attainable concentration, if no oxygen is used $[M_{O2}\,L^{-3}]$, here assumed to be $10\,gO_2\,m^{-3}$.
S_{O2} = measured concentration of dissolved oxygen $[M_{O2}\,L^{-3}]$

In order to keep this example simple the transport of oxygen by influent and effluent of the reactor is neglected. The resulting material balance equation becomes:

$$\frac{dS_{O2}}{dt} = oe + r_{O2} = k_la \cdot (S_{sät} - S_{O2}) + r_{O2} \qquad (13.8)$$

r_{O2} = oxygen consumption (its value is negative, because O_2 is consumed) $[M_{O2}\,L^{-3}\,T^{-1}]$.

The diurnal variation of oxygen consumption r_{O2} was measured during a typical day; the result is shown in Fig. 13.29.

In the control loop a delay between a change of the controlled variable S_{O2} and the adaptation of the corresponding k_la value appears. It is composed of a dead time $T_t = 3$ min and a delay of sixth order with $\theta_{total} = 6$ min. This delay is caused by the response time of the oxygen electrode, the power electronics of the blower, the adaptation of the air flow rate (blower speed), the compression of air in the transfer line, and the ascent of the gas bubbles in the reactor.

13.7.2 System Performance Without Control

Table 13.6 shows the code for the simulation of the system in Fig. 13.28. Figure 13.29 shows the simulated oxygen concentration S_{O2} which results if the aeration intensity is kept constant at $k_la = 3\,h^{-1}$. The concentration varies between 0 and $7.2\,gO_2\ m^{-3}$, the aeration between 13:00 and 15:00 is not sufficient to cover the requirements; the k_la value would have to be increased, which would result in an even larger variation of the oxygen concentration. Obviously the setpoint of $w = 2\ gO_2\ m^{-3}$ can only be attained, if the performance of the blowers is adjusted to the requirements in a control loop.

Table 13.6 Code for Berkeley Madonna for the simulation of the system in Fig. 13.28

```
{Time-dependent oxygen requirement}
rO2 =#rO2m(time)    ; Read in the measured oxygen requirements gO2 m^-3 h^-1
{Model  for  the  reactor,  with  aeration,  kla = Position,  controlled  member,
SO = controlled variable}
SOsat = 10              ; Saturation concentration for dissolved oxygen gO2 m^-3
kla = 3                 ; Control member, intensity of aeration, h^-1
init SO = 5             ; Oxygen concentration, gO2 m^-3
d/dt(SO) = kla*(SOsat – SO) – rO2
```

This code describes the behavior of the system; in addition the code for the simulation of a PID controller must be included (Table 13.3)

13.7.3 Parameters of a PID Controller

In Sect. 13.6.6 two methods that yield suitable parameters for the tuning of a PID controller are described. These methods are used here.

Experiments with an Unstable Control Loop

In the simulation we can define the disturbance of the system (here the oxygen requirement r_{O2}) as constant (here with an average value of $20\ gO_2\ m^{-3}\ h^{-1}$), and with the help of a proportional controller find the critical gain at which the control loop starts to oscillate. In this case study, initial oscillations are totally attenuated with $K_P < 1.69$ and remain stable with $K_{P,crit} = 1.69$ (Fig. 13.30). An oscillation develops with a period of $T = 0.54$ h. With the aid of Table 13.4, suitable parameters for the adjustment of the controller can be found (Table 13.7, nos. 1–3). The frequency or period of the oscillation depends on the time constants of the system.

In a real system it is not possible to keep the disturbance constant, but here too the gain of a proportional action controller can be increased until in the real experiment the control loop becomes unstable (Fig. 13.31). Because the distur-

k_la in h^{-1}, S_{O2} and Signal in $g\ O_2\ m^{-3}$

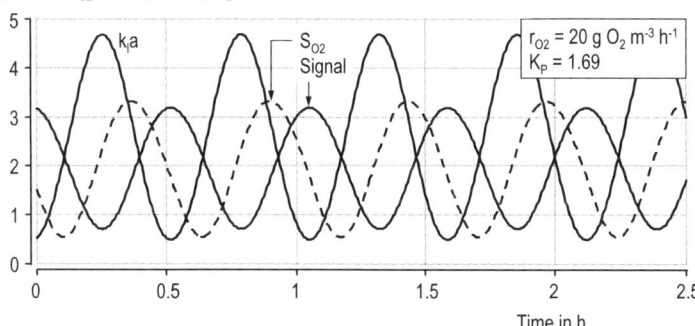

Fig. 13.30 Unstable control loop with constant oxygen requirement. The gain K_P of a proportional action controller was raised until an oscillation with constant amplitude resulted. The real oxygen concentration S_{O2} is not available for the control; only the retarded signal (with an approximate delay of 9 min) can be obtained

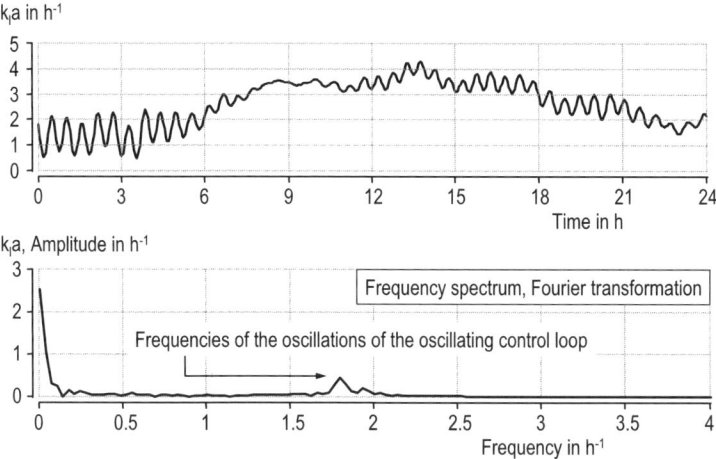

Fig. 13.31 *Above:* variation of the controlled k_la value with variable load (r_{O2} after Fig. 13.29) and a gain of the proportional action controller of $K_P = 1.7$. *Below:* the spectrum of the same time course after Fourier transformation

bance varies, no narrowly defined frequency of the oscillation results. For $K_P = 1.7$, it is a range of periods between 0.45–0.6 h (derived from Fourier transformation, Fig. 13.31, *below*). With this narrow range suitable parameters for the controller may be obtained.

Analysis of a Transient Step Response

The simulation of a transient step response is shown in Fig. 13.32. Around 06:00 the k_la value is increased by 20% with an inactive automatic controller. The simulated oxygen concentration reacts immediately. In the loop there is, however, a delay of some minutes until the reaction (new k_la value) takes place. From the size of the jump in the position ($\Delta y = \Delta k_la$) and the jump in the controlled variable ($\Delta x = \Delta$Signal) as well as the delays T_u (dead time plus apparent dead time) and T_g (transitory period, Fig. 13.32) the proportionality $K_S = \Delta x / \Delta y = 2.95$ is obtained. With the aid of Table 13.5 suitable parameters for the different automatic controllers can now be derived. These are shown in Table 13.7 (controllers nos. 4–6). In practice this experiment is difficult to accomplish with strongly variable load; periods with rather stable conditions must be found.

Minimization of the Variance of the Offset

The goal of control is to minimize the variance of the offset of the controlled variable e without running the risk of the control loop to become unstable. Both, in the simulation (here very simply) and in reality (with lengthy experiments), it is possible to obtain the variance of the controlled variable from an integral over e^2. With the adjustment of the parameters of the controller it is possible to minimize this variance. In Table 13.7, the results of this procedure are presented based on the simulation and automatic minimization of the variance (automatic controllers nos. 7–9).

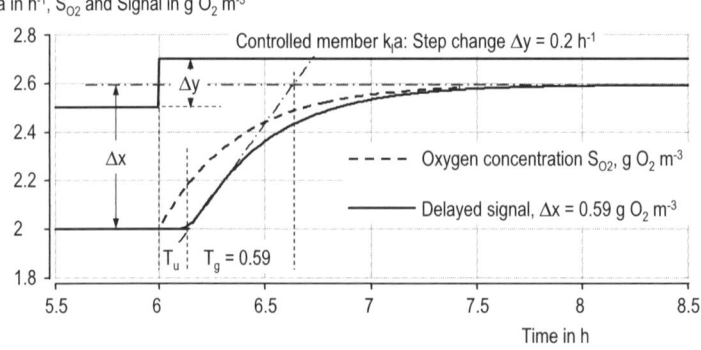

Fig. 13.32 Simulation of a step change of the k_la value and the resulting change of the signal, with inactive automatic controller but with consideration of the delay of the signal

The minimum variance usually results from fast controllers, i. e., high values of the transfer coefficients (gains), in particular of K_P. Such automatic controllers are directly below the stability limit. This becomes clear with control loop no. 7 in Table 13.7. The identified gain K_P amounts to 1.68. Figure 13.31 shows that this controller becomes instable with $K_P = 1.7$, with drastically increased variance. For practical application, the automatic controllers identified here with minimal variance must be made slower, i. e., the identified gain values must be reduced. This results in increased operational variance.

Choice of Suitable Parameters

The automatic control loops 1–9 in Table 13.7 characterize different situations:

For P controllers nos. 1, 4, and 7 an average oxygen concentration m_{SO2} that clearly deviates from the setpoint of $w = 2$ gO_2 m^{-3} results. This is the consequence of the permanent offset of a proportional action controller and becomes visible in Fig. 13.33, where proportional controller no. 10 cannot maintain the desired value w. The range of the measured concentrations is very wide (Fig. 13.35).

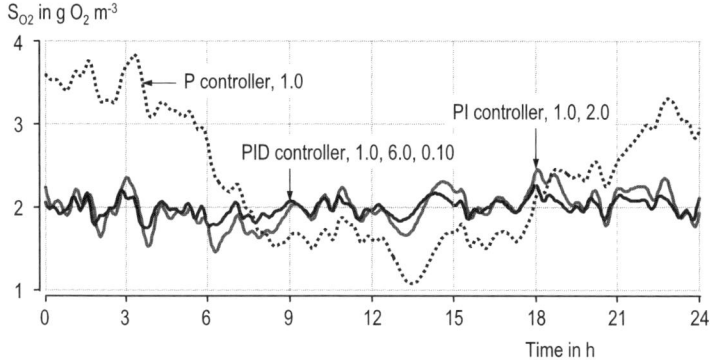

Fig. 13.33 Time course of the oxygen concentration for the controllers nos. 10–12 (Table 13.7)

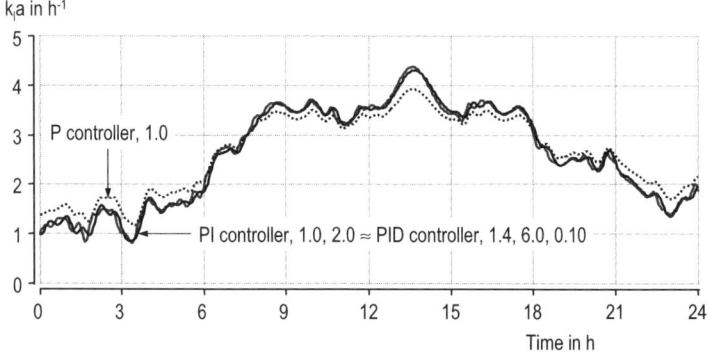

Fig. 13.34 Time course of the controlled position of the k_la value for the controllers nos. 10–12

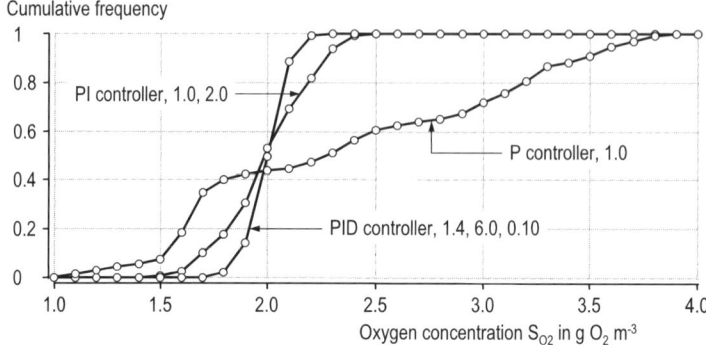

Fig. 13.35 Cumulative frequency of the oxygen concentrations for the automatic controllers nos. 10–12 (Table 13.7)

Table 13.7 Characterization of different controllers. m_{SO2} = average value of the resulting oxygen concentration S_{O2}, s_{SO2} = standard deviation of S_{O2}; < 5% and > 95% are the appropriate percentiles

No	Type	K_P	K_I	K_D	m_{SO2}	s_{SO2}	< 5%	> 95%
Result without control								
0	none	–	–	–	3.05	2.58	0	6.9
Parameters from the unstable controller, $K_{P,krit}$ = 1.69, T_{krit} = 0.54								
1	P	0.93	–	–	2.37	0.82	1.3	3.8
2	PI	0.84	1.87	–	1.99	0.23	1.6	2.3
3	PID	1.12	4.15	0.076	1.99	0.13	1.8	2.2
Parameters from transient step response, $\Delta k_l a$ = 0.2, ΔS_{O2} = 0.59, T_u = 0.14, T_g = 0.49								
4	P	1.19	–	–	2.30	0.67	1.4	3.4
5	PI	1.07	2.32	–	1.99	0.19	1.6	2.3
6	PID	1.42	5.07	0.10	1.99	0.11	1.8	2.2
Parameters from minimizing the variance of the offset								
7	P	1.68	–	–	2.23	0.53	1.5	3.2
8	PI	1.08	3.67	–	1.99	0.15	1.7	2.3
9	PID	1.61	16.1	0.17	2.00	0.07	1.8	2.2
Parameters chosen for Fig. 13.33 and Fig. 13.34								
10	P	1.00	–	–	2.34	0.77	1.3	3.7
11	PI	1.00	2.00	–	1.99	0.20	1.6	2.4
12	PID	1.40	6.00	0.10	1.99	0.10	1.8	2.2
Oscillating PID controller								
13	PID	2.00	6.00	0.10	2.00	0.25	1.5	2.5

The PI controllers nos. 2, 5, 8 and 11 can on average eliminate the offset. Smallest remaining variance of the controlled variable is obtained from the PID controllers nos. 3, 6, 9 and 12. With a standard deviation of only 0.07 g_{O2} m^{-3} the control loop nos. 9 achieves an excellent result, however, very close to the stability limit.

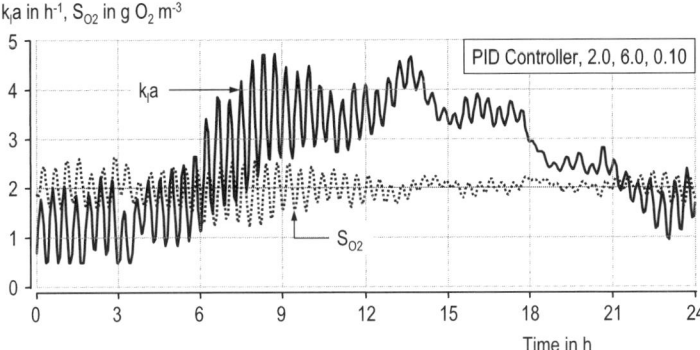

k$_l$a in h^{-1}, S$_{O2}$ in g O$_2$ m^{-3}

Fig. 13.36 Oscillations of the aeration system with an increased gain of the proportional member (controller no. 13 in Table 13.7)

A detailed analysis of the results of simulation for the automatic control loops nos. 1–9 and further simulation runs resulted in the parameter values selected for control loops nos. 10–12, which result in a small variance and sufficient stability contingency. The behavior of these controllers is characterized in Fig. 13.33. The differences in the controlled position of the k$_l$a value (Fig. 13.34) of the three control loops are small; they are, however, significant for the result. A small increase of the transfer coefficient K$_P$ is sufficient to induce oscillations in the aeration system (Fig. 13.36).

Concluding remarks

The extensive example introduced here is to point out some problems and characteristics of automatic controllers. The results correspond to observations that may also apply to practice. In reality, however, additional aspects would have to be considered, e. g., a single blower can hardly cover the whole necessary range of required capacity. Therefore, it will be necessary to control several machines in parallel. Frequently a different control strategy is applied: the required air is controlled with glider valves (control members) from a common collector. The blowers are then controlled in such a way that the pressure in this common collector remains constant. This procedure may help to control the aeration of several parallel treatment systems.

13.8 Fuzzy controllers

The term fuzzy means indistinct or vague. If we ask an experienced operator of a treatment plant for his control strategies in order to fulfill certain discharge requirements, he will frequently not be able to provide us with deterministic, one-to-

*one answers, but rather with verbal, possibly quite vague, descriptions of his ex-
perience like: "If X is large, then I choose Y to be high; if X is small, then I reduce
Y". The experience of the operator is very valuable, but in his decision X may not
be the only variable to be considered; rather many different additional observations frequently of rather subordinated nature may or may not have to be considered. Fuzzy automatic controllers try to convert such indistinct, verbal control
information into a mathematically manageable form.*

Here fuzzy control is explained on the basis of an example. The goal is to introduce the most important terms from the field of fuzzy control and not to provide
the basis to design in detail such automatic controllers.

13.8.1 Example of a Fuzzy Controller

A wastewater treatment plant operator is responsible for an activated sludge system. Daily he decides how much excess sludge (Q_{ES}) has to be removed from the
activated sludge tank. His most important information is the current activated
sludge concentration X_{AS}, which he measures daily. In addition, from experience
he knows that the weekday, the season, and the status from yesterday, etc. are to
be considered, however, with minor weight.

Based on the experience of the operator, we plan to automate the surplus sludge
removal. For the moment we only want to consider the aspects that the operator
considers to be important, especially the dependence on X_{AS}. An interview of the
chief operator results in:

*My goal is to maintain X_{AS} as close to 3 kgTSS m^{-3} as possible. If X_{AS} becomes
too large, the danger exists that the secondary clarifier will lose sludge; if X_{AS}
becomes too low, the nitrification performance is endangered. Specifically:*

- If X_{AS} is approximately 3.0 kg m^{-3}, then I adjust Q_{ES} to approximately 100 m^3
  d^{-1}. Approximately 3.0 kg m^{-3} means in the range 2.8–3.2 kg m^{-3}; this corre-
  sponds to normal operation.
- If X_{AS} is high, I set Q_{ES} to its maximum value of 200 m^3 d^{-1}. X_{AS} is definitely
  high at values over 3.3 kg m^{-3}, sometimes, however, already at 3.1 kg m^{-3}.
- If X_{AS} is low, I set Q_{ES} to its minimum value of only 20 m^3 d^{-1}. X_{AS} is definitely
  low at values below 2.6 kg m^{-3}, rarely, however, already below 2.9 kg m^{-3}.
- All intermediate values in the range $Q_{ES} = 20 - 200$ m^3 d^{-1} are possible.

All these statements are quite vague. We now want to convert them into a well-
defined control strategy. We will first derive a catalog of control rules:

Rule 1: IF X_{AS} = high THEN Q_{ES} = high
Rule 2: IF X_{AS} = low THEN Q_{ES} = low
Rule 3: IF X_{AS} = normal THEN Q_{ES} = normal
Rule 4: All values of Q_{ES} between low and high are possible

Or with numeric values:

Rule 1: IF $X_{AS} > 3.1$–$3.3\,kg\,m^{-3}$ THEN $Q_{ES} = 200\,m^3\,d^{-1}$
Rule 2: IF $X_{AS} < 2.9$–$2.6\,kg\,m^{-3}$ THEN $Q_{ES} = 20\,m^3\,d^{-1}$
Rule 3: IF $X_{AS} = 2.6$–$3.3\,kg\,m^{-3}$ THEN $Q_{ES} = 100\,m^3\,d^{-1}$
Rule 4: Q_{ES} may take all values between 20 and $200\,m^3\,d^{-1}$

Now we can draw the so-called membership functions for the control variable activated sludge concentration X_{AS} (Fig. 13.37). A membership function for X_{AS} indicates with which relative weight the statement "X_{AS} is small" or "X_{AS} is normal" is *true*. If the current value is $X_{AS,act} = 2.7\,kg\,m^{-3}$, then this value belongs to the range of normal concentrations with a weight of $p = 0.25$, and with a weight of $p = 0.67$ it is a member of the low concentrations (see Fig. 13.39, *above*). Overall it will rather be classified as low, however, occasionally (e. g., with $0.25/(0.67 + 0.25) = 27\%$ probability) also as normal. We say that the truth value of $X_{AS,act}$ being normal is 0.27.

As for the control variable X_{AS} membership functions are also derived for the position of the control member, here Q_{ES}. A possible interpretation of the information of the operator is given in Fig. 13.38.

The membership functions may either be explicitly named (LOW, NORMAL, HIGH) or often standardized abbreviations are used:

- N for negative (here low)
- NU for null (nominal value, here normal)
- P for positive (here high)

In addition, the name of the membership function expresses whether it describes an input (a controlled variable) or an output (the position of a control member).

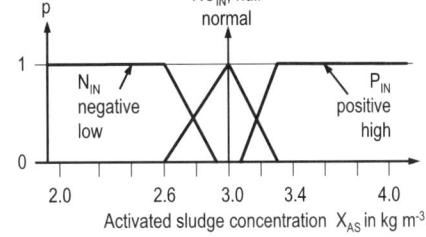

Fig. 13.37 Membership functions of the control variable activated sludge concentration X_{AS} for the terms "low", "normal", and "high"

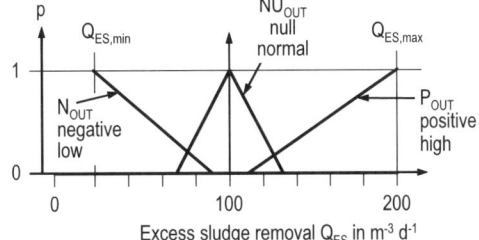

Fig. 13.38 Membership functions for the control member excess sludge removal Q_{ES} for the terms "low", "normal", and "high"

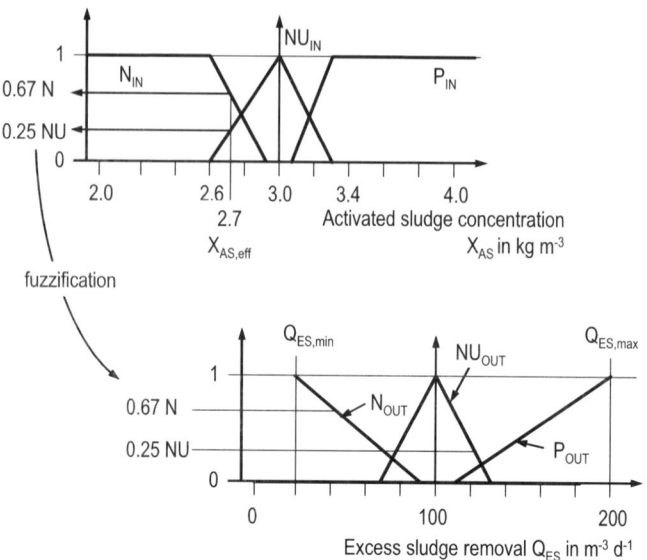

Fig. 13.39 Fuzzification of the current activated sludge concentration $X_{AS,act}$

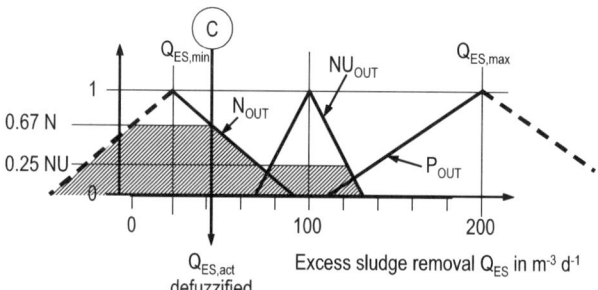

Fig. 13.40 Defuzzification according to the centroid method requires drawing all membership functions symmetrically. Subsequently, the centroid C of the shaded area is computed which results in the position of the control member Q_{ES}

Now we must interpret the current value of the controlled variable $X_{AS,act}$ in terms of the membership functions in Fig. 13.37, a process called fuzzification.

The procedure is introduced in Fig. 13.39. The value $X_{AS,act} = 2.7$ kg m^{-3} belongs with $p = 0.67$ to N_{IN}, with $p = 0.25$ to NU_{IN}, and with $p = 0$ to P_{IN}. According to rule 1 a value of X_{AS} that is a member of P_{IN} leads to a position of Q_{ES}, which is a member of P_{OUT} (respectively, N_{IN} to N_{OUT} or NU_{IN} to NU_{OUT}). Thus, we must give the membership function N_{OUT} a weight of $p = 0.67$ and NU_{OUT} a weight of $p = 0.25$. From these weights we now must derive the position Q_{ES}. There are several possibilities for this *defuzzification* process. The centroid method is shown in Fig. 13.40. A simpler possibility is the so-called singlet method, in which the weighted mean (weight p) of the maximum values of the individual membership functions (a singlet) is used. For the case example this results in:

$$Q_{ES} = 0.67 \cdot 20 + 0.25 \cdot 100 + 0.00 \cdot 200 = 41 \text{ m}^3 \text{ d}^{-1}.$$

Fuzzy control has the advantages that it can implement many different control rules quite easily, it can be based on empirical know-how, and it does not require large computation power. It is increasingly used in the control of wastewater treatment plants that must rely on empirical know-how to some extent.

Example 13.30: Analyzing your own fuzzy logic

Try to design a fuzzy controller for the adjustment of the quantity of water and the water temperature of the shower described in Sect. 13.1.1. Consider the possibilities that the shower was just used by your partner, that you want to refresh yourself after a hot summer day, etc.
You are organizing a grill party in the evening, but you must go shopping at noon. You want to consider the weather and all refusals of the invited guests. Formulate the fuzzy logic that leads you to the decision for how much you will buy or whether you shall call off the party.

13.8.2 Why Fuzzy Control?

Fuzzy controllers may be set up based on the relationship of the semantic variables (membership functions), which then only have to be parameterized for a particular application. This makes their implementation rather inexpensive. The know-how of one plant may therefore be put to beneficial use in a different plant.

It remains to be seen to what extent the mentioned advantages of fuzzy control prove to be valid in water technology, where the flow schemes and performance requirements change from plant to plant.

Chapter 14
Time Series Analysis

A time series consists of a series of values of a variable at successive times in regular intervals. Time series are the bases for the treatment of many engineering questions. In the analysis of time series we try to uncover regularities in the succession of the individual measurements and to derive information from these regularities. In addition, we will use time series analysis to develop stochastic models for the simulation of variables.

14.1 Time Series

In urban water management, we use many time series which we obtain over long time periods with entirely different frequencies.

Examples of time series are:

- The development of the population in the distribution area of a waterworks, which is frequently compiled annually
- the daily production of water in a water supply enterprise
- the measurement of a flow rate in a sewer in 1-min intervals
- the measurement of a pollutant concentration on line in 10-min intervals or daily in a mixed sample
- the measurement of the rain intensity in 1-, 5- or 10-min intervals
- etc.

All these time series have in common that they express the condition of a (measured) variable in regular intervals; they are the basis for many statistical evaluations and allow us to derive important information about these variables which serve us in designing and analyzing systems and plants.

14.2 Stationary Time Series

A time series is *stationary* if its expected value is not subject to a trend in the course of time and the dispersion of the individual values remains constant over time. *Nonstationary* time series are subject to a trend and/or the dispersion of the individual values is not independent of time.

We *test for stationarity* of a time series by obtaining the average value and the variance of the first half of the time series and comparing it with similar results of the second half.

Example 14.1: Stationary and nonstationary time series

The development of the population of a city over the years is an nonstationary time series, the expected value typically increases (or decreases) continuously.
The treatment performance of a wastewater treatment plant is frequently a stationary time series. Its value changes from day to day; its expected value and its dispersion remain, however, constant over longer time periods.

Example 14.2: Test for stationarity of a time series

Figure 14.1 shows the elements of a (simulated) time series over 360 days. *Is this series stationary?*
For the test, the time series is split in the middle and average and standard deviation of these partial time series are obtained with the following code in BM:

```
{The time series is read in as #x(i)}
STARTTIME = 0
STOPTIME = 0          ; Initialization is sufficient
DT = 1                ; Irrelevant, STARTTIME = STOPTIME
low[1..180] = #x(i)   ; Elements of first half of time series
high[181..360] = #x(i) ; Elements of second half of time series
m_low = arraymean(low[*]) ; Average of low        Result:  3.05 ± 0.09
```

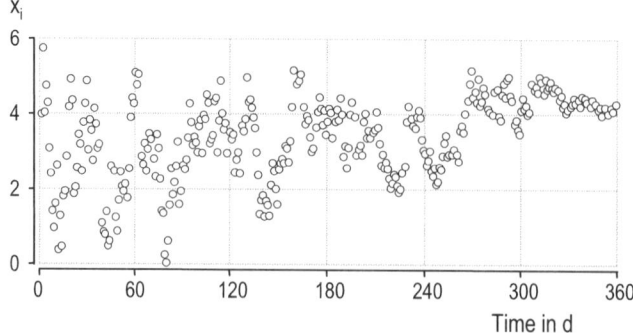

Fig. 14.1 Example of an instationary time series

m_high = arraymean(high[*])	; Average of high	3.82 ± 0.06
sig_low = arraystddev(low[*])	; Standard deviation of low	1.19
sig_high = arraystddev(high[*])	; Standard deviation of high	0.79

The results indicate a significant linear positive trend in the data; the average values of the two series are different.

The variance of the two partial series decreases with time, the standard deviation of the higher elements is significantly smaller than that of the lower elements.

This time series is not stationary!

14.3 Case study: Yearly Variation of the Temperature

The following time series serves as a uniform basis for the introduction of the concepts of time series analysis. It is based on the daily measured temperature of the wastewater in the influent of a wastewater treatment plant.

The characteristic of the temperature time series in Fig. 14.2 is the periodic variation in the yearly cycle which we may expect reliably. Likewise we expect that the water frequently has an above-average temperature today if it was rather warm yesterday. However, we hardly have a reason to assume that during the approximately 3 years over which the time series extends itself, the temperature follows an increasing or a decreasing trend or that the variation of the temperature changes. In addition to these expected processes, some other processes play a role but cannot be reconstructed any more, and we must explain them with stochastic processes.

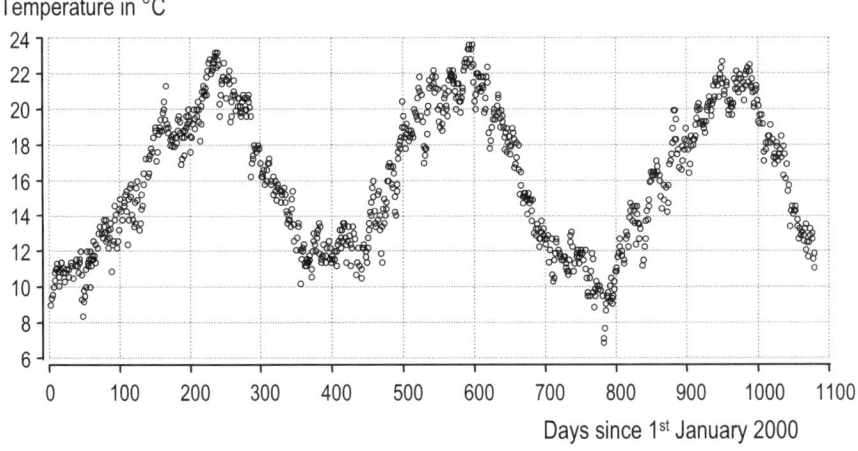

Fig. 14.2 Time series of the wastewater temperature in the influent of a large wastewater treatment plant in Switzerland, daily measurements at 0900 hrs (1079 successive measured values, no missing value)

In the analysis of this time series we recognize some structure in the data which permits us afterwards to identify models for the processes that explain or at least describe the behavior of the temperature. Based on such an analysis we can then answer specific questions, like:

- Which temperature range do we have to expect in a certain season?
- Can we make a good prediction of the temperature tomorrow if we know the temperature today?
- Which succession of low temperatures is possible? Such a succession determines the possible performance of the biological treatment during cold periods.

In this chapter we will discuss simple possibilities for the identification of structures and models in time series.

14.4 Conventional Statistical Characterization

A conventional statistical characterization of a time series does not consider the temporal sequence of the measured values; the relationship between two consecutive values is lost.

A purely statistical characterization of the time series in Fig. 14.2 could cover the following elements known from descriptive statistics; their evaluation does not consider the succession of the individual measurements:

- A histogram, as shown in Fig. 14.3 provides information about the frequency of specific measured values within a fixed temperature interval. In the example, a bimodal distribution results as a consequence of the pronounced yearly variation which we cannot model with a theoretical standardized distribution.
- The cumulative frequency of the measured values in Fig. 14.4 highlights which fraction of the measurements do not exceed a certain value. If we want to select a temperature value where less than 20% of the daily values fall below, we would have to select 12°C.

Caution: The time series in Fig. 14.2 is subject to large periodic variation, which leads to the bimodal distribution. Since this series does not cover three full years (which would require 1096 elements rather than 1079), the histogram and the cumulative distribution are not representative of the data of three full cycles. Depending on the question asked, we may want to use only two full years or extend the series to three full years, possibly based on a stochastic model to be developed.

Table 14.1 summarizes several typical statistical characteristics. For the design of a biological process we gain little from these values; in addition, we cannot answer the questions raised in Sect. 14.3.

Fig. 14.3 Histogram of the individual measured values of the temperature in Fig. 14.2

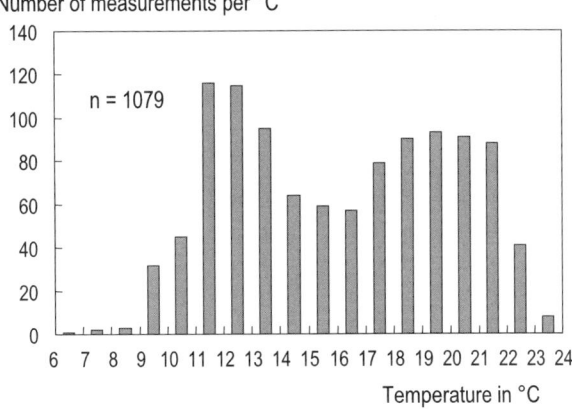

Number of measurements per °C

Fig. 14.4 Cumulative frequency of the measured values of the temperature in Fig. 14.2

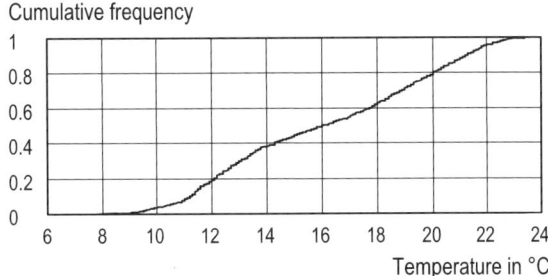

Cumulative frequency

Table 14.1 Statistical characteristic numbers of the time series in Fig. 14.2

Parameter	Symbol	Value	Units
Number of measurements	n	1079	–
Frequency of the measurement	f	1	d^{-1}
Expected value	$E(T)$, $\overline{T}$, m_T	16.06	°C
Median (50% value)	$T_{50\%}$	16.0	°C
Minimum	T_{min}	6.9	°C
Maximum	T_{max}	23.6	°C
Standard deviation	SD, s_T	3.87	°C
Variance	$Var(T)$, s_T^2	14.94	$°C^2$

14.5 Moving Average

With the moving average we compute the local trend of the time series by smoothing short-term variations in an averaging process. Depending upon the question, we select a different number of sequential elements for the smoothing process or select a weighting procedure of the measured values.

We differentiate between the arithmetic and the geometric moving average.

14.5.1 Arithmetic Moving Average

In the arithmetic moving average (abbreviated to moving average) all averaged measured values in the regarded window have the same weight.

The moving average at time i, averaged over five measured values, can be computed with Eq. (14.1). The time i lies in the center of the arithmetically averaged measured values:

$$g_{5,i} = \frac{x_{i-2} + x_{i-1} + x_i + x_{i+1} + x_{i+2}}{5} \quad \text{with} \quad i = 3 \ldots n-2 \tag{14.1}$$

$g_{5,i}$ = moving average over five measured values at time i
x_i = measured value at time i
n = number of measured values [−]

With Eq. (14.1) the moving average value is located in the center of the averaged range; this is possible only if an odd number of values is averaged. With an even number, the first and the last measurement in the range obtain the weight ½. Sometimes the result is assigned to the time of the last measured value in the averaged interval, which results in a delay. However, the averaging permits an even number of measured values.

For a general number λ of averaged measured variables (λ = length of the average) Eq. (14.1) becomes:

$$g_{\lambda,i} = \frac{\sum\limits_{k=i-(\lambda-1)/2}^{i+(\lambda-1)/2} x_k}{\lambda} = \frac{x_{i-(\lambda-1)/2} + \ldots + x_i + \ldots + x_{i+(\lambda-1)/2-1} + x_{i+(\lambda-1)/2}}{\lambda} . \tag{14.2}$$

If the result is referred to the time of the last measurement, then

$$g_{\lambda,i} = \frac{\sum\limits_{k=i-\lambda+1}^{i} x_k}{\lambda} = \frac{x_{i-\lambda+1} + x_{i-\lambda+2} + \ldots + x_{i-1} + x_i}{\lambda} . \tag{14.3}$$

In a program the moving average from Eq. (14.3) can be obtained from (see Example 14.3):

$$g_{\lambda,i} = \frac{g_{\lambda,i-1} \cdot \lambda - x_{i-\lambda} + x_i}{\lambda} \quad \text{with} \quad i \geq \lambda . \tag{14.4}$$

Figure 14.5 shows the moving average of length $\lambda = 7$, located in the center of the averaged section, for a small window of the time series in Fig. 14.2. Clearly the moving average smoothes the measured values, i.e., it reduces the high-frequency short-term variations; the larger the length λ, the larger this effect becomes, and thereby some information may be lost.

For the dimensioning of biological wastewater treatment plants, moving average values of the load or the temperature are frequently used. The length λ of the

Temperature in °C

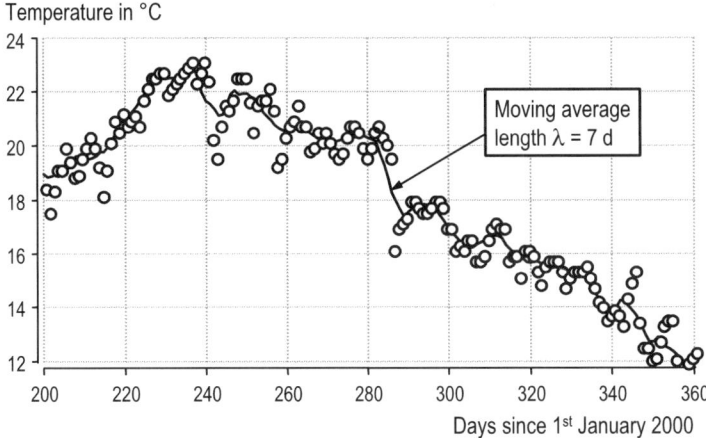

Fig. 14.5 Moving average of the temperature. Cutout from Fig. 14.2, averaged over $\lambda = 7$ days and located in the center of the section represented

average may approximately correspond with the mean solids retention time. Here the length λ of the averaging process has a meaning defined by the problem.

Example 14.3: Computation of the arithmetic moving average with Berkeley Madonna

The following code computes the arithmetic moving average of a complete time series over the length λ (lambda). In analogy to Eq. (14.2), the result is located in the center of the averaged period. The code handles even and uneven λ, because BM interpolates linearly between data points.

```
{Time series is read as #T(time) from data file, tested}
STARTTIME = 4                    ; Beginning time of computation, > DT*λ/2
STOPTIME = 1000                  ; End time of computation, < DT*(n − λ/2)
DT = 1                           ; Time step for time series in Fig. 14.2
lamda = 7                        ; Length of moving average, even or uneven
Selec[1..lamda] = delay(#T(time + DT*(lamda + 1)/2),DT*i)
                                 ; Choosing the values to be averaged
MA = arraysum(Selec[*])/lamda    ; Obtaining the moving average in center
```

14.5.2 Geometric Moving Average

In the geometric moving average, the individual measured values obtain an exponentially decreasing weight that corresponds to a geometric series of weights. This simulates, e.g., washing out of microorganisms from biological treatment systems, and corresponds to a convolution of the time series with an exponential distribution.

In biological treatment processes the memory of the past load of the plant decreases exponentially over time. In an activated sludge system, the mean residence time of the microorganisms corresponds roughly to the solids retention time (today typically >10 d). The activated sludge is completely mixed and individual microorganisms are washed out of the plant according to an exponential distribution (Eq. (7.20)). The geometric moving average characterizes this situation by using a geometric series of weights in the moving average.

An infinite geometric series has the form:

$$a_0, a_1, a_2, ..., a_\infty$$

where $a_i = a_{i-1} \cdot p$ or $a_i = a_0 \cdot p^i$ with $0 < p < 1$.

The sum of all elements of this series amounts to:

$$\sum_{i=0}^{\infty} a_i = a_0 \cdot \frac{1}{1-p} \quad \text{or with } a_0 = 1: \quad \sum_{i=0}^{\infty} a_i = \sum_{i=0}^{\infty} p^i = \frac{1}{1-p} = \lambda . \tag{14.5}$$

Therefore

for $a_0 = 1$ one has $p = \dfrac{\lambda - 1}{\lambda}$. \hfill (14.6)

For the computation of the geometric moving average of the time series X with elements x_i, we rely on the geometrically decreasing weights of earlier elements of the time series. Considering Eq. (14.5) to obtain λ the result is

$$g_{g,\lambda,i} = \frac{\sum_{k=-\infty}^{i} \left(x_k \cdot p^{i-k} \right)}{\sum_{k=-\infty}^{i} p^{i-k}} = \frac{\sum_{k=-\infty}^{i} \left(x_k \cdot p^{i-k} \right)}{\lambda} \tag{14.7}$$

$g_{g,\lambda,i}$ = geometric moving average with length λ at time i
x_k = element k of time series
λ = sum of the weights which corresponds to the length λ in the arithmetic moving average [T]
p   = common ratio of the geometric series

The geometric moving average is based on an infinite number of weights; the computation of individual moving average values can thus become very complex. The recursive computation described below minimizes this expenditure. One has

$$g_{g,\lambda,i} = \frac{\sum_{k=-\infty}^{i} \left(x_k \cdot p^{i-k} \right)}{\lambda} = \frac{x_i + \sum_{k=-\infty}^{i-1} \left(x_k \cdot p^{i-k} \right)}{\lambda} = \frac{x_i + p \cdot \sum_{k=-\infty}^{i-1} \left(x_k \cdot p^{i-1-k} \right)}{\lambda}$$

$$= \frac{x_i}{\lambda} + p \cdot g_{g,\lambda,i-1} \tag{14.8}$$

Temperature in °C

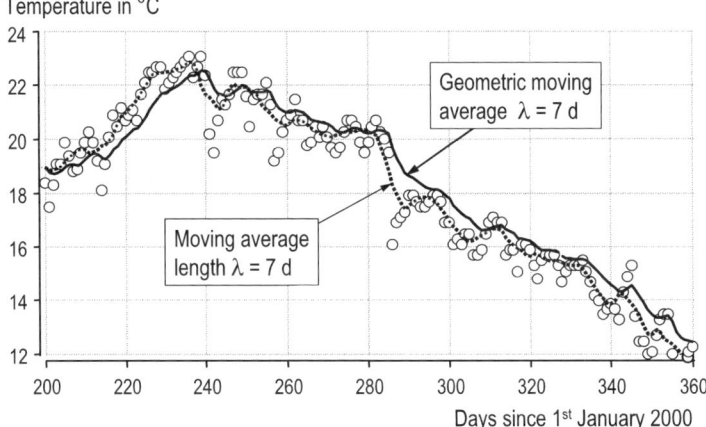

Days since 1st January 2000

Fig. 14.6 Subset of the geometric moving average of the time series in Fig. 14.2, with a length of 7 days and located at the end of the section represented. The arithmetic moving average of equal length is given for comparison

With Eq. (14.6) this yields

$$g_{g,\lambda,i} = \frac{x_i + g_{g,\lambda,i-1} \cdot (\lambda - 1)}{\lambda} = (1-p) \cdot x_i + p \cdot g_{g,\lambda,i-1}. \qquad (14.9)$$

With Eq. (14.9), the geometric moving average at time i is obtained recursively from the moving average for time $i-1$ and the current value of the time series x_i. There remains the problem of obtaining the starting value $g_{g,\lambda,0}$. Since empirical time series are always finite, we must estimate this value pragmatically. A suitable approach is

$$g_{g,\lambda,0} = x_0 \qquad (14.10)$$

or an improved estimate, if there is reason to deviate from Eq. (14.10). A poor starting value will be degraded asymptotically over the time period of 2λ to 3λ.

Equation (14.9) with the initial value after Eq. (14.10) is suitable for programming (Example 14.5). In Fig. 14.6, the geometric moving average of length $\lambda = 7$ is compared with the equivalent arithmetic moving average for a subset of the time series in Fig. 14.2. The value is located at the end of the averaged period, therefore the values react rather late to changes in the trend. The comparison with the arithmetic average shows differences of $\pm 1.5°C$ in the time series, which is very significant for the design of a biological process.

Example 14.4: Comparison of the weights of the values in moving average procedures

What are the weights of the elements of a time series in the arithmetic and in the geometric moving average of length $\lambda = 3$?

Fig. 14.7 Comparison of the weighting of the measured values in the arithmetic and geometric moving average ($\lambda = 3$)

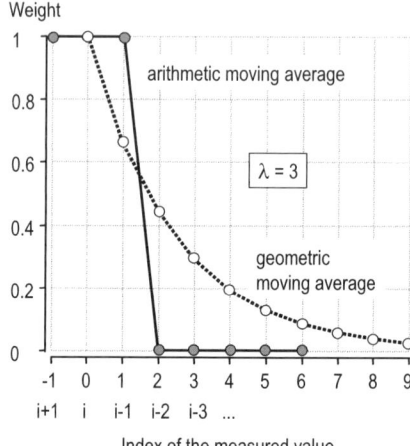

Figure 14.7 compares the weights of these two moving averages. Obviously large differences exist in these weights. Both series must be standardized with the sum of the weights λ.

Example 14.5: Computation of the geometric moving average with Berkeley Madonna

The following code computes the geometric moving average of length λ (lambda) for a time series of duration STOPTIME. The result is computed with Eq. (14.9) and is independent of the initial value T_0 only for larger values of time.

```
{The time series is read in as #T(time)}
STARTTIME = 0          ; Beginning of computation
STOPTIME = 365         ; End of computation
DT = 1                 ; Time step, applicable for time series in Fig. 14.2
lambda = 7             ; Length of the average
T = #T(time)           ; Reading the time series
init Gg = #T(0)        ; Geometric moving average, initialization with T₀
next Gg = (Gg*(lambda−1) + #T(time))/lambda
                       ; Reduce old sum, add new value, standardize
```

14.6 Trend Lines

Trend lines follow time series over a larger time range with a functional description, frequently polynomials are used. The parameters of the functions are identified, e.g., with the aid of the minimization of sum of the squared deviations. Spreadsheet programs offer this possibility as built-in functions.

A linear trend corresponds to the regression line between time and the elements of the time series according to:

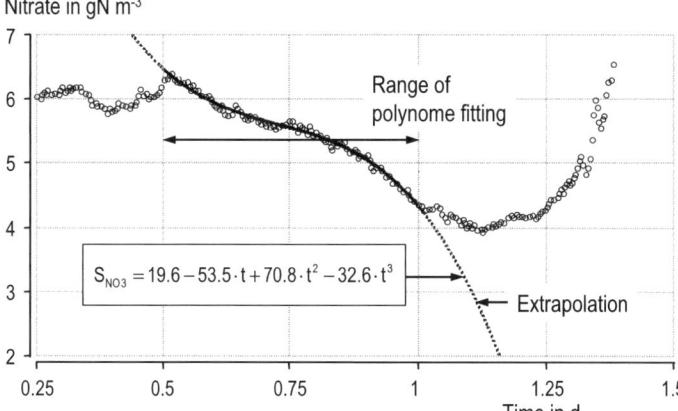

Fig. 14.8 Adjustment of a polynomial of third degree to the nitrate concentration in the effluent of an activated sludge system (on line measurement, $\Delta\tau = 5$ min). The polynomial was identified within the range of $0.5-1$ d (12 h) and cannot be used for any extrapolation from this range

$$P_1(t) = b_0 + b_1 \cdot t \; . \tag{14.11}$$

Polynomials of higher order have the form

$$P_n(t) = b_0 + b_1 \cdot t + ... + b_n \cdot t^n = \sum_{i=0}^{n} b_i \cdot t^i \; . \tag{14.12}$$

In Berkeley Madonna polynomials can be adapted to a time series by identifying the $n+1$ parameters (option *Curve Fit*). If these functions have no mechanistic background, they are suitable for interpolation, but do not, however, permit any extrapolation, as is illustrated in Fig. 14.8.

Example 14.6: χ^2 test for the polynomial fit in Fig. 14.8

The electrode that was used to measure the nitrate concentration in the treated effluent (Fig. 14.8), has a reproducibility of the result (deviation from the average value) which we estimate with a normally distributed variable with a standard deviation of $\sigma = 0.05$ gNO$_3$–N m^{-3}.

Fitting a polynomial of third degree to the measured nitrate concentrations within the range $0.5-1.0$ days in Fig. 14.8 results in $\chi^2 = 182$. With 145 measured points and four estimated parameters, 140 degrees of freedom remain.

We must reject the hypothesis that the deviation between measurement and polynomial is normally distributed with $N(0,0.05)$ at the 95% level of significance ($\chi^2 > 169$). With an estimated standard deviation of $\sigma = 0.06$ gNO$_3$–N m^{-3}, the result would be highly significant ($p = 22\%$).

The number of sign changes of the difference $NO_{3,\text{measured}} - NO_{3,\text{polynomial}}$ is 37. This is clearly lower than the expected value of 72 and points to the fact that some autocorrelation is contained in this data which is not captured by the polynomial.

Example 14.7: Fitting a polynomial to another function in Berkeley Madonna

For the adjustment of a function to data, BM makes the *Line Fit* option available. This function minimizes χ^2, which is determined exclusively from locations at which data is available. If we must adapt a computed time series (with elements at the distance of the time step DT), then we must use the Optimize option, which minimizes a computed test variable. A possibility is:

f=.....   ; Computed time series which is fitted to a polynomial
                                           (or other function)
Pf=a+b*time+c*time^2   ; Polynomial to be fitted to the original function f
a=1 b=1 c=1   ; Initial values of polynomial parameters
INIT chi2=0   ; χ^2 which is to be minimized
NEXT chi2=chi2+(f–Pf)^2

χ^2 obtained from this code must still be standardized with the standard error of the function f, before a χ^2 test becomes meaningful. In order to test the polynomial fit, we can also count the number of sign changes of the residuals according to:
INIT SignChange=0
NEXT SignChange=if (f–Pf)*Delay((f–Pf),DT)<0 THEN SignChange+1
ELSE SignChange

14.7 Removing a Trend

Frequently statistical characteristics of data are not very meaningful if their time series are subject to a trend. It may help to separate the trend from the time series and then to analyze the residuals.

Possibilities to remove a trend from a time series x_i and to produce a time series of residuals z_i are:

- Subtraction of the trend line P_i from the time series x_i; the residuals z_i do not contain the identified trend any longer ($z_i = x_i - P_i$);
- Subtraction of the moving arithmetic average from the time series results in a new time series free of trend ($z_i = x_i - g_{\lambda,i}$).
- The formation of differences between two sequential elements of the time series ($z_i = x_i - x_{i-1}$) results in a trend-free time series.

Figure 14.9 provides an example of data which have their trend removed by forming differences of two successive elements of the time series. Figure 14.10 is based on the same data, but the trend is removed by using a moving average. The scale of the residuals is clearly different, in addition the residuals from the moving average demonstrate a strong autocorrelation (see Sect. 14.10). The time series of the residuals can now be analyzed independently based on their statistical characteristics which provide information for possible stochastic models.

Fig. 14.9 Measured temperature (Fig. 14.2) and times series corrected for a trend by taking the difference between two consecutive elements

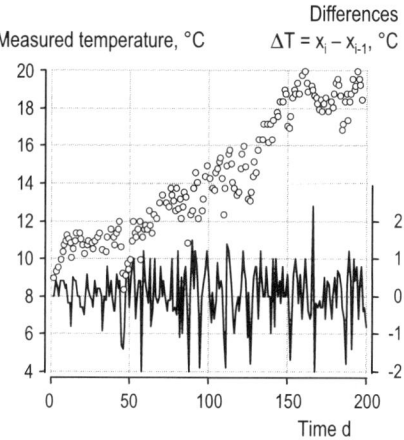

Fig. 14.10 Moving average of the temperature with length of 7 days and time series corrected for trend by taking the difference between the data and the moving average

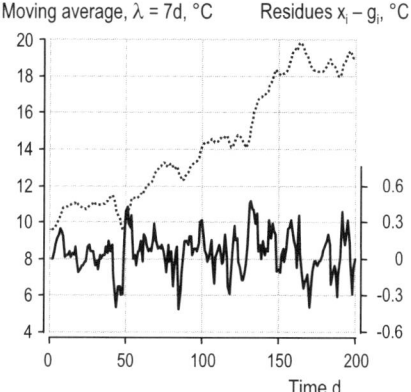

14.7.1 Correcting for the Average Value

Many methods of time series analysis become more transparent if the expected value of the elements vanishes ($E(x) = 0$ or $\mu_x = 0$).

If the trend is removed from a time series, then usually the expected value of the residuals disappears because the trend was identified accordingly. If no trend removal takes place, it might be helpful to correct a time series for its average value according to:

$$z_i = x_i - \mu_x .$$ (14.13)

The new time series z_i makes it easier to obtain information on autocorrelation (Sect. 14.10).

14.8 Logistic Growth

The model of logistic growth is frequently used if processes approach a saturation level, as is, e. g., the case with an increasing population that will finally occupy the entire available settlement area. Since this model is based on a logical argument (saturation), there is the danger that new arising processes will only be recognized rather late.

Exponential growth of a population is not sustainable; it results in an exploding increase of the population (see Fig. 14.11):

$$\frac{dx}{dt} = \mu \cdot x \quad \text{with the solution} \quad x = x_0 \cdot \exp(\mu \cdot t)$$

x = size of the population in a geographically defined region
x_0 = population at time $t = 0$
μ = growth rate $[T^{-1}]$
t = time $[T]$

In many natural and urban systems, we observe that initially the population increases exponentially, but with time the growth rate decreases and the population approaches the carrying capacity of the system. This situation is modeled in a simple manner with the logistic growth model:

$$\frac{dx}{dt} = r \cdot x \cdot \left(1 - \frac{x}{K}\right) \tag{14.14}$$

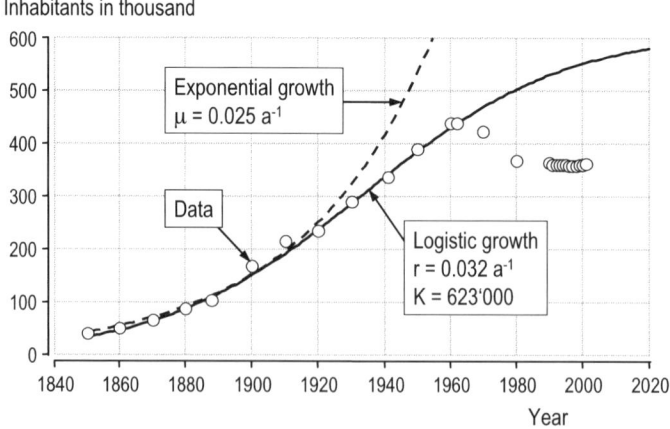

Fig. 14.11 Development of the population in the geographical area of today's city of Zurich. The model of logistic growth was fitted to the data from 1850 to 1960; that for exponential growth to the data from 1850 to 1920

Equation (14.14) has the analytical solution:

$$x(t) = \frac{K \cdot x_0}{x_0 + (K - x_0) \cdot \exp(-r \cdot t)}$$

x   = size of the population [#] with its initial value x_0
r   = maximum growth rate $[T^{-1}]$
K   = carrying capacity of population of the system [#]

With increasing population, the growth, which is initially exponential, slows as the growth rate is reduced linearly and drops to zero as the carrying capacity K is reached.

In Fig. 14.11 the logistic growth model is applied to the population of the geographic area that today forms the city of Zurich. Over more than 100 years this model follows the development excellently. After 1960, however, new social processes led to a development that was not foreseeable by this simple model (mobility, automobile, smaller families, rapid increase of income, etc.).

The logistic growth model was for a long time the most frequently applied model for the prediction of the population trend. The development of some long-lived public infrastructures (water supply, solid waste incineration, waste water treatment, etc.) was based on this model and resulted in overcapacity that may still cause problems today.

The logistic growth model describes *nonstationary* time series whose expected value is not constant over time.

14.9 Discrete Fourier Transformation

Discrete Fourier transformation decomposes time series into a sum of cosine and sine functions with gradually increasing frequency. Here we use this analysis especially in order to identify periodically repeating contributions to time series.

The function f(t), which is characterized by a time series x_i with the $N + 1$ elements x_0, x_1, x_2,..., x_N at intervals of Δt can be transformed into the sum of trigonometric functions g(t) (Fourier polynomial), according to

$$g(t) = \frac{a_0}{2} + \sum_{i=1}^{n} \left(a_i \cdot \cos\left(2 \cdot \pi \cdot \frac{i}{T} \cdot t \right) + b_i \cdot \sin\left(2 \cdot \pi \cdot \frac{i}{T} \cdot t \right) \right) \tag{14.15}$$

$$a_i = \frac{2}{T} \cdot \int_0^T x(t) \cdot \cos\left(2 \cdot \pi \cdot \frac{i}{T} \cdot t \right) \cdot dt \tag{14.16}$$

$$b_i = \frac{2}{T} \cdot \int_0^T x(t) \cdot \sin\left(2 \cdot \pi \cdot \frac{i}{T} \cdot t \right) \cdot dt \tag{14.17}$$

$$i = 0,1,...,n \qquad n \leq \frac{N}{2}$$

a_i, b_i =   Fourier coefficients, $a_0 = \mu_x$, $b_0 = 0$
T =   time interval covered by the time series $(t(x_N) - t(x_0) = N \cdot \Delta t)$
$g(t)$ =   Fourier polynomial

In total Eq. (14.15) contains $2n + 1$ unknown parameters (a_0, ..., a_n and b_1, ..., b_n) which can be computed from the $N + 1$ elements of the time series. With increasing number n of the functions, the fit to the time series becomes ever more exact until, with $n = N/2$, the fit is perfect.

Equation (14.16) represents a weighted mean of the function f(t) where the weights are defined by the cosine within the range T, and Eq. (14.17) represents the equivalent sine-weighted average of this function. i/T is the frequency of the harmonious oscillations. $a_0/2$ corresponds to the arithmetic mean of the function f(t) in the range T. Example 14.8 demonstrates how a simple function can be transformed into a Fourier series with the aid of Eq. (14.15).

The two trigonometric functions in Eqs. (14.16) and (14.17) can be combined into a single cosine with the observable amplitude c_i and an additional phase shift ϕ_i

$$a_i \cdot \cos\left(2 \cdot \pi \cdot \frac{i}{T} \cdot t \right) + b_i \cdot \sin\left(2 \cdot \pi \cdot \frac{i}{T} \cdot t \right) = c_i \cdot \cos\left(2 \cdot \pi \cdot \frac{i}{T} \cdot t + \phi_i \right)$$

$$c_i = \frac{a_i}{|a_i|} \cdot \sqrt{a_i^2 + b_i^2}, \quad \phi_i = \arctg\left(-\frac{b_i}{a_i} \right)$$

$$(14.18)$$

With Eq. (14.18) Eq. (14.15) becomes:

$$g(t) = \frac{c_0}{2} + \sum_{i=1}^{n} \left(c_i \cdot \cos\left(2 \cdot \pi \cdot \frac{i}{T} \cdot t + \phi_i \right) \right).$$

Example 14.8 demonstrates how the parameters of a Fourier decomposition can be obtained and how accurate the result finally is.

Example 14.8: Decomposition of a function based on Fourier transformation

For a defined function, the Fourier coefficients can be obtained with Berkeley Madonna from Eqs. (14.16) and (14.17).

From Eq. (14.16) it follow that $\dfrac{da_i}{dt} = \dfrac{2}{T} \cdot f(t) \cdot \cos\left(2 \cdot \pi \cdot \dfrac{i}{T} \cdot t \right)$.

We obtain the amplitude for a frequency from Eq. (14.18).
The following code covers two periods of length T. In the first period the parameters are identified; in the second period the Fourier polynomial is computed:

```
METHOD Euler          ; Integration with Euler is accurate
STARTTIME = 0         ; Beginning of integration
STOPTIME = 4          ; Twice the length of the time series
DT = 1/50             ; Time step
T = 2                 ; Length of the time series
f = graph (time) (0,25) (0.5,35) (0.7,7.5) (0.8,27.5) (1.3,30) (1.5,45) (1.6,25)
                      ; Time series example, see Fig. 14.13
```

n = 25   ; Number of coefficients

init a[0..n] = 0   ; Amplitude of cosine i

init b[0..n] = 0   ; Amplitude of sine i

d/dt(a[0..n]) = if time < T then 2*f*cos(2*pi*i*time/T)/T else 0

                            ; Eq. (14.16), integrated until time = T

d/dt(b[0..n]) = if time < T then 2*f*sin(2*pi*i*time/T)/T else 0

                            ; Eq. (14.17), integrated until time = T

c[0] = a[0]/2   ; Arithmetic mean

c[1..n] = sqrt(a[i]^2 + b[i]^2); Amplitude for frequency i, Eq. (14.18)

{the following computation is only valid when time > T and therefore a_i and b_i are known}

m = 9   ; Degree of the Fourier polynomial

g[0] = if time < T then 0 else c[0]

g[1..m] = if time < T then 0 else g[i−1] + a[i]*cos(i*time*2*pi/T) + b[i]*sin (i*time*2*pi/T)   ; Fourier polynomial, recursively

Fourierpolynomial = g[m]   ; Fourier polynomial of degree m

df = delay(f,T)   ; Time series delayed for second period for comparison

Figure 14.12 shows the discrete-frequency spectrum c_i of the original function shown in Fig. 14.13. The first nine amplitudes (i/T < 5) describe a large part of the variation. Higher frequencies have only small amplitudes. Figure 14.13 compares the original function with the ninth-degree Fourier polynomial. The adjustment is already quite good; the polynomial with 50 degrees would result in a perfect fit. The computation with a time step of DT = 1/50 during one period of T = 2 results in N = 100 data points plus the initial value.

Berkeley Madonna makes the discrete frequency spectrum (analogous to Fig. 14.12) based on a so called fast Fourier transformation FFT available. Any time dependent function or time series can be transformed. The computation is

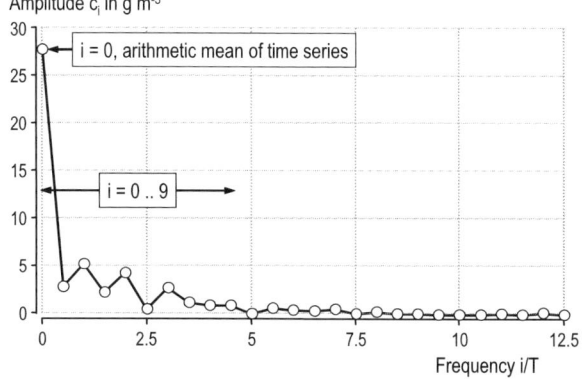

Fig. 14.12 Discrete frequency spectrum (amplitudes of the cosine function after Eq. (14.18)) for the original function in Fig. 14.13

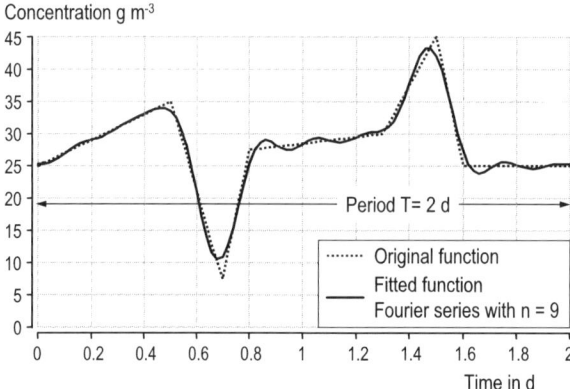

Concentration g m⁻³

Fig. 14.13 Example of an original time series and its image in form of a Fourier polynomial of ninth degree. The frequency spectrum of this function is shown in Fig. 14.12

based on efficient methods which for large n are much faster than the code in Example 14.8.

14.10 Autocorrelation, AR(1) Model

If the current value of a time series correlates with the preceding ones, we speak of autocorrelation. We know that the weather, for example, is autocorrelated: if it was sunny yesterday, then the sun will be out today with a large probability. Random numbers are not autocorrelated, because they do not depend on the other elements of the time series: if I throw a 6 with a die, I do not have a greater chance of throwing a high number in the next trial.

With the trend, the average value, and the periodic contributions we identify deterministic, reproducible aspects of a time series. The residuals of a time series after the subtraction of these deterministic contributions are seemingly caused by random, stochastic processes. We ask: *can we recognize further deterministic contributions in these residuals?*

Example 14.9: Time series analysis versus population analysis

In statistics we deal with large populations of values of one or several variables. Typically these values contain some deterministic relationship and some stochastic contribution. Statistic methods support us in the separation of these two fractions.

If the individual elements of a population are a time series, then the exact sequence of the elements is of interest and we can extract information from this sequence: *we analyze the time series* and may identify autocorrelation.

If the sample is composed of elements of a large population without any temporal relationship between the individual values (the sequence of drawing the elements is totally irrelevant). We infer with the help of *statistical methods* from the sample to the population: *we analyze the population* and may obtain its statistical characteristics.

14.10.1 Autoregressive Models

An autoregressive model makes a prediction for the expected value of a new element based on the past elements of a time series.

In an autoregressive model the current value of a time series is composed of:

- a deterministic contribution based on the values of past elements
- an additional contribution caused by a purely stochastic process (noise)

$$x(t_i) - \mu_x \equiv \tilde{x}(t_i) = \sum_{j=1}^{p} \alpha_j \cdot \tilde{x}(t_{i-j}) + \eta_i \tag{14.19}$$

$x(t_i)$	=	elements of a stationary time series
μ_x	=	average of the elements of the time series $x(t_i)$
$\tilde{x}(t_i)$	=	elements corrected for the average of the time series, $\mu_{\tilde{x}} = 0$
α_j	=	weight of the regression with lag j
η_i	=	stochastic process contribution to element $x(t_i)$, noise

Equation (14.19) describes an autoregressive model of order p (an AR(p) model). Here we will only discuss AR(1) models (p = 1) and assume the time series to be stationary and corrected for the average value. Simplifying, we abbreviate $\tilde{x}(t_i)$ to x_i. Equation (14.19) then becomes

$$\text{AR(p):} \quad x_i = \sum_{j=1}^{p} \alpha_j \cdot x_{i-j} + \eta_i , \tag{14.20}$$

$$\text{AR(1):} \quad x_i = \alpha_1 \cdot x_{i-1} + \eta_i . \tag{14.21}$$

The noise η_i summarizes the stochastic processes, i.e., the unpredictable part of the time series. We assume here η_i to correspond to *Gaussian white noise* (white noise is derived from the picture of a television screen without a signal), i.e., a stationary stochastic process whose elements are normally distributed with expected value $\mu_\eta = 0$ and standard deviation σ_η. Its elements are entirely independent of each other. We designate the ith element of such a process by η_i. It is impossible to distinguish a sample of η_i from a time series of η_i.

There are two limiting cases of an AR(1) process:

- $\alpha_1 = 0$: the AR(1) process becomes pure Gaussian white noise, $x_i = \eta_i$. This process is stationary and there exists no relationship between the individual elements of the time series.
- $\alpha_1 = 1$: the process becomes a random walk (see Sect. 4.2.3). Its variance is not stationary; the standard deviation of the elements increases with the root of time. Its expected value is stationary, $\mu_x = 0$.

For values of α_1 within the range

$$0 \leq |\alpha_1| < 1 \qquad (14.22)$$

the AR(1) process is stationary; the larger the regression weight α_1, the larger the deterministic fraction of the process becomes. For $\alpha_1 > 1$ the expected value μ_x is not stationary and increases exponentially with time.

A valuable array of parameters for the characterization of the degree of auto-correlation of a time series is the autocorrelation function $\rho(k)$:

$$\rho(k) = \frac{N}{N-k} \cdot \frac{\sum\limits_{i=k}^{N} x_i \cdot x_{i-k}}{\sum\limits_{i=0}^{N} x_i^2} . \qquad (14.23)$$

The autocorrelation function $\rho(k)$ computes the coefficient of correlation r be-tween the current element x_i and the element x_{i-k} (k time steps in the past). In the literature one finds instead of Eq. (14.23) also a relation without the first factor (N/(N−k)). For large numbers of elements the differences are not of importance. Here we use Eq. (14.23).

Example 14.10: Simulation of an AR(1) process

The following code was used to generate the time series in Fig. 14.14:

```
STARTTIME = 0          ; Beginning of simulation
STOPTIME = 1000        ; End of simulation, total 1 + STOPTIME / DT elements
DT = 1                 ; Time step
a1 = 0.9               ; Regression weight α₁
sigh = 1               ; Standard deviation of white noise σₙ
init x = 0             ; Initializing time series, could be chosen randomly
next x = a1*x + normal(0,sigh,8)   ; time series xᵢ, seed = 8 makes it reproducible
```

Example 14.11: Computation of the autocorrelation function with BM

The following code allows the autocorrelation function $\rho(k)$ to be obtained for a broad range of values of k. The time series to be analyzed must be corrected for its average value, $\mu_x = 0$.

```
{The time series to be analyzed is read in as data: #x(time), tested}
STARTTIME = 0          ; Beginning of computation
STOPTIME = 1000        ; End of computation
DT = 1                 ; Time step of data series
x = #x(time)           ; Reading of elements of time series
{Computation of covariance function ρ(k)}
N = (Stoptime – Starttime)/DT + 1   ; Number of elements in time series
ord = 100              ; Highest order of autocorrelation to be computed, < N/4
```

init Cov[0..ord] = 0   ; i > 0 covariance, i = 0 variance of elements
next Cov[0..ord] = if time − Starttime > = i*DT then Cov[i] + x*delay(x,i*DT)
                        else 0
rho[0..ord] = if time = Stoptime then N/(N−i)*Cov[i]/Cov[0] else 0
                  ; Covariance function $\rho(k)$

Example 14.12: Autocorrelation of white noise

White noise is defined by a time series where the individual elements are independent of each other, thus no autocorrelation exists. The autocorrelation function is: $\rho(k) = 1$ for $k = 0$ and 0 for $k > 0$.

Below we will discuss the properties of a simulated time series that was based on an AR(1) model and obtained from the code in Example 14.10. The statistical characteristics of the time series are summarized in Table 14.2. Since an AR(1) model contains a stochastic contribution (η_i), obtaining the same sequence again is generally not possible. The code in Example 14.10 always supplies the same sequence because the stochastic function was made reproducible with the aid of a seed. The model is

$$x_i = 0.9 \cdot x_{i-1} + \eta_i (\sigma_\eta = 1).$$

Figure 14.14 shows the first 500 elements x_i of this simulated time series. The autocorrelation is clearly visible: a large value is followed by a second large value, etc. If the elements were independent of each other ($\alpha_1 = 0$), we would expect $v_x = 0.5$ sign changes per element. The observed number of sign changes is $v_x = 0.17$, which is a strong indication of autocorrelation. Figure 14.20 shows the relation between the expected number of sign changes and the regression weight α_1 of an AR(1) time series (here empirically compiled based on a large number of stochastic simulation runs). For $\alpha_1 = 0.9$ we expect a value of $v_x = 0.15$; the deviation is not significant.

Figure 14.16 shows the correlation between the elements x_i and their immediate predecessor x_{i-1}. The obvious correlation is called autocorrelation with lag 1 (time shift between the correlated elements is 1 time step). In contrast to a normal correlation where pairs of data from different series are plotted, the autocorrelation

Table 14.2 Computed statistical characteristics of the simulated time series x_i and the associated stochastic process η_i (see Fig. 14.14 and Example 14.10)

Parameter	Value	Comment
N	1001	Number of elements simulated and analyzed
m_x	0.00	Empirical average value of all elements x_i
s_x	2.31	Empirical standard deviation of all elements x_i
s_η	0.99	Empirical standard deviation of stochastic contribution η_i
$\rho(1)$	0.90	Empirical autocorrelation coefficient with lag 1 (x_i to x_{i-1})
v_x	0.17	Empirical number of sign changes per element of time series x_i
v_η	0.51	Empirical number of sign changes per element of time series η_i

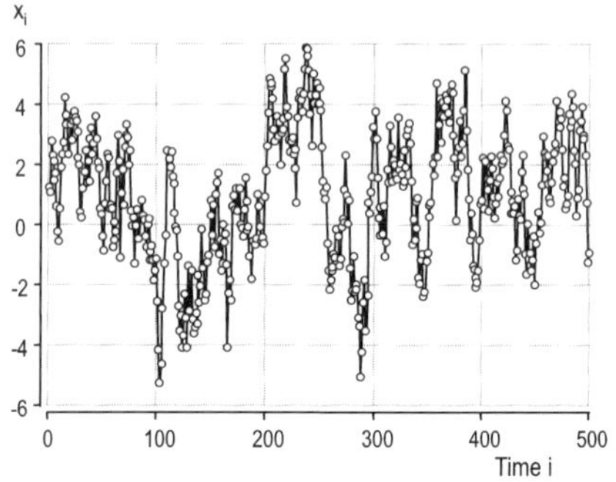

Fig. 14.14 Simulated time series x_i.
AR(1) model with $\alpha_1 = 0.9$, $\sigma_\eta = 1$,
$N = 500$

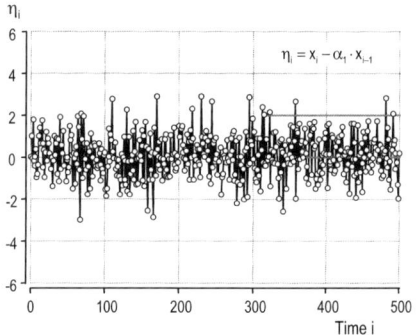

Fig. 14.15 Stochastic process η_i (white
noise) isolated from the series x_i in
Fig. 14.14 with Eq. (14.24): $\eta_i = x_i -$
$0.9 \cdot x_{i-1}$

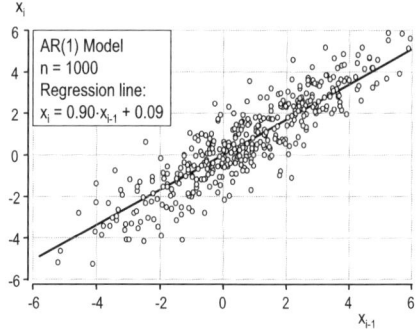

Fig. 14.16 Autocorrelation of the time
series x_i with lag 1

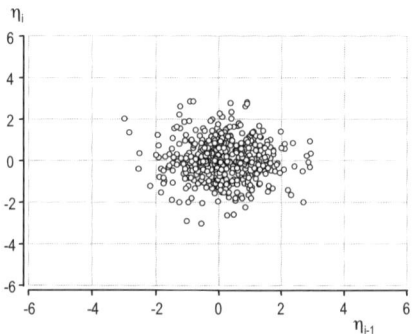

Fig. 14.17 Lacking autocorrelation of
the time series of the stochastic process η_i

relates elements of a single time series which are temporarily shifted. The slope of the autoregression line is equal to the autocorrelation coefficient with lag 1, $\rho(1)$ and its intercept should disappear, if the elements x_i are corrected for their average value, i. e. $\mu_x = 0$. For data which follows an AR(1) model, the autocorrelation coefficient $\rho(1)$ is equal to the regression strength α_1 of the AR(1) model.

With the slope of the autoregression in Fig. 14.16, the regression strength α_1 is known. This allows isolating the stochastic fraction η_i of the time series according to:

$$\eta_i = x_i - \alpha_1 \cdot x_{i-1} . \tag{14.24}$$

The time series η_i is shown in Fig. 14.15, it does not contain any autocorrelation. The number of sign changes per element of η_i amounts to $v_x = 0.51$, very close at the expected value of 0.5. In addition, the comparison of the time series x_i and η_i clearly shows the decrease of the standard deviation of the elements of x_i to η_i: $\sigma_x > \sigma_\eta$. The relationship between these two parameters is given by (see Fig. 14.20):

$$\sigma_x = \frac{\sigma_\eta}{\sqrt{1 - \alpha_1^2}} \quad \text{or} \quad \sigma_\eta = \sigma_x \cdot \sqrt{1 - \alpha_1^2} . \tag{14.25}$$

The autocorrelation function $\rho(k)$ of a time series that corresponds to an AR(1) model consists of a geometrical series according to

$$\rho(k) = \alpha_1^k \quad \text{with} \quad k \geq 0 . \tag{14.26}$$

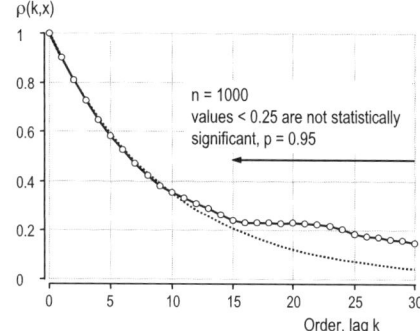

Fig. 14.18 Autocorrelation function $\rho(k)$ based on 1000 simulated elements of the time series x_i. *Dotted line* indicates the ideal result for AR(1) with $\alpha_1 = 0.9$

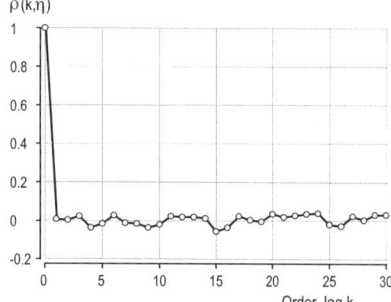

Fig. 14.19 Autocorrelation function $\rho(k)$ based on 1000 isolated values of the time series η_i

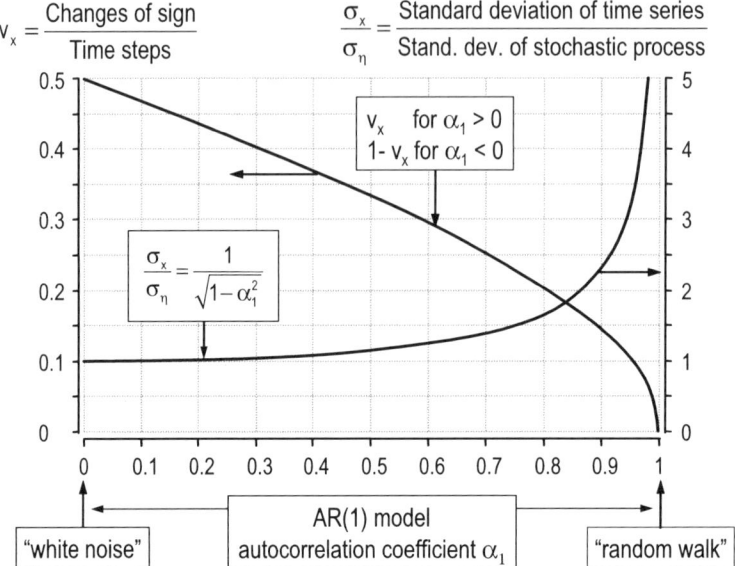

Fig. 14.20 Relationship between the number of expected sign changes v_x (empiric) and the standard deviation σ_x of the elements x_i of an AR(1) time series (computed) and the regression weight α_1 as well as the standard deviation of the stochastic process σ_η. For negative values of α_1 applies: Number of sign changes $= 1 - v_x$. Assumption: η_i corresponds to Gaussian white noise

Since the computation of the autocorrelation function of a time series is quite easy (see Example 14.11), it may be used as an indication whether a time series reasonably follows an AR(1) model. Figure 14.18 shows the autocorrelation function for the time series x_i which corresponds over wide ranges to the theoretically expected values (Eq. (14.26)). As shown in Fig. 14.19, the time series η_i does not contain any autocorrelation; this is an indication of the fact that this series corresponds to white noise. Additionally, the elements of both time series x_i and η_i could be tested for normal distribution, e. g., graphically as shown in Fig. 14.24 or with a specific test.

14.10.2 Summary on AR(1) models

We characterize a time series which we want to model with an AR(1) model, by compiling the following information:

- Correcting the time series regarding trends, periods and average value. Subtracting these corrections and obtaining the residuals x_i which we want to model;
- Analysis of the stationarity of the time series x_i: Is the standard deviation of the first half of the time series equal to the standard deviation of the second half?

- Statistical characterization of the time series: Average value (should be $=0$), standard deviation σ_x, sign changes v_x and correlation function $\rho(k)$. If $\rho(k)$ does not point to an AR(1) model, more complex models may have to be considered which are not discussed here (e. g. ARIMA);
- Graph of the autocorrelation with lag 1 (similar to Fig. 14.16) and identification of the regression line. The slope corresponds to the regression weight α_1;
- Isolating the residuals η_i and examination of this new time series: Does it fulfill the requirements of Gaussian white noise (autocorrelation function $\rho(k)$, sign changes v_η, normal distribution)? The standard deviation σ_η and α_1 are the two parameters of the AR(1) model.

14.10.3 Identification of an AR(1) model

Figure 14.21 shows the COD in 1 min intervals as measured with a new online electrode in the primary effluent of a wastewater treatment plant. The optical electrode reacts very fast to changes in the wastewater. During the 5 hours of measurements that are available, the results show a clear trend which is captured with a polynomial of second degree. Figure 14.21 shows the residuals R_i after correcting the time series for average and trend.

Do these residuals still contain structure that characterizes the process (real variation of the COD concentration), or do they primarily describe the stochastic characteristics of the electrode?

In Fig. 14.22 the strong autocorrelation of the residuals R_i is identified. If this autocorrelation is separated from the residuals, there remains a time series η_i

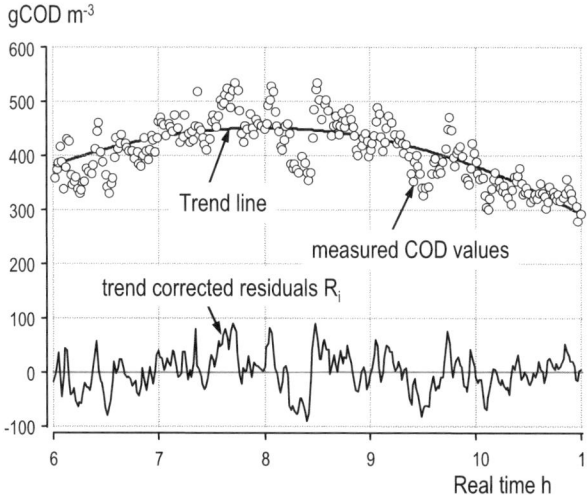

Fig. 14.21 Online measurement of the COD in the primary effluent in 1-min intervals, trend line (second-order polynomial) and trend-corrected residuals of the COD

R_i in gCOD m^{-3}

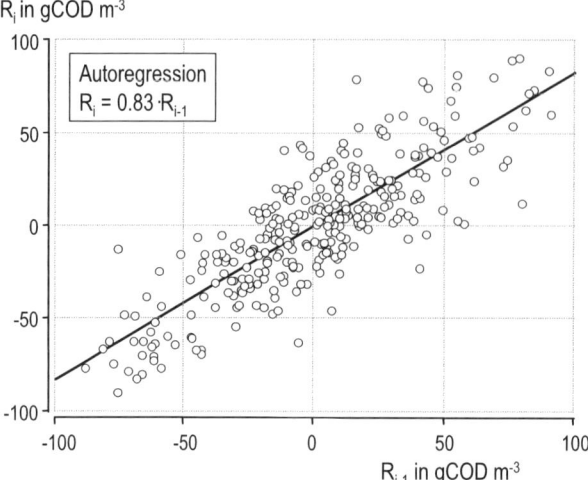

Fig. 14.22 Autoregression of the trend corrected residuals R_i of the COD in Fig. 14.21

η_i in gCOD m^{-3}

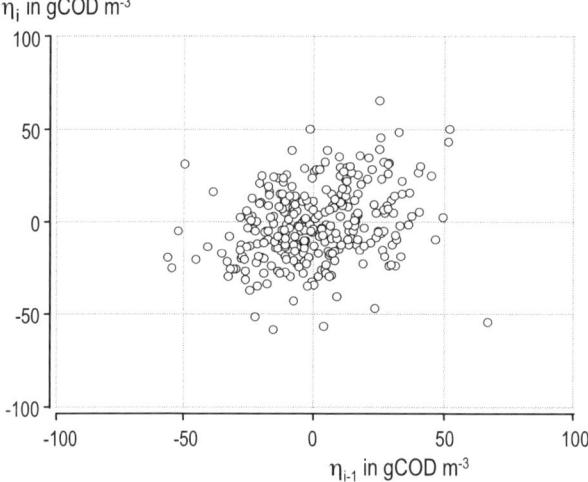

Fig. 14.23 Lacking autocorrelation of the stochastic fraction η_i of the AR(1) model of the residuals R_i

(noise) which is not autocorrelated any more (Fig. 14.23). The elements of η_i are in addition approximately normally distributed (Fig. 14.24), thus we conclude that η_i corresponds to Gaussian white noise (also the autocorrelation function $\rho(k)$ which is not shown points to white noise). The number of sign changes of the residuals R_i amounts to $v_R = 0.19$ per element; from v_R and from Fig. 14.20 we obtain a value of $\alpha_1 = 0.82$, which agrees well with the autocorrelation coefficient $\rho(1) = 0.83$ as identified from the data (slope of regression line in Fig. 14.22).

Fig. 14.24 Normal Q–Q plot of the stochastic residuals η_i (tests for normal distribution)

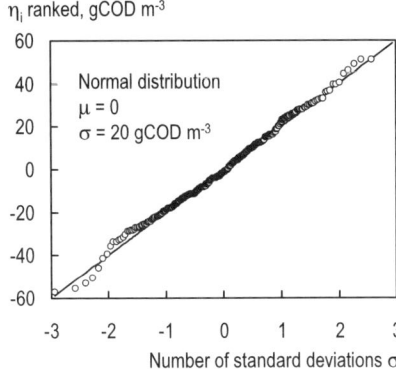

The number of sign changes of the noise η_i is $v_\eta = 0.41$, which is statistically significantly lower than the expected value of 0.5; this points to the fact that the identified AR(1) model does not seize all structural components. The relationship of the standard deviations $\sigma_P/\sigma_\eta = 35.5/20.0 = 1.8$ agrees well with the theoretically expected value from Eq. (14.25).

The residuals R_i therefore contain information which is influenced by past measurements, i.e., the trend line does not seize all deterministic processes. An AR(1) model is suitable to model the residuals; the remaining noise is approximately white and Gaussian. The stochastic contribution from the electrode is most probably smaller than the identified noise ($\mu_\eta = 0$, $\sigma_\eta = 20$ gCOD m^{-3}), because this noise includes the purely stochastic contribution of the real concentrations which cannot be assigned to the noise of the electrode.

Example 14.13: Code for Berkeley Madonna for the analysis of the time series in Fig. 14.21

```
{The time series is read in as COD(time), tested}
STARTTIME = 360              ; Beginning of simulation at 06:00
STOPTIME = 660               ; End of simulation at 11:00
DT = 1                       ; Time step 1 Min
n = (Stoptime − Starttime)/DT ; Number of measurements − 1
COD = #COD(time)             ; Reading the time series
p = a0 + a1*time + a2*time^2 ; Polynomial correcting for trend and average
a0 = −671  a1 = 4.68  a2 = −0.00488; Fitted parameters of the trend line
R = COD − p                  ; Trend corrected residual Ri of COD
R_1 = delay(R,1)             ; R_{i−1}
sR = b0 + b1*R_1             ; Autoregression line
b0 = 0.0  b1 = 0.825         ; Fitted parameters of autoregression
init chi2 = 0
next chi2 = chi2 + (sR−R)^2  ; Test variable to fit autoregression
h = R − 0.83*R_1             ; Correction for autoregression, ηi
h_1 = delay(h,1)             ; η_{i−1}
init sig2R = 0
```

next sig2R = sig2R + R^2/n   ; Variance σ^2 of the residuals
init vR = 0
next vR = if R*R_1 < 0 then vR + 1/n else vR   ; Sign changes of R_i, v_R
init sig2h = 0
next sig2h = sig2h + h^2/n   ; Variance σ^2 of η_i
init vh = 0
next vh = if h*h_1 < 0 then vh + 1/n else vh   ; Sign changes of η_i, v_η

14.11 Case study

In this case study we analyze the time series of the temperature of the wastewater in Fig. 14.2 and derive a stochastic model for the temperature which we will then apply to questions which cannot be answered without stochastic modeling.

14.11.1 Task, Question

A biological wastewater treatment plant is to be extended. In the influent the temperatures in Fig. 14.2 were measured. The future plant must reach a defined performance at temperatures above 10°C. Since temperatures below 10°C reduce the nitrification performance over a long period of time (nitrifiers are partially washed out), we pose the following question: *How many days in series do we have to expect with temperatures below 10°C?*

14.11.2 Procedure

Since the measurements over only about 3 years do not permit to derive a statistically secured statement, we will try to derive a model of the time course of the temperature which includes as much as possible the structural components of the measured time series. We can then analyze the model over a very long period (e. g. 1000 years) and evaluate the results statistically.

We will follow the following steps:

- Correcting for trend;
- Fourier analysis;
- Analysis of the residuals for autocorrelation;
- Analysis of the remaining residuals for Gaussian white noise;
- Deriving a stochastic model;
- Simulation of 1000 yearly variations of the temperature and counting the periods with low temperature;
- Provide a cumulative frequency distribution in order to answer the initial question.

14.11.3 Trend Line

A linear regression over the entire time period of the measured temperatures (Fig. 14.2) results in:

Temperature $= 15.2°C + 0.0016 \cdot$ time in days.

Since the time series is subject to strong periodic variations, we must examine whether we capture entire periods (oscillations) with the time series, or if we introduce an artificial linear trend. Three years cover 1095 days; with 1079 data points we miss 16 days which here fall into winter, where temperatures are far away from the linear regression line. The regression does not identify a true trend because the missing data are systematically low and at the end of the time series. There is no cause to hypothesize a linear trend in the data, we conclude that trend correction is not required.

The trend analysis would be more meaningful after the elimination of the yearly variation. Here we do without this step because we do not have a reason to expect a trend.

14.11.4 Fourier Transformation

A Fourier transformation of the data with the help of the FFT routine of Berkeley Madonna results in Fig. 14.25. The picture is a cutout of the spectrum of amplitudes of the discrete frequencies which are derived from the data ($f_{max} = 1\ d^{-1}$, $f_{min} = 1/1079\ d^{-1}$ plus the average value with $f = 0$). Only the average value and the yearly variation have amplitudes which are clearly larger than the dispersion of the measured values. The spectrum hardly justifies considering frequencies larger than $f = 1/365\ d^{-1}$. As an interesting detail, a weekly variation is identified with the frequency $f = 1/7\ d = 0.143\ d^{-1}$ and an amplitude of $0.3°C$.

As a result we can identify a yearly variation and illustrate this with only one cosine function with the frequency of $f = 1/365\ d^{-1}$ and an average value. Since the Fourier transformation of the 1079 measured data points does not result in exactly this frequency, we identify the parameters of this function again (BM), which results in

$$T_{YV,i} = 16.0 + 5.17 \cdot \cos\left(\frac{2 \cdot \pi \cdot i}{365} + 2.51\right) \tag{14.27}$$

T_{YV} = yearly variation of the expected value of the temperature [°C]
i   = index of the day in the time series, starting 1st Jan. 2000 [–]

Equation (14.27) describes the determined, periodic fraction of the temperature development. We assume that the residuals R_i correspond to a stochastic process which we will analyze further

$$R_i = T_i - T_{YV,i} = T_i - 16.0 - 5.17 \cdot \cos\left(\frac{2 \cdot \pi \cdot i}{365} + 2.51\right)$$

R_i = residual of the temperature on day i, corrected for yearly variation [°C]

Figure 14.26 shows the time series of the residuals R_i; no obvious periodic structures remain, however, there is a positive autocorrelation which expresses itself in particular in the small number of sign changes of this time series (counting results in $v_R = 0.15$ sign changes per element). There is no indication of nonstationarity ($\sigma_{R,i<500} \approx \sigma_{R,i>500}$) or trend. We analyze the time series of the residuals for autocorrelation.

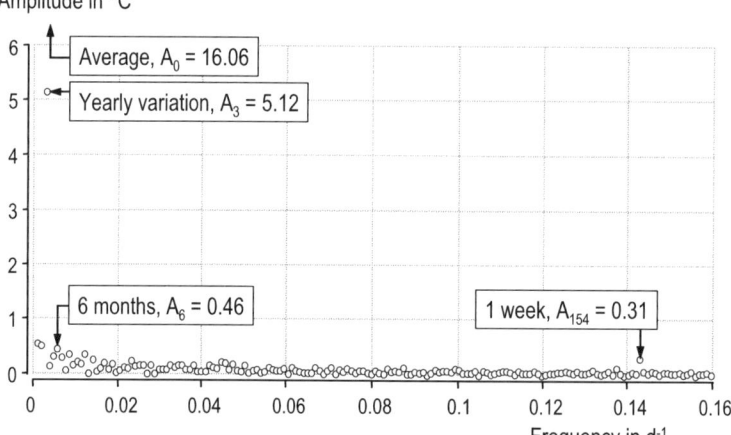

Fig. 14.25 Frequency spectrum of the measured temperatures (FFT, Berkeley Madonna)

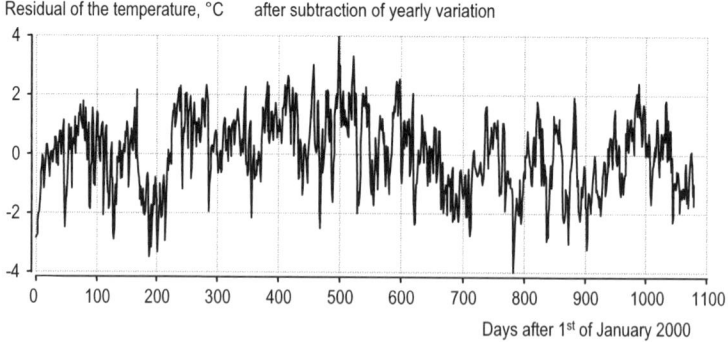

Fig. 14.26 Residuals of the temperature after subtraction of the yearly variation according to Eq. (14.27)

14.11.5 Analysis of the Residuals: AR(1) Model

As a model for the autocorrelation, we assume that with a large probability a high temperature of yesterday leads to an increased temperature today. An appropriate AR(1) model would have the form

$$R_i = \alpha_1 \cdot R_{i-1} + \eta_i \left(\sigma_\eta \right). \tag{14.28}$$

Figure 14.27 shows the autocorrelation of the residuals of the temperature R_i for lag 1; with $r = \alpha_1 = 0.85$, this correlation is significant. From $\sigma_R = 1.25°C$ and the identified $\alpha_1 = 0.85$, Eq. (14.25) results in $\sigma_\eta = 0.65$. This value agrees with the value computed from the elements of the time series η_i. We obtain

$$R_i = 0.85 \cdot R_{i-1} + \eta(0.65°C). \tag{14.29}$$

Based on Eq. (14.29), we can further correct the temperature residuals for auto-correlation; the remainder should correspond to a random process $\eta_i(\sigma_\eta)$

$$\eta_i = R_i - \alpha_1 \cdot R_{i-1}$$

η_i = temperature residuals after elimination of the yearly variation and the auto-correlation with lag 1 [°C]

Figure 14.28 shows the time series of the temperature residuals η_i. We now examine whether this time series corresponds sufficiently to the model of Gaussian white noise: the residuals produce $v_\eta = 0.41$ sign changes per element with an expected value of 0.5. This deviation is statistically highly significant and points to remaining structural model errors. Visual inspection reveals that the negative residuals are larger than the positive ones – the distribution of the residuals is skewed to the left. In Fig. 14.29, the distribution of the residuals η_i is compared with the distribution of Gaussian white noise. Over a wide range the agreement is quite good. In the range of $\pm 2 \cdot \sigma_\eta$ (95% of the values) the deviation is $<0.25°C$. Figure 14.30 shows the autocorrelation function of the time series η_i; its difference

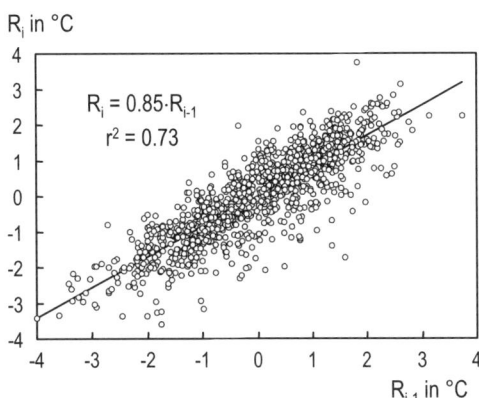

Fig. 14.27 Autocorrelation of the temperature residual R_i after subtraction of the yearly variation

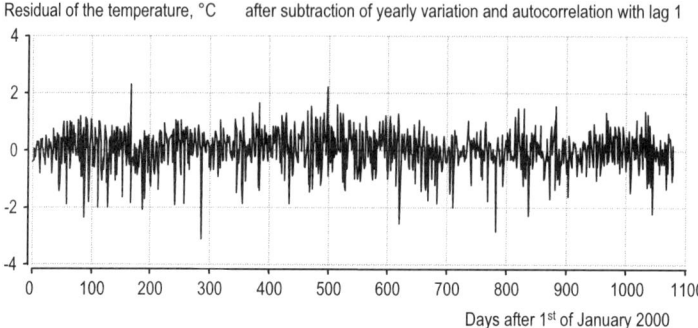

Fig. 14.28 Time series η_i of the stochastic temperature residuals after elimination of the yearly variation and the autocorrelation with lag 1

Fig. 14.29 Comparison of the distribution of the corrected residuals η_i with the normal distribution of Gaussian white noise

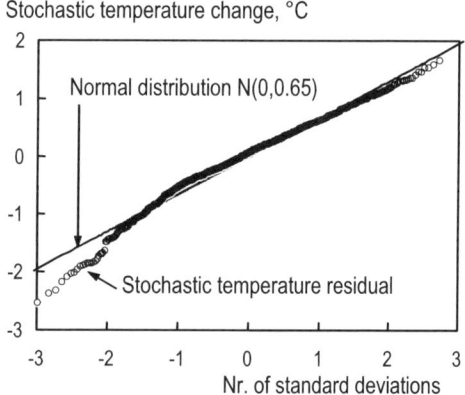

Fig. 14.30 Autocorrelation function $\rho(k)$ for the stochastic residuals η_i

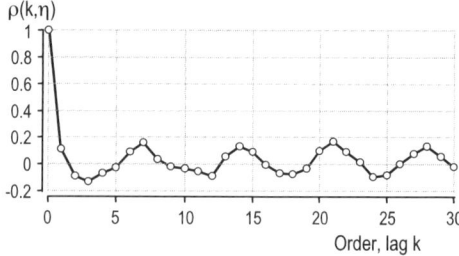

from zero is not statistically significant. (The autocorrelation function $\rho(k)$ has a period of 7 days which stems from the weekly pattern which was already identified in the Fourier transformation, Fig. 14.25).

A further, independent possibility to derive the parameters of the AR(1) model is to identify the sign changes v_R and the standard deviation σ_R of the time series R_i and then to derive from Fig. 14.20 the properties of the AR(1) model. With $v_R = 0.15$ and $\sigma_R = 1.25°C$ we obtain:

- from Fig. 14.20: $\alpha_1 = 0.89$
- from Eq. (14.25): $\sigma_\eta = 0.57°C$
- and therefore instead of Eq. (14.29):

$$R_i = 0.89 \cdot R_{i-1} + \eta_i (\sigma_\eta = 0.57°C). \tag{14.30}$$

14.11.6 Synthesis

Here the elements of the stochastic models are put together and the model is then used to answer the original question based on a long term simulation with the model.

The model for the stochastic simulation of the time series in Fig. 14.2 consists of a yearly variation and an autoregressive model AR(1)

$$T_i = T_{YV,i} + R_i ,$$

$$R_i = \alpha_1 \cdot R_{i-1} + \eta_i (\sigma_\eta) ,$$

$$T_i = \mu_T + A_{YV} \cdot \cos\left(2 \cdot \pi \cdot \frac{i}{365} + \phi \right) + R_i . \tag{14.31}$$

In Fig. 14.31, the data are compared with one run of the stochastic simulation. In addition the yearly variation of the expected temperature is drawn as well as the range of $\pm 3 \cdot \sigma_R$. For the 1079 data points or simulated values, we expect from normally distributed residuals R_i that approximately three values would lie outside this range. Both for the data and for the simulation this estimation is correct.

Thus, with Eq. (14.31) we have a model available which has a statistical behavior close to the behavior of the temperature data. Its implementation is given in Example 14.16. We can now use this model to simulate a long sequence of years and analyze the periods with low temperature. Figure 14.2 provides the cumulative distribution of the length of these periods for the two parameter sets which were

Table 14.3 Parameters and results of the identified models (Fig. 14.20). The two independent parameter sets are underlined

Source	Yearly variation		Autocorrelation					Period $< 10°C$
	μ_T	A_{YV}	Figure	σ_R	α_1	σ_η	v_R	80%
Data	16.0		14.2	1.25			0.15	
Fourier transformation	16.0	5.17	14.25					
Correlation coefficient			14.27	1.25	0.85	0.65	0.22	10 d
Sign changes			14.20	1.25	0.89	0.57	0.15	13 d

Temperature in °C

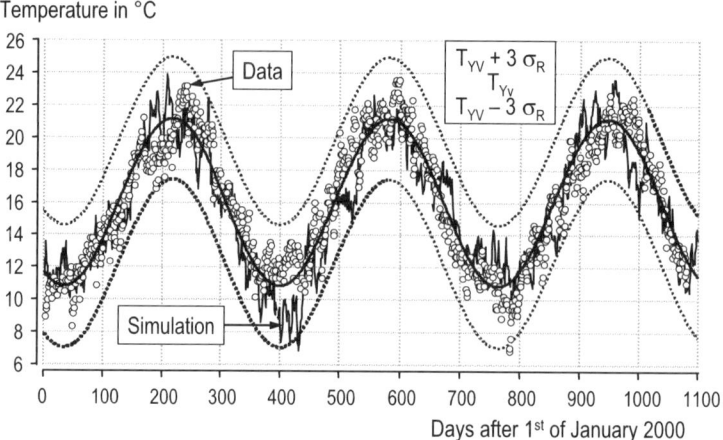

Days after 1st of January 2000

Fig. 14.31 Measured data and simulated temperature course (one realization as an example), Fourier polynomial for yearly variation (T_{YV}), Fourier polynomial $\pm 3\ \sigma_R$

independently identified (Table 14.3). If we are interested, e. g., in the length of a period of cold days which appears on average once every 5 years (80% value), the estimate is 10–13 days. In addition the model could also provide us with a few typical series of temperatures which we would expect in this plant and could use in a rather detailed design procedure.

The fact that there is no single best estimate (say 12 days) relates to second-order uncertainty: the two parameter sets were identified independently, one set gives priority to the variance of the residual and the other one prioritizes the sign changes or the fact that there might be additional processes that influence the data. The original data covers about 3 years however not three full winter periods. The results for these three years would be: 6, 0 and 14 days in series with $T < 10°C$. The estimate of 10–13 days once every 5 years provides us with more security than the simple evaluation of the original data. The model equations rely on the entire yearly pattern and do not only consider some of the temperature values below 10°C.

Example 14.14: Prognosis based on the AR(1) model

On day 150 of the year you measure a temperature of the wastewater of 16.2°C. *Which temperature do you expect tomorrow (day 151)?*
The yearly variation with Eq. (14.27) results for day 150 in an expected value of the temperature of 17.9°C. The residual amounts to $R_{150} = 16.2 - 17.9 = -1.7°C$. According to Eq. (14.29), this residual is degraded to $0.85 \cdot (-1.7) = -1.4°C$. The stochastic process η_i has the expected value 0, therefore the best estimate for the residual $R_{151} = -1.4°C$. The expected temperature on day 151 is $T_{YV,151} = 18.0°C$ and thus the estimated value for the 151st day becomes $T_{151} = 16.6°C$ with a standard error of $\sigma_\eta = 0.65°C$. Without the AR(1) model and only based on the Fourier polynomial the estimate would be $T_{YV,151} = 18.0°C$ with $\sigma_R = 1.25°C$, since no

Fig. 14.32 Simulated values for the maximum number of days in sequence in any one year, during which the wastewater temperature is lower than 10°C. The computation is based on 1000 yearly simulation runs with the stochastic model in Eq. (14.31) and the two identified parameter sets (Table 14.3)

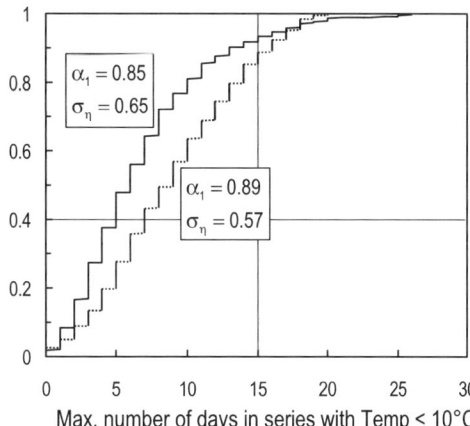

Cumulative frequency of the years

$\alpha_1 = 0.85$
$\sigma_\eta = 0.65$

$\alpha_1 = 0.89$
$\sigma_\eta = 0.57$

Max. number of days in series with Temp < 10°C

autocorrelation with the previous day would be considered and the information of T_{150} could not be used.

Example 14.15: Code for the analysis of the time series

The following code was used to analyze the temperature time series (tested):

```
STARTTIME = 0
STOPTIME = 1079            ; Length of the time series, d
DT = 1                     ; Time step between elements, d
T = #T(time)               ; Reading the time series
Ttrend = a + b*time        ; Determine the trend line with the option Curve
                             Fit
a = 15.2      b = 0.0016   ; Result
TYV = Tavg + ATemp* (cos(2*pi*Time/365 + Phase))
                           ; Identify the yearly variation after FFT
Tavg = 16.0   ATemp = 5.18   Phase = 2.51        ; Result
R = T - TYV                ; Subtract the yearly variation to obtain R_i
R_1 = delay(R,DT)          ; Delay R_i by 1 time step
init sign = 0              ; Count the sign changes in the residuals
next sign = if R*R_1 < 0 then sign + 1/1078 else sign
                           ; Result: v_R = 0.15 changes per element
AR = Kappa + a1*R_1        ; Identify autoregression AR(1)
Kappa = 0.00     a1 = 0.85 ; Result
init chi2 = 0   next chi2 = chi2 + (AR-R)^2
                           ; Test variable for the identification of AR(1)
init sig2R = 0             ; Variance of R_i
next sig2R = sig2R + R^2/1078 ; Result: σ²_R = 1.55
h = R-AR                   ; η_i residual after subtraction of autocorrelation
h_1 = delay(h,DT)          ; Delay of η_i by 1 time step
```

```
init sign_h=0                    ; Count the sign changes of ηi
next sign_h=if h*h_1<0 then sign_h+1/1078 else sign_h
                                 ; Result: v_h=0.41
init sig2h=0                     ; σ_η^2, variance of η_i
next sig2h=sig2h+h^2/1078        ; Result: σ_η^2=0.43
```

Example 14.16: Code for the simulation of the time series

The following code has been used in order to simulate the temperature time series over 1000 years:

```
STARTTIME=250              ; 250 for cold period, 1 for sign changes
STOPTIME=600               ; 600 for cold period, 1000 for sign changes
DT=1                       ; Time step 1 d
sigh=0.57                  ; Standard deviation of noise η_i
a1=0.89                    ; Regression strength of autoregression, α_1
init AR1=normal(0,sigh)    ; Simulated residual R_i, AR(1) model
next AR1=a1*AR1+normal(0,sigh)
T=16+5.17*cos(2*pi*time/365+2.51)+AR1
                           ; AR(1) model plus yearly variation
init sign=0                ; Count sign changes
next sign=if AR1*delay(AR1,1)<0 then sign+1 else sign
init Days=0                ; Count days with low temperature in series
next Days=if T<10 then Days+1 else 0
init Period=0              ; Obtain longest period with low temperature
next Period=max(Period,Days)
```

Chapter 15
Design under Uncertainty

The design of systems and plants is frequently based on the simulation of the expected behavior of these systems. Since these computations build on parameters subject to different errors, uncertainties, and lacking information, the model predictions are subject to uncertainty. Techniques such as Monte Carlo simulation help quantifying the resulting uncertainties in the design and to consider them in our decisions. As a result we obtain the probability that a planned system will be a success or a failure.

15.1 Dealing with Uncertainty

Engineers, and here in particular civil and environmental engineers, typically design prototypes that cannot be tested in their function and performance before they are built, e. g., as pilot plants or physical models in the laboratory. Historically we used our experience and gradually introduced innovations, continuously enlarging the range of experience and application (*adaptive design*). Today we increasingly try to innovate based on models and simulation and their prediction of the performance of the planned system. While building on the experience with the operation of existing plants includes the resulting variability of the performance, we must specifically deal with this aspect when using mathematical models.

Figure 15.1 shows schematically a plant with the load (input) which we want to convert into a product (output) in a plant with its typical performance. In a wastewater treatment plant, the wastewater and its pollutants correspond to the *load*. As a *product* we receive the treated wastewater, which has to satisfy discharge requirements. Treatment processes within the plant provide that the required *performance* of the plant is obtained under a variety of loading conditions (dry and wet weather, peak and low loading conditions). Depending upon the design procedure, we will include capacity reserves that give us sufficient confidence that the

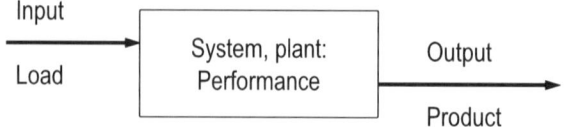

Fig. 15.1 Schematic description of a system or plant in operation

design of the plant can provide the required performance. This additional capacity is chosen, based on the following reasoning:

- We proceed from a high design load: treatment plants are frequently dimensioned for a pollutant load which we assume that in the distant future will not be exceeded during at least 85% of the days (6 out of 7 days).
- Or we may proceed from increased requirements for the product, e. g., we formulate discharge requirements that exceed the requirements of the receiving water and that are based on rare loading conditions. We will then design the plant such that we can expect that the quality requirements can well be reached. A partial failure of the plant does therefore not lead to ecological disaster.
- Alternatively we estimate the performance of the plant based on our experience with other, similar designs. In this experience many problems which can develop are already contained, e. g., that the plant is not optimally operated; or that the plant is subject to weekly load variation which we do not consider explicitly; or that an industrial company affects the waste water composition in an unknown way; etc.
- An alternative estimation of the performance that is increasingly used today is based on the simulation of the expected behavior of the plant with deterministic models which include the relevant processes as reliably as possible. We simulate a variety of loading situations, analyze the results and derive a robust and economical design.

Example 15.1: Discharge requirements are specified, based on rare situations

In the Swiss ordinance for wastewater discharge of 1976 (which is not valid any more) for many pollutants the limiting concentrations were specified for Q_{347}, where Q_{347} is the flow of receiving water that is exceeded on average during 95% of the days. Many treatment plants were then designed such that the required performance to maintain receiving water quality could be achieved in the winter with 10°C during 80% of the days. In addition, this performance was specified for an increased wastewater quantity and pollutant load that was expected in the distant future.

The probability that all these assumptions would coincide is extremely small and led to significant performance reserves in the plants that were built. These reserves are used in the meantime to provide additional performance (denitrification, biological phosphorus removal, etc).

Example 15.2: Load, performance, and product of a bridge

Traffic with a certain frequency and distribution of weight under different weather conditions (load) is to be led safely across an obstacle (product). The bridge (the plant or system) is built with a carrying capacity (performance) that provides a sufficiently large security in fulfilling the demanded performance.

Here the security is ensured by the interaction of many factors: as the design load, heavy vehicles and extreme wind influences are considered. For the performance of the concrete and the steel, only a fraction of the full potential strength of these materials is used. This results in a construction on whose security we (the users) rely beyond any doubt. Such buildings have a failure probability which is very small (but not nil) – a careful civil engineer is in the course of his professional career hardly so unfortunate that he is responsible for the design of a bridge with insufficient performance (at least, if we do not consider aging and corrosion).

This high security (small failure probability) stands contrary to the frequency of the failure of systems that environmental engineers have to design. Society is not willing, and it would hardly be justifiable, to design wastewater treatment plants in such a way that receiving waters are overloaded only once every 1000 years. *Thus, in the environmental engineering sciences we must deal with relatively high probabilities of insufficient system performance.*

15.2 Variation and Uncertainty

Variation refers to the different loads of a plant that occur in reality. We know that this variation takes place, but frequently we cannot forecast when the load takes which value or even which extent the variation has.

Uncertainty refers to variables from which we accept that they have constant values which is in particular true for parameters. We can, within a range, indicate with which probability these variables assume a certain, fixed value. Uncertain variables do not vary; they have a specific value which, however, we do not know exactly. Uncertainty characterizes our ignorance about the value of a variable.

Variation can be determined experimentally by a series of observations. The different values that we obtain are real (but of course subject to measuring error) and have a probability distribution. We speak of *objective probability*, which can be determined by repeated measurements. An increase in the number of measurements does not reduce the variation, but it gives us more confidence in the observed variation. We characterize the variation, e. g., with a probability distribution of possible values.

Uncertainty results from a different cause: few, poor or lacking experiments, unknown disturbances, problematic model structure, etc. Although a parameter for a particular situation has only one value, we indicate a probability distribution of values that we perceive as possible. We speak of *subjective probability*, which may be strongly affected by subjective expert opinion. Additional investigations and experiments can typically reduce this uncertainty.

In the design of plants we must differentiate between variation and uncertainty. The variation of the load leads to the choice of different, relevant design loads. Our uncertainty over the behavior of the plant leads to a conservative design and thus frequently to expensive extra performance capacity. While a plant must master the given variation of the load or the variable disturbances, we can possibly reduce the uncertainties by additional investigations with appropriate time and cost.

Example 15.3: Variation and uncertainty in load and performance of a bridge

The bridge in Example 15.2 is exposed to a *variable load*. Heavy and light vehicles with high and low frequency during strong and weak wind load cross the bridge; its performance must suffice in all combinations of loading conditions, thus we must consider different load conditions in the design phase.

Once built, the bridge exists in a certain form, with a certain maximum carrying capacity (performance). This performance depends on the quality of the concrete and the steel as well as on the shape and design of the bridge. We can estimate the probability that the bridge reaches a certain performance; without destroying the bridge; we will, however, never know the exact value.

Example 15.4: Variation and uncertainty in wastewater treatment

A treatment plant must be able to deal with an array of different loads. The temperature varies throughout the year; the amount of wastewater and the pollutant load depends on the weather conditions and varies throughout the week. We must know this variation and consider it in design and operation.

The performance of a nitrifying activated sludge plant depends heavily on the maximum growth rate of the nitrifiers. This is a constant that we can measure for a certain situation, however, only inaccurately. If we are willing to perform expensive and careful experiments, we can reduce the uncertainty and thus realize the plant with smaller reserve capacity. Not considering the delay and the cost of the extra experiments, this usually leads to a more economical design.

Variation refers to really occurring, different loads of a plant; uncertainty characterizes the lack of our knowledge about the value of a parameter. In both cases Monte Carlo simulation is used in order to characterize the effects on the expected result. Since for each possible set of parameters each combination of the varying loads is also possible, a nested MC simulation is applied as illustrated in Fig. 15.2. This two-dimensional Monte Carlo simulation is computationally very costly: 1000 different loads and 1000 sets of parameters require 1 million computations. Today this is only possible for rather simple models. If variation and uncertainty are not differentiated, then the cost of computation is drastically reduced; we lose, however, the possibility to evaluate these two aspects separately; we can no longer

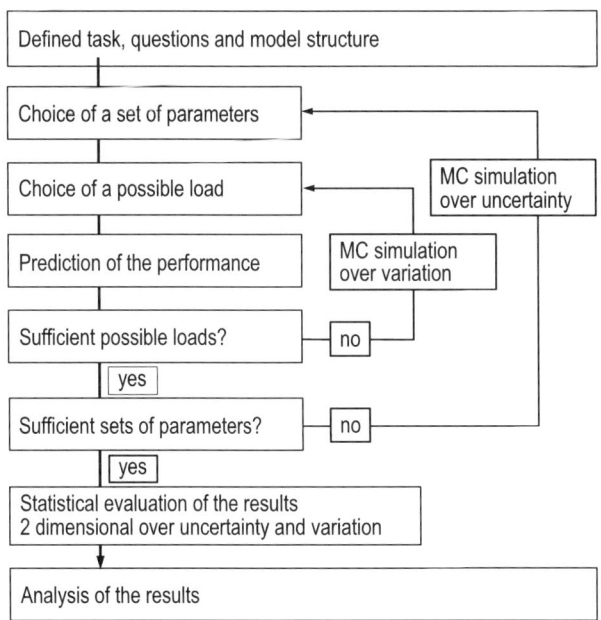

Fig. 15.2 Progression of a two-dimensional Monte Carlo simulation with consideration of parameter uncertainty and variable load

differentiate between the effect of variation (which we cannot reduce) and the effect of uncertainty (which we can reduce) on our result.

Example 15.5: Response of a fish population to a toxic spill

This example is similar to Example 12.16 but here the response of an entire population of fish will be discussed rather than only an individual animal.

Assume that fish react as follows to the exposition (integral of concentration over time) of a pollutant: The tolerated exposition E_T up to an unacceptable damage of an individual is normally distributed: $E_T \sim N(\mu_{ET}, \sigma_{ET}) = N(750\,\text{g min m}^{-3},\ 50\,\text{g min m}^{-3})$.

After an accident the concentration of the toxic pollutant in a river has the following time course of the concentration: $C = C_0 \cdot e^{-k \cdot t}$. Since the observation of the accident is unreliable, the pollutant load can only be indicated inaccurately. The following estimated values are available:

C_0 = evenly distributed in the range of $5 < C_0 < 15\,\text{g m}^{-3}$

k = normally distributed with expected value $\mu_k = 0.02\,\text{min}^{-1}$, $\sigma_k = 0.002\,\text{min}^{-1}$.

What is the probability that more than 20% of the fish population will be damaged?

As opposed to Example 12.16 this question must differentiate between uncertainty (the time course of the pollutant concentration) and variability (the tolerable expo-

Probability, that a given fraction will not be exceeded

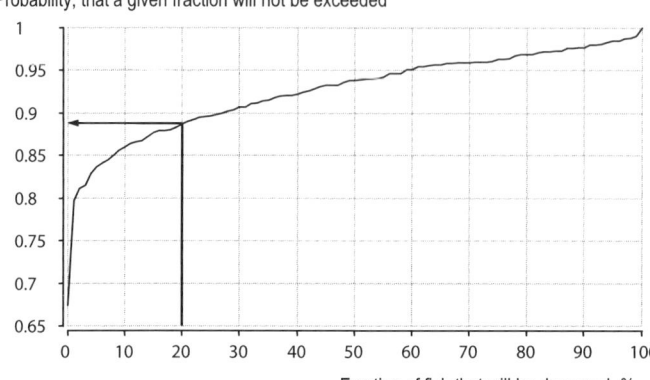

Fraction of fish that will be damaged, %

Fig. 15.3 Result from a two-dimensional Monte Carlo simulation: the cumulative probability distribution for the fraction of damaged fish in Example 15.5

sure of different animals). This results in a two-dimensional MC simulation, which is implemented in the following BM code. First the tolerable exposure of 1000 individual fish is computed (here with a seed to obtain the same 1000 fish in all runs) and then the computations are repeated 1000 times with different time courses of the pollutant concentration. At the end the result is analyzed in a cumulative probability distribution. The mean value from 1000 simulations provides the cumulative distribution shown in Fig. 15.3. Less than 20% of the individual fish will be affected with a probability of approximately 90%.

The code for BM is as follows:

```
METHOD RK4               ; Integration with fourth-order Runge–Kutta
STARTTIME = 0            ; Beginning of the simulation
STOPTIME = 500           ; End of simulation, min
DT = 1                   ; Time step, min
DTout = 500              ; The final result is sufficient
n = 1000                 ; Number of individual fish
init Et[1..n] = normal(750,50,i)
next Et[1..n] = Et[i]    ; Random choice of tolerable exposures
                         ; Describes variability of fish and is equal to the inner
                         MC loop

init C0 = random(5,15)
next C0 = C0             ; Random choice of initial pollutant concentration g m³
init k = normal(0.02,0.002)
next k = k               ; Recession constant for pollutant concentration, min⁻¹
C = C0*exp(−k*time)      ; Time course of pollutant concentration
init E = 0    d/dt(E) = C ; Exposure is equal for all fish
Damage[1..n] = if E > Et[i] THEN 1 else 0
                         ; Check whether damage occurs
```

FractionDamaged = arraymean(Damage[*])
                              ; Fraction of damaged fish
init cumulative[0..100] = 0   ; Cumulative distribution of FractionDamaged, %
next cumulative[0..100] = if FractionDamaged < i/100 then 1 else 0

This code is repeated 1000 times (*Batch Run*) and the mean of cumulative is plotted against [i].

The code has significant potential for optimization.

15.3 Case Study

The following didactical case study serves as an introduction to probabilistic design. The case is artificial and deliberately kept simple. It will point out some problems and a possible procedure. After the description of the task follows a deterministic design procedure, and afterwards it will be shown what additional information a probabilistic procedure may yield.

15.3.1 Task

A reactor for the disinfection of drinking water with the aid of ozone is to be designed. It shall reduce the concentration of active oocysts of cryptosporidium by a factor of 100 during at least 95% of the time. This goal shall be reached with a probability of 90%. If the goal is not reached (10% chance) the plant will be retrofitted at extra cost.

Surface water may contain oocysts (resistant resting cells) of *Cryptosporidium* (pathogenic protozoa), which frequently occur in extremely low concentrations ($<1\ m^{-3}$). Their analysis is rather complex and expensive; however, they endanger the consumer of drinking water, because only a single cyst can lead to illness. Depending upon the quality of the raw water, therefore, one specifies by which factor the treatment must reduce their concentration. A performance control approach is not possible as the analysis of *Cryptosporidia* is too inaccurate and too costly. *Here we proceed from the realistic task that in the water from a lake which is well protected, the concentration of cysts is to be reduced in a first disinfection step with ozone (O_3) by a factor of 100.*

Unfortunately in the course of ozonation some carcinogenic disinfection by-products may be produced. For this reason the maximum concentration of ozone must be kept as low as possible; here it is to be limited to $1\ gO_3\ m^{-3}$.

We use the simple kinetic model shown in Table 15.1. Ozone reacts with the organic fraction of the water matrix (dissolved organic carbon, DOC) and thereby disintegrates in a first-order reaction; thus its concentration decreases continuously. Disinfection is proportional to the ozone concentration S_{O3} and to the concentration of cysts C_C. For the two parameters of the model (k_{O3}, k_D), we find in

Fig. 15.4 Schematic representation of the ozonation reactor: a cascade with six equal reactors in series. Ozone is added to the first and fourth reactor. Ozonation is controlled, so that a constant ozone concentration can be expected in compartments 1 and 4

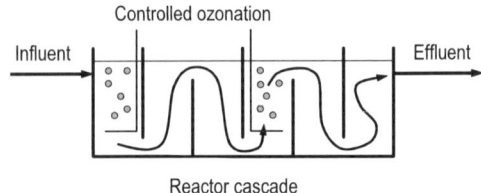

Controlled ozonation

Influent   Effluent

Reactor cascade

Table 15.1 Kinetic model for the degradation of ozone and the disinfection of drinking water (S_{O3} = ozone concentration, C_C = concentration of cysts)

Process	S_{O3} $gO_3\,m^{-3}$	C_C $\#\,m^{-3}$	Process rate ρ
Decay of ozone	-1		$k_{O3} \cdot S_{O3}$
Disinfection		-1	$k_D \cdot C_C \cdot S_{O3}$

Table 15.2 Kinetic parameters for the disinfection of oocysts of *cryptosporidium* with ozone (realistic literature values, see Gujer and von Gunten 2003)

Parameter	Temperature 5°C	10°C	Units
k_{O3}	52	84	d^{-1}
k_D	230	550	$m^3\,g^{-1}O_3\,d^{-1}$

the literature the values listed in Table 15.2. The temperature of lake water varies in the yearly cycle within the range 5–10°C; this has a large effect on the disinfection processes.

Ozone is produced locally and is added to the water via gas exchange. A disinfection reactor is planned, as shown in Fig. 15.4. In a cascade of six reactors, ozone is added to the first compartment. Ozonation is controlled so that an average concentration of $S_{O3,1} = 1\ gO_3\,m^{-3}$ is reached (measured). Since ozone decays along the reactor, additional ozone is added in the fourth compartment based on a second control loop.

Task/Questions

How large must the ozonation reactor be built in order to reduce the concentration of the cysts even under unfavorable temperature conditions by a factor of 100 ($C_C/C_{C,0} < 0.01$) during at least 95% of the time (variation) with a probability of at least 90% (uncertainty)?

Here the probability of 90% stands for the uncertainty that the design does not allow to reach the goal; i. e., if 100 plants with possibly a different purpose and definitely in different locations were built based on the same risk profile, we expect that 90 of them would provide the expected performance. We select a small value for this probability if we are ready to accept the risk that the plant has to be upgraded. A high value of this probability means that we are risk averse and accept a large probability that we will have to build large and expensive capacity reserves. Here we accept a probability of 10% that the plant must be retrofitted after it has been placed in operation. However, retrofitting will then be based on reliable operational experience which may permit an economical solution to be found.

The parameter 95% of the time describes the variation of the performance under variable conditions. The goal is reached if at an unfavorable temperature less than 5% of equally likely operating conditions result in a concentration of cysts larger than 1% of the influent concentration.

15.3.2 Variation

The absolute load of the plant with cysts is not known explicitly. The expenditure to analyze a sufficient number of samples is too large (looking for <1 cyst m^{-3}). Instead a goal is given, in the sense of a risk reduction by at least a factor of 100, i.e., that the efficiency of the plant should, independent of the load with cysts, be in excess of 99% removal.

Here variation results because the kinetics of disinfection is temperature dependent (Table 15.2). At high temperatures ozone decay is fast (k_{O3}), slowing disinfection. However, the increased temperature accelerates disinfection (k_D). Thus, it is not evident *a priori* whether low or high temperatures are overall rate determining for disinfection.

The temperature in the intake of lake water varies in the course of the year. This variation is very slow in comparison with the hydraulic residence time of the ozonation reactor. Thus, we must consider this variation, but we can evaluate the operating conditions based on an instantaneous steady state with constant temperature (see Sect. 10.3.4). Since the temperature range of 5–10°C is given, we can include this variation without problems by determining the critical temperature with a deterministic model.

The two parameters of disinfection (k_{O3}, k_D) depend on the pH value and the content of organic materials in the lake water (expert knowledge) as well as on the temperature; these two further effects are not considered in the model (Table 15.1). These two state variables too are subject to very slow variation in comparison to the hydraulic residence time, which again allows the plant to be considered at steady state. However, these two parameters lead to variation even at a given temperature. For individual steady-state computations we can regard these parameters as constant. The extent of the variation must be based on experience and expert knowledge, because no measurements exist and obtaining such measurements would delay the construction of the plant.

The control of the ozone concentration in the first and fourth reactor results in a variation of the concentration around the set point of 1 gO$_3$ m^{-3}. The course of this variation can contribute to the behavior of the reactor. The variation itself can be obtained from measurements in similar plants or from detailed simulation.

15.3.3 Uncertainty

If we assume that the model in Table 15.1 describes the kinetics of disinfection correctly, i. e., that there are no structural uncertainties, there remain the following uncertainties in our computation:

- The *hydraulic model of the reactor*, based on the model of the reactor in Fig. 15.4 as a cascade of six equal stirred tank reactors. However, complete mixing of the individual compartments is not guaranteed without installing mixing units. The flow through parts of the reactor may be insufficient, leading to dead volume.
- *The influent* is a measured value, adjusted to 10,000 $m^3\,d^{-1}$. This value is subject to systematic deviations due to the possibility of incorrect calibration of the measurement device.
- The control of the *ozone concentration* is based on a measurement that is again subject to measurement error. The repeatability of the measurement is clearly better than the variation that arises from the controller. A lack of accuracy of the measurement as a consequence of erroneous calibration may, however, lead to systematic deviations.
- Even at fixed temperature, pH and DOC concentration, determination of the two kinetic parameters k_{O3} and k_D is subject to considerable experimental error. Here our estimate of the uncertainty of kinetic parameters is based on expert knowledge. We assume that uncertainty is much smaller than variation and can therefore be neglected. Thus, for this simulation we regard only the variation. The parameter uncertainty becomes of importance if we want to predict the performance of the plant for a specific, well-defined operating condition.

Example 15.6: The allocation of variation and uncertainty depends on the model

Whether a variable varies or its value is uncertain depends on the underlying model. In the present case study we do not consider the influence of the pH value and the organic load of the water in the estimation of the kinetic parameters; this leads to the variation of these parameters. We could, however, extend the model and include as further state variables pH and DOC (dissolved organic carbon). In this way we could capture the variation of k_D and k_{O3} in the model and thus only the uncertainty from the experimental determination of the additional model parameters would remain.

Increasing separation of variation and uncertainty entails increasingly more complex models; thereby the procurement of the necessary information becomes ever more complex and more expensive, but our uncertainty in the prediction of an instantaneous value is (may be) reduced. Here we would have to make available additional time series of pH and DOC of the raw water, at the cost of additional expense and time. Moreover, we would have to compile a model that permits us to predict the influence of these state variables on kinetics.

Example 15.7: Measurement error and product quality

In the present case study drinking water that is expected to contain at least 100 times fewer *Cryptosporidia* spores than the raw water during 95% of the time is produced. This goal is derived from a risk analysis. The analysis of *Cryptosporidia* spores is subject to considerable error. The question arises of whether analytical error has to be considered as an additional stochastic process in the design of the plant?

A high count of *Cryptosporidia* due to measurement error does not cause an extra risk; this is an argument in favor of basing evaluation on expected values. However analysis of these organisms is expensive and occurs at low frequency. Thus expected values are subject to a wide confidence interval. Controlling the performance of the plant is not easy in the short term.

The water works have to fulfill legal requirements, which are formulated based on analytical results. Typically these requirements do not allow that measured concentrations exceed required limits with a frequency close to 50% but rather in only 5% or even 1% of the samples. Fulfilling such requirements requires us to consider measurement error as an additional stochastic process.

The conclusion is that, if our focus is on reduction of risk, we do not have to include measurement error as an additional process; with a focus on fulfilling legal requirements and performance control we must consider measurement error. Here ozonation is only one of several treatment steps in series – thus our focus is on risk reduction, indicating that measurement error should not be included in the evaluation of a single treatment step. It must however be included as an additional stochastic process if the performance of the entire treatment chain is evaluated experimentally.

15.3.4 Representation of Variation and Uncertainty

In a deterministic view we proceed from a fixed value or a time series of the variables. In stochastic modeling we regard uncertain and time dependent variables as random variables, whose probability distributions must be available.

The uncertainty in obtaining or estimating model parameters does not permit us to make a clear statement about the absolute value of a parameter. However, we can obtain the probability that a parameter is contained in a given interval of values. The probability distribution of the parameter values can be obtained, e. g., from experiments (see Sect. 12.5) or it results from *expert opinion* based on experience with similar situations and processes and/or careful theoretical evaluations. For variables subject to variation, we can usually obtain the probability distribution of their value experimentally by repeated observations. Here we select the following representation of parameter uncertainties, measurement errors, and stochastic variations:

$$p = \mu_P \cdot f_U \cdot f_V \qquad\qquad (15.1)$$

p = value of a parameter which we use in our computations
μ_P = (expected) value of a parameter, as it would be used in a deterministic design procedure
f_U = dimensionless factor for the characterization of uncertainty. We must indicate its probability distribution. Frequently this factor is normally distributed, with expected value of 1 and given standard deviation $N(1, \sigma_{fU})$ or evenly distributed with upper and lower bounds of the distribution
f_V = dimensionless factor similar to f_U, however, for the characterization of variation

With the factors f_U and f_V, we express the uncertainty or the variability of the parameter P in dimensionless form. In Table 15.3, parameters with their uncertainties and variations are represented for the case study of the ozonation.

Example 15.8: Reduction of parameter uncertainty

The uncertainty of parameters does not have a fixed value. The higher the number of carefully performed experiments we make, the more exactly we can determine the model parameters, however, the larger the costs of experimentation.

Let us assume the uncertainty of a parameter value which is determined in a single experiment to be normally distributed with $f_U \sim N(1, 0.25)$. Each experiment costs € 2,500.

What would be the cost of reducing the expected relative error of this parameter to 10%, or 5%?

The mean error decreases with the root of the number of experiments. Thus, we obtain for $\sigma_{Parameter} = 0.1$:

$$n = \left(\frac{\sigma_{Experiment}}{\sigma_{Parameter}} \right)^2 = \left(\frac{0.25}{0.1} \right)^2 = 7 \text{ experiments with costs of € 17,500.}$$

For $\sigma_{Parameter} = 0.05$ we obtain $n = 25$ with costs of € 62,500.

Obviously we can decrease the uncertainty with additional costs; whether the time and the financial expenditure are worthwhile must be decided on a case-by-case basis.

Table 15.3 Expected values, uncertainty, and variation of the model parameters, variable disturbances, and measured values for the case study of ozonation of drinking water (values based on literature, experience, and expert judgment)

Expected values				
No.	Parameter	Temperature	Symbol	Value	Units
1	Influent	–	Q	10,000	$m^3 d^{-1}$
2	Reactor volume	–	V	to be defined	m^3
3	O_3 controlled	–	$S_{O3,1}, S_{O3,4}$	1	$gO_3 m^{-3}$
4	Ozone decay	5°C	k_{O3}	52	d^{-1}
		10°C	k_{O3}	84	d^{-1}
5	Disinfection	5°C	k_D	230	$m^3 g^{-1} O_3 d^{-1}$
		10°C	k_D	550	$m^3 g^{-1} O_3 d^{-1}$

Parameter uncertainty			
No.	Parameter	Symbol	Distribution	Rational
1	Influent	$f_{U,Q}$	normal(1, 0.05)	Measurement error, MID
2	Reactor volume	$f_{U,V}$	random(0.8, 1.0)	Model uncertainty, dead space
3	O_3 measured	$f_{U,SO3}$	normal(1, 0.05)	Measurement error
4	Crypto measured	$f_{U,C}$	normal(1, 0.25)	Measurement error
5	Ozone decay	$f_{U,kO3}$	random(0.9, 1.1)	Experimental uncertainty
6	Disinfection	$f_{U,kD}$	random(0.9, 1.1)	Experimental uncertainty

Variation			
No.	Parameter	Symbol	Distribution	Rational
1	Influent	$f_{V,Q}$	1	Required production
2	Reactor volume	$f_{V,V}$	1	Accurate construction
3	O_3 control	$f_{V,SO3}$	random(0.9, 1.1)	Variation in control loop
4	Ozone decay	$f_{V,kO3}$	random(0.7, 1.3)	Variation, pH, DOC
5	Disinfection	$f_{V,Kd}$	random(0.8, 1.2)	Variation, pH

1 The influent is constant but subject to an expected measuring error of 5%. This data may originate from the supplier of the measuring instrument or detailed experiments.
2 The reactor volume is to be computed; uncertainty lies in the fraction of the available reaction volume. For the computation with uncertainty a fraction of 0–20% is considered as dead space (based on experience; once in operation this may be obtained from a residence time distribution).
3 The ozone concentrations in the first and fourth reactor are controlled via ozonation. An expected systematic measuring error of 5% (calibration error, based on experience) results and a range of ±10% of the set point in the form of a slowly varying offset of the controlled variable (based on observation in operation or simulation).
4 Measurement of *Cryptosporidium* is quite inaccurate. In our analysis we will however not include this inaccuracy (see the discussion in Example 15.7).
4,5,6 The constants for the decay of ozone and for disinfection are subject to experimental error even at well-defined temperature, pH, and DOC concentration. In addition, they vary as a consequence of the variation of pH and DOC (literature and expert knowledge).

Discussion of Uncertainty and Variation in the Case Study

Table 15.3 shows the uncertainties and variations of the parameters of the model. The expected values correspond to the setting of the task or are taken from the literature (Table 15.2). The uncertainties and variations are based on suppliers' information, experiments and expert opinion (experience). There is no reason to expect correlations between the first three parameters. The variation of the kinetic parameters k_{O3} and k_D could correlate, if e. g., the pH value affects both variables. Since we do not have any detailed information on such correlation, we proceed here from uncorrelated parameter values.

Calibration errors for the probes which measure $S_{O3,1}$ and $S_{O3,4}$ could also be correlated if standard solutions were produced with erroneous concentrations. Since measuring errors are smaller than variation of S_{O3} such a correlation is not considered here (see also Example 15.9).

Example 15.9: Effect of calibration error on plant performance

The control of the ozone concentration in the first and fourth reactor compartment requires the signal of an online ozone electrode. Such electrodes must be recalibrated in regular intervals; this is done with standard solutions. Most probably both these electrodes will be calibrated in sequence. Therefore, if the required standard solutions are produced with an erroneous concentration, both electrodes will be subject to correlated systematic calibration error.

Calibration error will result in ozone concentrations which deviate from their ideal set point. The correlation of the calibration errors, will lead to correlated deviations of the signals of both electrodes. Therefore, if one electrode results in a low ozone concentration, the other electrode most probably will also lead to a low concentration.

Low ozone concentration will lead to poor disinfection performance, an effect which will be enhanced if ozone in both ozonated reactors is low simultaneously. Thus, correlated calibration error will lead to higher uncertainty about reactor performance than uncorrelated calibration error. (see also Example 12.11)

15.3.5 Deterministic Design

In the design and dimensioning of a plant we proceed primarily deterministically, i. e., uncertainties are not explicitly considered but only overall in the form of safety factors. The procedure used here is very simple but in principle correct: On the basis of a deterministic model the performance of the plant for different design loads is predicted, and the uncertainties are considered by adding some extra capacity for safety reasons.

As basis for the deterministic dimensioning we use a model of the plant. Here we select a dynamic model which is implemented in Berkeley Madonna (Table 15.4). We integrate this model forward in time until a steady state is reached (relaxation) and can then analyze the results.

Table 15.4 Code for Berkeley Madonna for the deterministic simulation of the ozonation and disinfection reactor. Parameter values correspond to run no. 2 in Table 15.5. The model is based on the dynamic computation of the stationary condition (relaxation)

{deterministic design – ozonation reactor, tested}
METHOD RK4   ; Integration with fourth-order Runge–Kutta
STARTTIME = 0   ; Beginning of simulation
STOPTIME = 0.2   ; Time to reach steady state, d
DT = 0.00025   ; Time step, d

{deterministic model parameters}
Vtot = 400   ; Total reactor volume, m^3
V = Vtot / 6   ; Volume of a single reactor, m^3
Q = 10000   ; Constant influent, $m^3 d^{-1}$
Ccin = 1   ; Relative concentration of cysts in influent
kO3 = 52   ; Reaction constant for ozone decay, d^{-1}
kD = 230   ; Reaction constant for disinfection, $m^3 g^{-1} d^{-1}$

{dynamic balance equations}
init SO3[1..6] = 1   ; Initial concentration for ozone, $gO_3 m^{-3}$
d/dt(SO3[1]) = 0   ; Controlled concentration in reactor 1
d/dt(SO3[4]) = 0   ; Controlled concentration in reactor 4
d/dt(SO3[2..3]) = Q*(SO3[i−1]−SO3[i])/V−kO3*SO3[i]
                           ; Ozone balance for reactors 2 and 3
d/dt(SO3[5..6]) = Q*(SO3[i−1]−SO3[i])/V−kO3*SO3[i]
                           ; Ozone balance for reactors 5 and 6
init Cc[1..6] = 0.001   ; Relative initial concentration of cysts
d/dt(Cc[1]) = Q*(Ccin−Cc[1])/V−kD*Cc[1]*SO3[1]
                           ; Balance for cysts in reactor 1
d/dt(Cc[2..6]) = Q*(Cc[i−1]−Cc[i])/V−kD*Cc[i]*SO3[i]
                           ; Balance for cysts in reactors 2–6
Ccout = Cc[6]   ; Effluent concentration of cysts

In a deterministic dimensioning we have different possibilities to deal with uncertainty and variation. In no case can we quantify, however, the uncertainty of the result explicitly. Possible methods are:

1. From the many possible load situations (variation) we determine the most unfavorable combination of the variables subject to variation (operating conditions and load of the plant), with which we then just satisfy the target situation.
2. We first dimension the plant (here the reactor volume), using expected values for all model parameters such that we just satisfy the requirements. We subsequently include margins of safety which ought to ensure that even under unfavorable conditions the goal can be achieved, here, e. g., by doubling the computed reactor volume. Based on parameter sensitivity (change of the parameters, section 12.3.2) we might even be able to rationalize the required enlargement of the computed volume.

3. As an alternative to the enlargement of the volume, we proceed from stricter requirements to the expected performance of the plant. Here, e. g., by increasing the required efficiency from $C/C_0 < 0.01$ to $C/C_0 < 0.001$. This performance must then be achieved with, e. g., an extreme load and best estimates for parameter values (expected values). A sensitivity analysis with unfavorable parameter combinations is a possible extension.
4. Alternatively we can proceed from an extreme design load and obtain the required volume to achieve the goal for an unfavorable combination of model parameters.

Method 1

Without further analysis, it is not clear whether disinfection is favored in the summer by increased temperatures and accelerated ozone decay or in the winter with slow ozone decay. A comparison of the necessary volume at 5 and 10°C (see runs 1 and 2 in Table 15.5) shows that the necessary volume becomes clearly larger at 5°C than at 10°C. The variation of the temperature is thus considered here, by selecting 5°C as the critical load. Thus, all further simulation runs will refer to 5°C, a temperature that is expected to last for several month per year.

Table 15.5 Model parameters and performance for different deterministic computations. Italic variables are given; underlined ones computed

Run No.	Temp °C	k_{O3} d^{-1}	k_D $m^3 g^{-1} d^{-1}$	Q $m^3 d^{-1}$	S_{O3} $g\,m^{-3}$	f_V –	V_{tot} m^3	C/C_0 –
1	10	84	550	10,000	1	1	151	0.01
2	5	52	230	10,000	1	1	400	0.01
3	5	52	230	10,000	1	1	800	0.0017
4	5	52	230	10,000	1	1	1016	0.001
5	5	86	138	10,825	0.83	0.8	1550	0.01

Method 2

For the computation of the necessary volume in steady state we must provide the mass balances for all six reactors both for the ozone concentration and for the cysts (Table 15.4). Here the balance equations are written in dynamic form and integrated forward until a steady state is reached. We obtain the required dimensions by determining the volume V_{tot} which allows reaching our goal. In Berkeley Madonna we can use the optimization routine and minimize the difference between computed and desired concentration in the effluent by varying V_{tot}:

$$\left|C_{C,out} - C_{C,Goal}\right| = \text{Min.}$$

The resulting, necessary total volume at 5°C is $V_{tot} = 400 \, m^3$ (see Table 15.5, run 2).

In order to be able to cope with unfavorable situations, we must include capacity reserves. A possibility is to *double* the computed minimum *volume* of $400 \, m^3$ to $V_{tot} = 800 \, m^3$ (the idea being to include a safety factor of two). Thus, an efficiency of 99.83% rather than 99% results for expected parameter values. The expected residual concentration is reduced from $C/C_0 = 0.01$ to 0.0017 by a factor 6 (run 3 in Table 15.5). Why we should choose a factor of two and not anything else is open for debate.

Method 3

A second possibility is to make the *requirement for disinfection* more stringent: From $C/C_0 = 0.01$ to 0.001. We can achieve this more stringent condition with a volume of $V_{tot} = 1016 \, m^3$ (run 4 in Table 15.5). In disinfection we tend towards rather large margins of safety, because the lack of sufficient disinfection endangers the consumers. If this were a wastewater treatment plant, we could choose to reduce the requirement from $C/C_0 = 0.01$ to 0.005. This would require $V_{tot} = 530$ rather than $400 \, m^3$.

Method 4

As an extreme loading situation we can select the most unfavorable parameter combination. Run 5 in Table 15.5 combines the most unfavorable parameter values (with normally distributed variables approximately 95% of the value, k_D small, k_{O3} large, Q large, etc.). With this combination a total volume of $V_{tot} = 1550 \, m^3$ is required to achieve the goal of $C/C_0 = 0.01$ without reserves.

Discussion

In a deterministic procedure, handling the uncertainty in our prediction is based on a rather arbitrary choice of safety margins which is usually based on subjective experience and priorities. The procedure does not permit a statement about the design reserves and the chances of not reaching the goal. Depending upon the choice of the reserves, an uneconomical but frequently successful or an insufficient design may result.

When entire plants are designed as prototypes, our norms do not usually specify required capacity reserves or safety margins. While, e. g., for buildings implicit or explicit safety factors can be specified, this is rather difficult for the individually planned performance of an entire plant. Depending upon the risk profile (willingness to accept a risk) of the owner of the new plant or the consulting engineer, lesser or larger precautionary capacity reserves are realized which strongly affect the economics of the design.

The computations for runs 3–5 in Table 15.5 include different margins of safety which cannot be quantified in detail. Since there is no standard or well-founded experience for such rather rarely built plants, it is in the discretion of the different owners or the advising engineer to choose the reactor volume to be built.

Example 15.10: Design standards for ozonation of drinking water in the US

The United States Environmental Protection Agency USEPA specifies in detail how ozonation reactors are to be designed, built, and operated in order to achieve a certain degree of disinfection. These specifications are, however, based on a series of laboratory scale experiments and theoretical analysis which include uncertainty. The resulting deterministic design procedure is therefore a short cut based on an uncertainty or risk analysis.

15.3.6 Uncertainty-Based Design

A design procedure based on stochastic calculations considers explicitly all quantifiable stochastic load variations and parameter uncertainties which we identify in the development of the deterministic model of the plant. With the help of Monte Carlo simulation, we determine the effects of variation and uncertainties on the performance of the plant. The result is an indication of the distribution of the probability that the plant can actually reach a specified performance.

A Procedure for Stochastic Calculations

The described case study is influenced both by variation of the load and by uncertainty of the values of the model parameters. The setting of the task requires us to make a statement on the frequency of exceeding the limiting value (variation, 95% values) as well as on the uncertainty of reaching the goal (90% probability). This requires a two-dimensional Monte Carlo simulation (Fig. 15.2): For many possible sets of parameters (here 1000), different load combinations (here 1000) are to be simulated. In order to carry out the resulting 1 million simulation runs as efficiently as possible, we must compute the steady state for each combination of load and parameter set at the smallest expenditure. For the deterministic computations we could use the relaxation procedure of the dynamic solution (asymptotic approximation of the steady state, Table 15.4), because only a few computations were necessary. Here we must employ 1 million computations, and thus computation, time becomes import. The algebraic equations which describe the steady state of the six reactors (dynamic equations without accumulation, $dC/dt = 0$) can be solved about 10 times faster with the appropriate routines. The code for the stochastic computations is therefore based on the root finder algorithm of Berkeley Madonna (Table 15.6 and Example 15.12).

Table 15.6 Code for the stochastic simulation of the ozonation reactor. The code is suitable for the inner loop of the two-dimensional Monte Carlo simulation in Fig. 15.2

```
STARTTIME = 1          ; Beginning of simulation
STOPTIME = 1000        ; End of simulation, number of steady states
DT = 1                 ; Time step, time = counter for inner loop over variation
Run = 1                ; Counter for outer loop, parameter plot, uncertainty
```

{Uncertainty: stochastic variables that are kept constant over time
Uncertainty is computed under the option parameter plot}

```
init fuQ = normal(1,0.05)    next fuQ = fuQ
                             ; f_{U,Q} is normally distributed and remains constant
init fuV = random(0.8,1)     next fuV = fuv
                             ; f_{U,V} is randomly distributed
init fuSO1 = normal(1,0.05)  next fuSO1 = fuSO1
                             ; Measuring error of S_{O3,1}, normally distributed
init fuSO4 = normal(1,0.05)  next fuSO4 = fuSO4
                             ; S_{O3,4} is independent of S_{O3,1}
fukO3 = 1      fukD = 1      ; Uncertainty of k_{O3} and k_D are neglected
```

{Variation: stochastic variables that are recomputed randomly after each time step}

```
fvSO1 = random(0.9,1.1)   ; SO3,1 variation due to control loop, evenly distribu-
                            ted, no uncertainty
fvSO4 = random(0.9,1.1)   ; SO3,4 variation due to control loop, evenly distribu-
                            ted, no uncertainty
fvkO3 = random(0.7,1.3)   ; Variation of k_{O3} is randomly distributed
fvkD = random(0.8,1.2)    ; Variation of k_D is randomly distributed
```

{Model Parameters}

```
Vtot = 500              ; Total reactor volume, m^3
V = fuV*Vtot / 6        ; Active volume of a single reactor, m3
Q = fuQ*10000           ; Influent, m^3 d^{-1}
kO3 = 52*fvkO3*fukO3    ; Reaction constant for ozone decay at 5°C, d^{-1}
kD = 230*fvkD*fukD      ; Reaction constant for disinfection, 5°C, m^3 g^{-1} d^{-1}
```

{Balance equations, solution for steady state with Root Finder. A total of ten unknowns}

```
SO1 = fvSO1  SO4 = fvSO4 ; S_{O3,1} and S_{O3,4} are controlled, remain constant
```

{here follow the ten stationary material balance equations in the form of the *Root Finder*}

```
Guess SO2 = 0.5  Limit SO2 <= 1   Limit SO2 >= 0
                 RootS SO2 = Q*(SO1–SO2)/V – kO3*SO2
Guess SO3 = 0.5  Limit SO3 <= 1   Limit SO3 >= 0
                 RootS SO3 = Q*(SO2–SO3)/V – kO3*SO3
Guess SO5 = 0.5  Limit SO5 <= 1   Limit SO5 >= 0
                 RootS SO5 = Q*(SO4–SO5)/V – kO3*SO5
Guess Sout = 0.5 Limit Sout <= 1  Limit Sout >= 0
                 RootS Sout = Q*(SO5–Sout)/V – kO3*Sout
```

Cin = 1   ; Relative influent concentration of cysts
Guess Cc1 = 0.5   Limit Cc1 <= 1   Limit Cc1 >= 0
                   RootS Cc1 = Q*(Cin−Cc1)/V − kD*SO1*Cc1
Guess Cc2 = 0.5   Limit Cc2 <= 1   Limit Cc2 >= 0
                   RootS Cc2 = Q*(Cc1−Cc2)/V − kD*SO2*Cc2
Guess Cc3 = 0.5   Limit Cc3 <= 1   Limit Cc3 >= 0
                   RootS Cc3 = Q*(Cc2−Cc3)/V − kD*SO3*Cc3
Guess Cc4 = 0.5   Limit Cc4 <= 1   Limit Cc4 >= 0
                   RootS Cc4 = Q*(Cc3−Cc4)/V − kD*SO4*Cc4
Guess Cc5 = 0.5   Limit Cc5 <= 1   Limit Cc5 >= 0
                   RootS Cc5 = Q*(Cc4−Cc5)/V − kD*SO5*Cc5
Guess Cout = 0.5   Limit Cout <= 1   Limit Cout >= 0
                   RootS Cout = Q*(Cc5−Cout)/V −kD*Sout*Cout

{Evaluation of the cumulative frequency of Cout, i stands for concentration and must be scaled}
init cumCout[0..250] = 0   ; Range of $C/C_0 = 0-0.025$, graphics: cumCout[i] against i
next cumCout[0..250] = if Cout < i/10000 then cumCout[i] + 1/Stoptime else cumCout[i]

{Search for that 95% value for a set of the uncertain parameters}
Count[0..250] = if cumCout[i] < 0.95 then 1 else 0
Cout95 = if time < Stoptime then 0 else Arraysum(Count[*])/10000 + 0.0001
                            ; 95% value of Cout

Example 15.11: Variation, time constants and static models

The variation of a variable disturbance occurs over a certain time horizon, i.e. in the present case study this means that the measured concentrations of cysts in the raw water correlate over a certain time strongly among themselves (autocorrelation; see Sect. 14.10). The concentration in the lake water will not jump from minute to minute over orders of magnitude. If we can assume that the concentration remains constant over the hydraulic residence time of the water in our reactors ($\theta_h \approx 2$ h) and that the hydraulic loading does not vary either, then we can rely on static computations; if, however, the concentrations or the flow vary rapidly, we cannot ignore the effects of these dynamics on the performance of the plant and must simulate it in all its detail.

In the biological treatment of wastewater the situation is entirely different: here the load changes significantly during a hydraulic residence time, so that a dynamic view is necessary (see Sect. 10.3.4).

Example 15.12: Stochastic simulation of the ozonation plant

For the case study of ozonation Table 15.6 gives a code that permits the simulation of the steady state of the plant for different loads. This code is suitable for the internal loop of the Monte Carlo simulation (variation) in Fig. 15.2. The outside loop (uncertainty) is obtained by calling this program 1000 times under the option parameter plot.

In this code the system variable *time* is used to count the number of steady-state computations. The uncertain parameters are initially selected randomly and are then kept constant for an entire run of the code. All variable parameters are changed randomly after each time step DT. Finally for each 1000 variations the 95% value of the computed performance C_{out} is determined.

The uncertain parameters are changed for each of the 1000 calls of the program. The variable *run* is increased from 1 to 1000 within the option parameter plot. For each call, the 95% value of the computations is stored. In the end in a separate routine the frequency distribution of these 95% values is determined, which requires the final results of the parameter plot to be written to a file that can then be sorted and analyzed (e. g., in Excel).

The code in Table 15.6 is not fully optimized for speed. This is not required here because the overall computation time of a two-dimensional MC simulation is still rather small (< 30 s). For larger models it may be required to optimize the code.

Figure 15.5 shows the probability with which we may expect to obtain a given disinfection performance with increasing reactor volume. The basis for the parameter uncertainty and the variation of the load is given in Table 15.3. We must interpret the resulting probability as follows: *considering all knowledge that the design engineer has at the time of the dimensioning of the plant, he expects that the design goal of the plant will be reached with the probability indicated. If many plants (in different locations) were designed based on the same method, then it should be possible to validate this probability.*

Cumulative frequency of the 95% value (uncertainty)

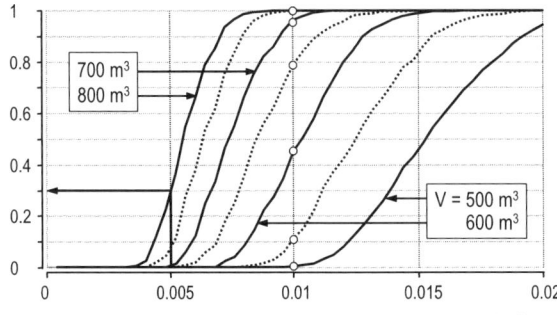

Residual fraction of remaining cysts, C/C_0, 95% value

Fig. 15.5 Cumulative frequency of the 95% value of the performance C/C_0 that may be expected from ozonation reactors with increasing total volume V in winter at 5°C. For each volume, 1000 Monte Carlo simulations with different parameter sets have been performed for each of 1000 different loads per volume. *Reading example*: with a reactor volume of 800 m^3 we can expect to obtain with a probability of 30% a disinfection performance of $C/C_0 = 0.005$ in 95% of the operating conditions at 5°C

Probability of $C/C_0 \leq 0.01$ in 95% of the samples

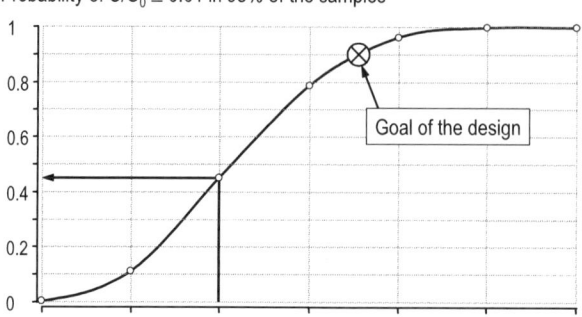

Total reactor volume, m³

Fig. 15.6 Probability that a plant with a given total volume will achieve a performance of $C/C_0 = 0.01$ during 95% of the time at 5°C. The circles correspond to the circles in Fig. 15.5. *Reading example:* with a total volume of the ozonation reactor of 600 m³ we can expect to reach the goal of $C/C_0 \leq 0.01$ during 95% of the time at 5°C with a probability of 45%

Figure 15.6 is extracted from Fig. 15.5 and shows, for increasing reactor volume V_{tot}, the probability that the goal of $C/C_0 < 0.01$ is achieved in 95% of the loading conditions. The task of designing a plant that is capable of delivering drinking water that fulfills the requirement ($C/C_0 \leq 0.01$) with at least 90% probability calls for a reactor volume of $V_{tot} \geq 675$ m³. With this volume, there is a 10% chance that the plant does not satisfy our expectations and therefore must be upgraded. Figure 15.6 also shows that with less than 5% extra costs (more volume) the risk that insufficient drinking water is produced ($C/C_0 > 0.01$) can be reduced from 10% (with a volume of 675 m³) to only 3% (with a volume of 700 m³). Considering these low extra costs, the owner decides to set the volume at 700 m³.

15.3.7 Operational Experience and Retrofitting of the Plant

Frequently the design must take place without a comprehensive knowledge of the local situation. However, once the plant is in operation, we receive secure information that allows our design assumptions to be reexamined.

After the construction of the plant, the operator realizes that the value of k_{O3} has been estimated too small with the information on variation given in Table 15.3. The observation of low ozone concentrations in the effluent and several laboratory experiments to obtain the value of k_{O3} reveal that at 5°C $f_{V,kO3}$ is evenly distributed within the range 1.1–1.3 and not, as was used as the basis for the design, the range 0.7–1.3. In addition, experiments for the determination of the residence time distribution indicate that only 80–90% of the built reactor volume is active and 10–20% of the volume must be regarded as dead [thus $f_{U,V} = random(0.8, 0.9)$ instead of $random(0.8, 1.0)$ as used for the design].

Cumulative frequency of the 95% value (uncertainty)

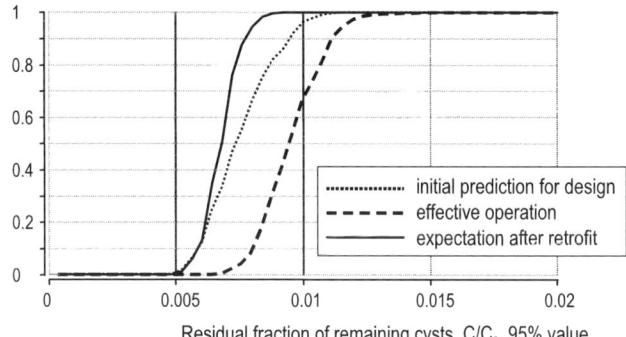

Residual fraction of remaining cysts, C/C_0, 95% value

Fig. 15.7 Probability that a reactor with a volume of $700\,\mathrm{m}^3$ reaches a certain performance (C/C_0, 95% value at 5°C). The original design is identical to the appropriate case in Fig. 15.5. The analysis of the effective operation considers the operational experience (a conditional probability distribution); the retrofit with mixing units reduces the dead volume and thereby reduces the risk of an insufficient performance

Both experimental observations are within the range of the uncertainty which the engineer used for his design, but both point to a reduced disinfection performance. The determination of the real performance of the plant is not possible because the determination of the small residual concentration of the oocysts of Cryptosporidium is too laborious, inaccurate, and expensive.

Considering the larger than initially expected risks, the owner decides to reexamine the design. With the reduced uncertainties based on operational experience, the remaining probability that the plant always produces a satisfactory quality of water is now only 67% rather than the expected 97% (Fig. 15.7).

In order to reduce the risk of producing insufficient drinking water quality to the level expected from the design, the owner decides to retrofit his plant as economically as possible. The engineer suggests installing mixing devices in the individual reactors and thereby eliminating the considerable dead volume. This very economical solution leads to an improvement of the performance that far exceeds that of the initial design (Fig. 15.7).

Clearly the design under uncertainty should be reevaluated once operational information becomes available and allows the quality of the uncertain information to be improved.

15.3.8 Critique of the Design Procedures

If engineers design a plant, they are conscious of the risk that the built plant may not live up to their expectations. However, there exist different possibilities to deal with this risk.

A deterministic design procedure does not allow one to obtain the probability with which a given goal will be achieved. Safety margins must be specified based on standards, legal requirements or experience. The fact that we lack understanding of the planned processes, that information may be unavailable, or that we have limited experience, etc. is not explicitly considered in the calculations. Overall economic optimization procedures and analysis of risk can hardly be used here. The results of the calculations are simple to communicate and are typically understood by the owners of the plant (who expects of course that the engineer should know all these details).

In a stochastic design procedure, we try to characterize our ignorance and to point out how this may affect the behavior of the planned systems. However, the identification of our ignorance and the quantification of the uncertainty are very subjective processes which, when analyzed themselves, are again very uncertain (*second-order uncertainty*). If we accomplish this work carefully and self-critically, Monte Carlo simulation provides us with the possibility to discuss the effects of our ignorance relative to the expected performance of the plant. To date, we have little experience in the interpretation of the results that we receive from such an investigation. In particular the owner will have problems dealing with the statement that the planned plant can fulfill his expectations only with, e. g., 90% probability and that there is a 10% chance that his new plant will have to be retrofitted after placing it in operation. From his point of view, he has hired and paid the consulting engineer to design a successful plant.

The forced, careful quantification of our ignorance has the advantage that we can judge when additional investigations are necessary and rewarding. At the same time we engineers become more conscious that not everything is feasible.

Given the ever-increasing computing power available, we will in the future see more and more stochastic design procedures. Learning to perform and deal successfully with such procedures may be a lengthy and possibly difficult task.

Today it is not clear who will carry the extra cost and financial risk of stochastic design procedures. Our customers will however happily accept the savings that are possible if such procedures are applied.

15.4 Second-Order Uncertainty

So far we have been dealing with uncertainty of parameter values and have assigned explicit values to this uncertainty. However, depending on how we obtain these values of uncertainty, even this uncertainty may be uncertain. We call this second-order uncertainty. Frequently this is neglected in uncertainty analysis.

In the previous case study on disinfection by ozonation the rate constants k_{O3} and k_D are rather uncertain (Table 15.3). An expert might be tempted to express his subjective uncertainty about these parameters in a rather fuzzy, uncertain way. An example is given in Table 15.7, where random distributions of the parameter values

Frequency of parameter value [-]

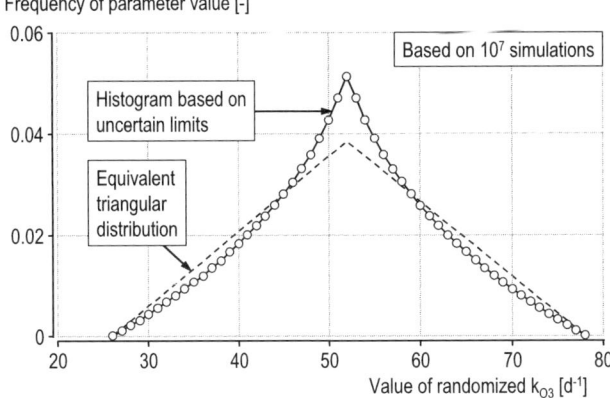

Fig. 15.8 Resulting distribution of k_{O3} due to the uncertain definition of the limits of the even random distribution of the parameter values (Table 15.7)

Table 15.7 Uncertain characterization of uncertainties, second-order uncertainty (see Table 15.3)

Variation				
No.	Parameter	Symbol	Distribution	Rational
4	Ozone decay	$f_{V,kO3}$	random(G_l,G_u)	Variation, pH, DOC
	Uncertainty	G_l	*random(0.5,1.0)*	*Lower bound of variation*
	about variation	G_u	*random(1.0,1.5)*	*Upper bound of variation*
5	Disinfection	$f_{V,Kd}$	random(G_{Dl},G_{Du})	Variation, pH
	Uncertainty	G_{Dl}	*random(0.75,1.0)*	*Lower bound of variation*
	about variation	G_{Du}	*random(1.0,1.25)*	*Upper bound of variation*

are specified but the boundaries of these distributions are again declared uncertain. We can of course include this second-order uncertainty in our analysis, but the procedure becomes ever less transparent (see Example 15.13).

Example 15.13: Uncertain definition of distributions may lead to unforeseeable results

In Table 15.7 the anticipated variation of the decay constant of ozone $f_{V,kO3}$ is formulated very vaguely. The upper and lower bounds of even random distributions are defined in an uncertain way themselves (second-order uncertainty). This is at first sight a cautious formulation of the expert. In the stochastic simulation this formulation has, however, the consequence that an entirely different probability distribution of the variable values of the parameter k_{O3} is obtained, as shown in Fig. 15.8 which was obtained from the following BM code:

```
STARTTIME = 1        ; Time is used to count the number of simulations
STOPTIME = 1E7       ; A total of 10^7 random values of k_O3 are determined
DT 1
```

DTout = 1E7   ; Output only after termination of the program
lo = random(0.5,1.0)   ; Lower uncertain limit of $f_{V,kO3}$
hi = random(1.0,1.5)   ; Upper uncertain limit of $f_{V,kO3}$
k = 52 * random(lo,hi)   ; Generation of stochastic values of k_{O3}, 52 = expected
                         value

{ Generation of a histogram of values of k_{O3} }
init h[26..78] = 0
next h[26..78] = if round(k) = i then h[i] + 1/STOPTIME else h[i]

Clearly it is more straightforward to define a triangular distribution with known properties rather than the nested combination of uncertain, even distributions.
The uncertain formulation of variation was introduced here for didactic purposes only and is not advised in practice.

Chapter 16
Problems

The following problems are arranged approximately in the sequence of this text. It is, however, not always possible to assign a specific problem exactly to an individual chapter.

Some examples require rather large sets of data. These are available from the homepage under www.sww.ethz.ch.

16.1 Composition Matrix and Conservation Equation

The use of the composition matrix and the conservation equation for elements and charge is just a systematic way to balance chemical reaction equations. Use this procedure to balance the following nitrification reaction:

$$1 \cdot NH_4^+ + \nu_{O2} \cdot O_2 \rightarrow \nu_{NO} \cdot NO_3^- + \nu_{H_2O} \cdot H_2O + \nu_{H^+} \cdot H^+$$

1. Write a stoichiometric matrix, indicating the units of all stoichiometric coefficients.
2. Write a composition matrix, using all relevant conservatives, and indicate the units of the composition factors.
3. Write down and solve all required conservation equations.
4. Repeat points 2 and 3 but use only the conservation of TOD and nitrogen, as is customary in many activated sludge models.

It is of course much simpler to balance this equation directly, but for microbial reactions and reactions involving group parameters such as COD this is frequently the most efficient approach.

16.2 Conservation of TOD

A sand filter treats the settled wastewater of a single house. Table 16.1 contains some average concentrations. Organic material (COD) contains some nitrogen at the redox level of $NH_4^+ - N$, $i_N = 0.04\,gN\,g^{-1}COD$. Ten percent of the organic material accumulates in the sand filter.

You should consider the processes of degradation of COD, Nitrification to NO_3 and denitrification to N_2.

Table 16.1 Wastewater characteristics in the influent and effluent of the soil filter (real data, organic material contains $i_N = 0.04\,gN\,g^{-1}COD$)

Material	Units	Influent	Effluent
Organic material	$gCOD\,m^{-3}$	320	32
Ammonium, NH_4^+	$gN\,m^{-3}$	82	6
Nitrate, NO_3^-	$gN\,m^{-3}$	0	51

1. How much nitrogen is lost due to denitrification?
2. How much oxygen must be delivered to this soil filter? How many cubic meters of air are required to treat $1\,m^3$ of wastewater if $1\,m^3$ of air contains roughly $300\,g\,O_2$?

16.3 Breakpoint Chlorination: Stoichiometry and Composition

Breakpoint chlorination consists of a sequence of reactions in which ammonium is oxidized with chlorine to mono-, di-, and trichloramine and finally to N_2. Table 16.2 provides the details of some of the reactions involved.

1. Derive a stoichiometric matrix for the reactions in Table 16.2.
2. In view of Table 16.2 what is the TOD of Cl?
3. Write the composition matrix for the stoichiometric matrix in question 1.
4. Write the full conservation equation for N and TOD for reactions 1, 4, and 6.

Table 16.2 Major reactions involved in breakpoint chlorination of water with low ammonium content

$NH_4^+ + HOCl$	$\rightleftarrows$	$NH_2Cl + H_2O + H^+$
$NH_2Cl + HOCl$	$\rightleftarrows$	$NHCl_2 + H_2O$
$NHCl_2 + HOCl$	$\rightleftarrows$	$NCl_3 + H_2O$
$2NH_2Cl + HOCl$	$\rightleftarrows$	$N_2 + H_2O + 3HCl$
$NH_2Cl + NHCl_2$	$\rightleftarrows$	$N_2 + 3HCl$
$3NH_2Cl + NCl_3$	$\rightleftarrows$	$2N_2 + 6HCl$

Fig. 16.1 Reaction scheme for anaerobic digestion of domestic sludge. Percentages indicate substrate flow (stoichiometrically) in the form of COD or CH_4 equivalents. Only the net flow of substrates (degradation minus biomass formed) through cell external pools are indicated. Number in circles identify different processes (Gujer and Zehnder, 1983)

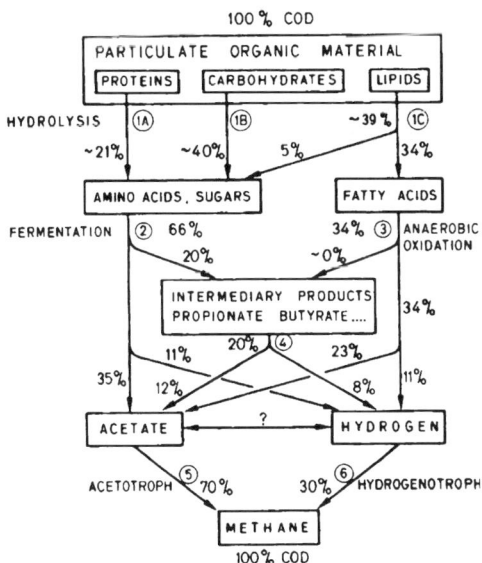

16.4 Deriving a Stoichiometric Matrix

Gujer and Zehnder (1983) suggested a reaction scheme for the degradation of sewage sludge in anaerobic digestion. It is reproduced in Fig. 16.1.

1. Write the stoichiometric matrix for the reaction scheme in Fig. 16.1.
2. Can you reconstruct the reaction scheme in Fig. 16.1 from the stoichiometric matrix or is any information lost in the transformation process in question 1?
3. What form of composition matrix corresponds to the reaction scheme in Fig. 16.1?
4. How has conservation of TOD been used in Fig. 16.1?

16.5 Mass Balance in the Steady State

A trickling filter is operated in steady state. You sample the influent and effluent and obtain the concentrations indicated in Table 16.3.

1. How much dinitrogen (N_2) is exchanged with the atmosphere?
2. How much oxygen is transferred into the wastewater?

Table 16.3 Wastewater characteristics in the influent and effluent of a trickling filter at steady state

Material	Units	Influent	Effluent
Dissolved oxygen	$gO_2\ m^{-3}$	2	6
Ammonium, NH_4^+	$gN\ m^{-3}$	20	2
Nitrate, NO_3^-	$gN\ m^{-3}$	0	15
Dinitrogen, N_2	$gN\ m^{-3}$	15	15
Organic material[1], COD	$gCOD\ m^{-3}$	200	80

[1] You may neglect the nitrogen content of the organic material

16.6 Ideal Reactors, Chemostats

Microbiologists use chemostats to grow microorganisms. A chemostat is nothing more than a CSTR at steady state.

You want to know which one of two microorganisms can be enriched in a chemostat under different operating conditions (flow rate, substrate concentration). For this, you write a program in BM and vary the operating conditions with the aid of a parameter plot (result versus parameter value).

The volume of the chemostat is $V = 1$ L.

The substrate concentration is to be varied in the range of $100\text{--}5000\ gCOD\ m^{-3}$ with $Q = 1\ L\ d^{-1}$.

The flow rate varies in the range of $Q = 0.1\text{--}10\ L\ d^{-1}$ with $S_{in} = 1000\ gCOD\ m^{-3}$.

The kinetic model for the two organisms is defined in Table 16.4. There is always enough oxygen present.

Table 16.4 Kinetic model of the two organisms which compete in the chemostat

Process	Oxygen S_{O2} gO_2	Substrate S_S $gCOD$	Organism A X_A $gCOD$	Organism B X_B $gCOD$	Rate ρ
Growth A	–	$-\dfrac{1}{Y_A}$	$+1$		$\mu_A \cdot \dfrac{S_S}{K_S + S_S} \cdot X_A$
Decay A	–		-1		$b_A \cdot X_A$
Growth B	–	$-\dfrac{1}{Y_B}$		$+1$	$\mu_B \cdot \dfrac{S_S}{K_B + S_S} \cdot X_B$
Decay B	–			-1	$b_B \cdot X_B$

Parameter values:

$\mu_A = 6\ d^{-1}$, $K_A = 5\ gCOD\ m^{-3}$, $Y_A = 0.67\ gCOD\ g^{-1}COD$, $b_A = 0.2\ d^{-1}$
$\mu_B = 12\ d^{-1}$, $K_B = 50\ gCOD\ m^{-3}$, $Y_B = 0.67\ gCOD\ g^{-1}COD$, $b_B = 0.6\ d^{-1}$

1. Set up the required mass balances.
2. Develop a solution for the steady state with $Q - 1\,L\,d^{-1}$ and $S_{in} = 1000\,gCOD$ m^{-3}. Comment on your results. (This can easily be solved algebraically.)
3. Develop a parameter plot that shows the composition of the biomass when you change the substrate concentration in the influent. How does this affect the substrate concentration in the effluent?
4. Develop a parameter plot that shows the composition of the biomass when you change the flow rate Q. How does this affect the biomass composition and the substrate in the effluent?

Hints for BM:

You can generate a parameter plot under the option *Parameters/Parameter Plot*. Choose the relevant parameter (S_{in}, Q) and analyze the final result of the state variables of interest.

Make sure that your simulations reach steady state.

16.7 Ideal Reactors, Plug Flow

Streeter and Phelps (1925) proposed a simple model that describes the degradation of BOD in very large rivers (they worked on the Ohio river). Their model assumes the river to be an ideal plug-flow reactor in steady state. Water quality is characterized by BOD and dissolved oxygen. The processes are degradation of BOD and reaeration through the free surface of the river.

The kinetic model is introduced in Table 16.5.

Table 16.5 Kinetic model of Streeter and Phelps for self-purification of rivers

Process	Oxygen S_O [g O_2 m^{-3}]	BOD C_S [g O_2 m^{-3}]	Process rate ρ
Degradation	-1	-1	$K_1 \cdot C_S$
Reaeration	$+1$		$K_2 \cdot (S_{sat} - S_O)$

For the Ohio river at 20°C Streeter and Phelps found the following parameter values:
$K_1 = 0.23\,d^{-1}$
$K_2 = 0.55\,d^{-1}$
$S_{sat} = 9\,gO_2\,m^{-3}$
At a particular location A they might have measured
$S_{O,A} = 7\,gO_2\,m^{-3}$ and $C_{S,A} = 24\,gBOD\,m^{-3}$.

1. Derive the mass balance equations for this particular model.
2. Draw up the length profile of the pollutant and the oxygen concentration along the river, starting at location A. Assume the mean flow velocity to be 1 m s^{-1}.
3. Provide an interpretation of your results.
4. What determines the location at which the dissolved oxygen reaches a minimum?
5. Is the model of an ideal plug-flow reactor justified for this application? Make reasonable estimates considering Table 4.4.

You have the choice of developing an analytical solution or to develop a solution in BM.

Hints for BM:

You can obtain the derivative of a state variable y by using $y' = f(y,p,t)$ rather than $d/dt(y) = f(y,p,t)$. y' can then be used and plotted as a variable. This is sufficient to answer question 5. Obtaining a solution for the turbulent plug-flow reactor is more involved, it requires the module *Boundary Value ODE* under *Model/Modules*.

16.8 Ideal Reactors, Sampling in Turbulent Flow

You want to establish the profile of the oxygen consumption along a 60-m-long aeration basin. To do this you first determine the turbulent diffusion coefficient in the basin with constant aeration in the entire tank (this requires the determination of a hydraulic residence time distribution). You then measure the oxygen concentration in four locations (influent channel, after 20 m and 40 m, and in the effluent).

The oxygenation rate is:

$$r_{ox} = k_la \cdot (S_{O,sat} - S_O).$$

Table 16.6 Characterization of the aeration tank at steady state

Parameter	
Total flow through aeration tank, Q	$4{,}800 \, \text{m}^3 \, \text{d}^{-1}$
Length, width, depth of the tank, L : W : H	$60 : 6 : 4 \, \text{m}$
Turbulent diffusion coefficient, D_T	$24{,}000 \, \text{m}^2 \, \text{d}^{-1}$
k_la value	$50 \, \text{d}^{-1}$
Oxygen concentration in the influent, $S_{O,in}$	$0.5 \, \text{gO}_2 \, \text{m}^{-3}$
Oxygen concentration after 20 m, $S_{O,20}$	$1.3 \, \text{gO}_2 \, \text{m}^{-3}$
Oxygen concentration after 40 m, $S_{O,40}$	$3.2 \, \text{gO}_2 \, \text{m}^{-3}$
Oxygen concentration in the effluent, $S_{O,out}$	$6.4 \, \text{gO}_2 \, \text{m}^{-3}$
Oxygen saturation concentration, $S_{O,sat}$	$10.0 \, \text{gO}_2 \, \text{m}^{-3}$

1. Develop the correct material balance equations for oxygen for the three sections of the aeration tank.
2. Estimate the rate of oxygen consumption in the three sections.
3. How could you improve this experiment?

16.9 Ideal Reactors, Disinfection

In a water treatment plant the raw water is to be disinfected with ozone. Since ozone is added with a gas exchange apparatus this introduces high turbulence. An example of a disinfection reactor is shown in Fig. 16.2.

Ozone is added to the first reactor compartment up to the constant concentration of $S_{O3,1} = 0.5 \, gO_3 m^{-3}$. The total flow is 10,000 $m^3 \, d^{-1}$.

1. You choose a total reactor volume of $V = 1000 \, m^3$. What fraction of organisms is disinfected if you choose a cascade of six equal reactors in series? (This may be solved algebraically, stepping from reactor to reactor).
2. What is the fraction of remaining organisms (and ozone), if you vary the number of reactor compartments from 2 to 30?
3. How would you design and operate the reactor if you would have to provide a disinfection performance of $C_{out}/C_{in} = 10^{-9}$.

Table 16.7 Kinetic model for disinfection with ozone. Ozone is subject to decay; the residual ozone interacts with the organisms

Process	S_{O3} $g \, O_3 \, m^{-3}$	X_B (bacteria) #/100 ml	Process rate ρ
Decay of ozone	−1		$k_{O3} \cdot S_{O3}$
Disinfection		−1	$k_D \cdot X_B \cdot S_{O3}$

Parameter values:
$k_{O3} = 10 \, d^{-1}$, $k_D = 1500 \, m^3 \, g^{-1} O_3 \, d^{-1}$

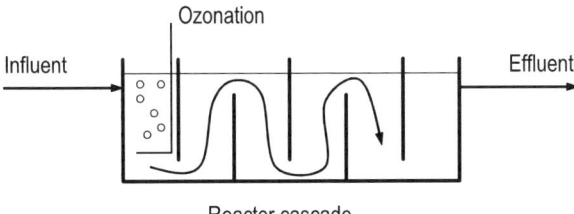

Reactor cascade

Fig. 16.2 Example of a disinfection reactor

Hints for BM:

You should introduce a variable (say n) that stands for the number of reactors. In BM you can then define the state variables and their mass balances in the form of array equations: init SO3[1..n] = 0.5 and d/dt(SO3[2..n]) = ... If you want to use a parameter plot, you must copy the effluent concentration onto a variable which is not an array: SO3out = SO3[n]. You can then plot the final value of SO3out.

The easiest way to keep the ozone concentration constant in the first reactor is to use: d/dt(SO3[1]) = 0.

16.10 Ideal Reactors, SBR

You want to model the rinsing process in a washing machine. The wet textiles contain 1 L of water. For rinsing, 5 L of fresh water are added in every cycle, which takes 1 min. After 3 min of tumbling, all the water is in equilibrium and contains an equal concentration of soil. After decanting 5 L of water, which takes again 1 min, the cycle starts again.

1. How does the concentration of soil in the rinse water decrease over time?
2. How many cycles are required to remove 99% of the soil?
3. How would you enhance your model, if the soil from within the textiles were released to the bulk of the water only during tumbling (3 min/cycle) in a reversible first-order process with a rate constant of $k_{forward} = k_{backward} = 0.5 \, min^{-1}$.
4. How could you improve the rinsing process? What are the parameters? How could you save water?

An analytical solution for questions 1 and 2 can easily be found. Here you are expected to write a program in BM and simulate the dynamics of this process.

Hints for BM:

At the end of *Equation Help*, BM tells you how to create periodic functions.

16.11 Residence Time Distribution, Cascade of CSTRs

You add 10,000 g of a tracer as a pulse to the influent of a reactor and observe the course of concentrations in the effluent as shown in Fig. 16.3. From visual inspection you assume that a cascade of CSTRs would be an adequate model of the entire reactor.

Concentration of tracer, g m⁻³

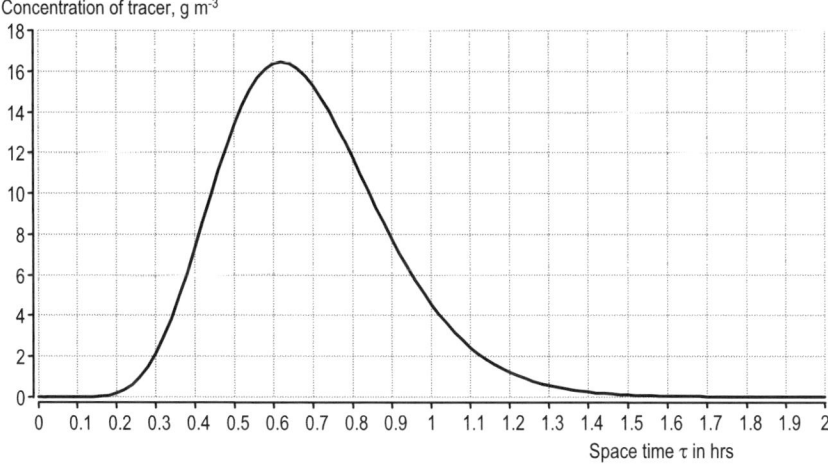

Fig. 16.3   Tracer concentration in the effluent of a cascade of CSTRs

1. What is the flow rate Q of water flowing through the reactor?
2. What is the volume V of the reactor?
3. How many reactors in series would you choose for the reactor model?

You may estimate the required information with approximations rather than an accurate and detailed analysis.

16.12 RTD, Reactor Model

In a longitudinal reactor a compound A is degraded in a first-order reaction according to $r_A = - k_A \cdot C_A$. The following parameters apply:

$$V = 200 \text{ m}^3, \ Q = 2400 \text{ m}^3 \text{ d}^{-1}, \ S_{A,in} = 50 \text{ g m}^{-3}, \ k_A = 60 \text{ d}^{-1}.$$

In order to characterize the reactor, you perform a tracer experiment where you add a pulse of tracer to the influent. The results are given in Table 16.8.

1. What is the effluent concentration $C_{A,out}$, if this reactor can be modeled as a cascade of CSTRs?
2. Is the amount of tracer in the effluent compatible with the experiment?
3. Are the mean hydraulic residence time θ_h and the standard deviation of the residence time distribution compatible with the operating conditions and the experiment?
4. If you choose to model the reactor as a turbulent plug-flow reactor with 25 compartments, would this affect the predictions of the performance?

There are several possible approaches to solve this problem: you may integrate the tracer curve with the aid of Simpson's rule, you can integrate the data in BM,

Table 16.8 Measured tracer concentration in the effluent after dosing a pulse of $E_T = 2500\,g$ of tracer at time t_0

$t-t_0$ [min]	$S_T\ [gT\ m^{-3}]$
0	0
40	4.8
80	11.0
120	10.0
160	6.3
200	3.2
240	1.4
280	0.5
320	0.2

or you can fit the tracer curve to a simulated curve. The same is true for the reactor performance. Try at least two of these options.

Simpson's equation for equidistant samples has the form:

$$\int_{a}^{b} f(x)\,dx = \frac{h}{3}\cdot(y_0 + 4y_1 + 2y_2 + 4y_3 + \ldots + 2y_{n-2} + 4y_{n-1} + y_n).$$

Hints for BM:

A Dirac pulse can be added to a plant either via the Pulse function (see *Equation help*) or in the form of an initial condition (init SA = EA / V).

The number of reactors in a cascade must be an integer. You may choose n = round(n_ident) when you try to identify it in *Curve Fit* or *Optimize*. Alternatively you may want to work with a slider.

The integral of d/dt(x) = #C(time)*Q will be based on linear interpolation between the individual data points #C(time).

16.13 RTD, Activated Sludge Tank

One type of activated sludge tank is based on high internal recirculation, as shown in Fig. 16.4. You want to find the hydraulic residence time distribution of a

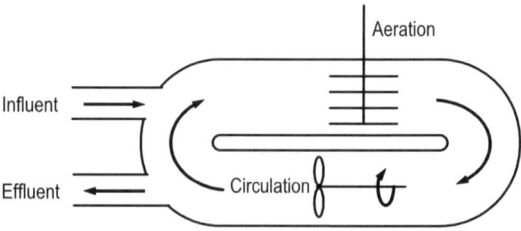

Fig. 16.4 Schematic of an activated sludge reactor with high internal recirculation

planned reactor with a volume of 5,000 m^3. The influent flow rate is 10,000 m^3 d^{-1}; the return sludge flow rate, which is delayed for 1 h in the secondary clarifier, is also 10,000 m^3 d^{-1}. The internal recirculation induces a flow of 120,000 m^3 d^{-1}.

1. Develop a model of this reactor based on a cascade of CSTRs.
2. Determine the hydraulic residence time distribution of the reactor and compare it to a single CSTR (do not forget the return sludge).
3. What is the mean hydraulic residence time of this reactor? How does it compare to the mean that you obtain from the simulated RTD? What is the reason for the difference?
4. Compare the performance of the reactor with internal recirculation and a comparable CSTR for a first order reaction with k=24 d^{-1} (neglect a possible reaction in the secondary clarifier).

Hints for BM:

You can delay the tracer concentration in the return sludge with the delay function of BM.

16.14 RTD, Flow Rate and Dispersion in a Sewer

A sewer is 1500 m long and does not have any additional influents from the side. At time t=0 you add 1 g of a conservative tracer dye to the influent of the sewer and you observe the concentration in the effluent as shown in Fig. 16.5.

1. How much water is flowing in the sewer?
2. What is the mean flow velocity in the sewer?
3. How large do you estimate the dispersion coefficient in this sewer?

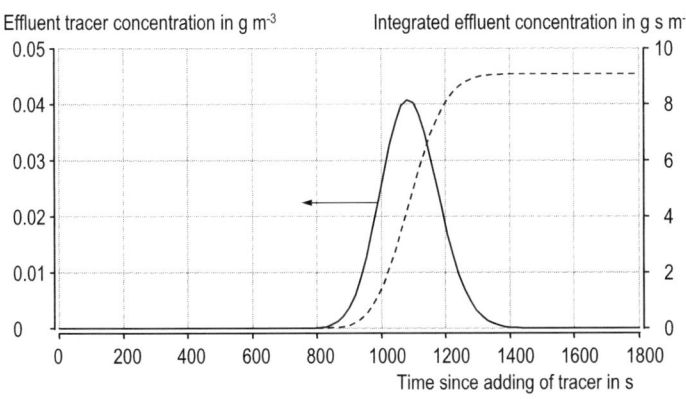

Fig. 16.5 Effluent tracer concentration after addition of 1 g of an inert tracer at time 0

16.15 Modeling a Sewer

You want to model the transport of pollutants in a sewer with the aid of a cascade of CSTRs. You estimate the dispersion coefficient based on literature information (Rieckermann et al., 2005) as $D_D = 0.15\,\mathrm{m}^2\,\mathrm{s}^{-1}$ (Table 4.4).

1. What length of the sewer will you model with a single CSTR?
2. If you would choose the model of a turbulent plug-flow reactor, what would be reasonable model parameters?

16.16 RTD, Disinfection Reactor

Derive the hydraulic residence time distribution for the ozonation/disinfection reactor in Sect. 16.8. Compare the results for 2, 6, and 30 reactors in series.

16.17 RTD, Additivity of τ_m and σ^2

Derive Eqs. (7.25) and (7.26) based on the statistical properties of a single reactor.

16.18 RTD, Turbulent Plug-Flow Reactor

A turbulent plug-flow reactor is 50 m long and has a cross section of $20\,\mathrm{m}^2$. The total flow through the reactor is $Q = 500\,\mathrm{m}^3\,\mathrm{h}^{-1}$. From experience you know that the turbulent diffusion coefficient induced by the aeration system will be on the order of $D_T = 300\,\mathrm{m}^2\,\mathrm{h}^{-1}$ (see Example 7.14).

1. What is the turbulence number N_T of this reactor?
2. If you want to model this reactor as a cascade of CSTRs, how many equal reactors in series would you choose?
3. If you model the reactor with 30 equal sections with back mixing (see Fig. 7.18), how large would you choose the back mixing R?
4. What is the difference in performance of the two reactor models for a first order process with a decay rate of $k = 2\,\mathrm{h}^{-1}$?
5. How does the performance of the two discretized reactor models compare with the numeric solution of the mass balance equation for the turbulent plug-flow reactor?
6. How do the RTDs of the two discretized models compare?

16.19 Heterogeneous Systems: Filtration

Ives (1960) proposed a model for a rapid sand filter: the concentration of particles C is reduced over the depth x of the sand filter according to:

$$\frac{\partial C}{\partial x} = -\lambda \cdot C \quad \text{where} \quad \lambda = \lambda_0 + a_1 \cdot \sigma - \frac{a_2 \cdot \sigma^2}{\sigma_{max} - \sigma},$$

where σ is the suspended solids that have accumulated locally in the filter $[g\,TSS\,m^3_{Filterbed}]$.

The head loss is approximated by:

$$\frac{\partial H}{\partial x} = k \cdot \left(\frac{\sigma_{max}}{\sigma_{max} - \sigma}\right)^{a_3}.$$

For a typical sand filter for tertiary wastewater treatment at a filtration rate of $v = 10\,m^3\,m^{-2}\,h^{-1}$ the following parameter values may be used (not calibrated): $\lambda_0 = 4\,m^{-1}$, $a_1 = 0.0005\,m^2\,g^{-1}\,TSS$, $a_2 = 0.0005\,m^2\,g^{-1}\,TSS$, $\sigma_{max} = 20{,}000\,gTSS\,m^{-3}$, $k = 0.25$, $a_3 = 4$.

The influent TSS concentration (secondary effluent) is $C_0 = 15\,gTSS\,m^{-3}$ and the filter bed has a depth of 1.5 m.

1. Develop a model for the removal of TSS over the depth of the filter with time.
2. Include a prediction of the head loss into your model.
3. How often does the filter have to be backwashed, if the maximal allowable head loss over the entire filter bed is 3 m? What is the amount of solids that has been accumulated in the filter bed when this limit is reached? How do the effluent concentration, the removed solids σ, and the head loss develop over time?

Hint for BM:

Look at the function arraysum under *Equation help*. It allows you to sum up the deposit of particles and the head loss over the depth of the filter bed.

16.20 Substrate Profiles in a Biofilm

In Sect. 9.4.3 the mass balance for a biofilm was derived as Eq. (9.12):

$$\frac{dj_S}{dz} = r_S, \text{ which together with Fick's law, Eq. (9.13) } j_S = -D_S \cdot \frac{dS}{dz}$$

results in $\dfrac{d^2S}{dz^2} = S'' = -\dfrac{r_S}{D_S}.$ $\hfill (16.1)$

Solving Eq. (16.1) requires two boundary conditions; one is the concentration at the surface of the biofilm $S_{S,surface}$ and the other is the derivative $S' = 0$ at the substratum (no substrate can diffuse into the substratum). A numeric solution will iteratively start at the substratum ($z = 0$) with $dS/dz = 0$ and a guess of a reasonable residual substrate concentration at the substratum. This guess will then be improved until the surface concentration is met. BM offers the option *Model/ Modules/Boundary Value ODE* to solve this problem.

The code has the form: init $S = S0$

init $S' = 0$

$S'' = -r/D$

1. Compute the concentration profile of ammonium over the depth of a nitrifying biofilm with the following kinetic expression:

$$r_{NH} = -\frac{\mu}{Y} \cdot \frac{S_{NH}}{K_{NH} + S_{NH}} \cdot \gamma$$

S_{NH} = Ammonium concentration
μ = $1\,d^{-1}$ = max. growth rate of nitrifiers
Y = $0.25\,gCOD\,g^{-1}N$ = nitrifier yield coefficient
K_{NH} = $1.5\,gN\,m^{-3}$ = saturation coefficient for ammonium
γ = $2000\,gCOD\,m^{-3}$ = packed density of nitrifiers
D_{NH} = $1.7 \cdot 10^{-4}\,m^2\,d^{-1}$ = molecular diffusion coefficient for ammonium

The biofilm is 0.5 mm thick and at the surface you measure $S_{NH} = 3\,gN\,m^{-3}$.

- How large is the resulting ammonium concentration at the substratum?
- How much ammonium can this biofilm nitrify? (There are two possibilities to obtain this result, one analyzes only the interface to the bulk water, and one integrates over the depth of the biofilm. Compare the results.)

2. Nitrification is an obligate aerobic process, therefore we must pay attention to the oxygen concentration. In addition the nitrate profile is of interest. An appropriate kinetic model is given in Table 16.9.

- What are the concentrations of the three compounds O_2, NH_4 and NO_3 at the surface of the substratum? What are the concentration profiles over the depth of the biofilm?
- Which compound is limiting the nitrification rate?
- What is the turnover (mass flux) of the three compounds?
- Can you derive an expression which allows you to obtain the nitrate and the ammonium flux when the oxygen flux is given?

3. The transport of substances from the bulk to the biofilm is hindered by a laminar boundary layer, which may be characterized with a mass transfer coefficient.

- How do the concentrations, the profiles, and the mass flux change if the boundary layer has a thickness of 50 µm.
- How large is the mass transfer coefficient across the boundary layer?

Assume equal values for the diffusion coefficients inside the biofilm and in the boundary layer.

Table 16.9 Kinetics for nitrification in a biofilm

Process	S_{O2} $gO_2\,m^{-3}$	S_{NH} $gNH_4\!-\!N\,m^{-3}$	S_{NO} $gNO_3\!-\!N\,m^{-3}$	ρ $gCOD\,m^{-3}\,d^{-1}$
Nitrification	-18	-4	$+4$	$\mu \cdot \dfrac{S_O}{K_O + S_O} \cdot \dfrac{S_{NH}}{K_{NH} + S_{NH}} \cdot \gamma$

$K_O = 0.5\,gO_2\,m^{-3}$	$D_{O2} = 2.1 \cdot 10^{-4}\,m^2\,d^{-1}$	$D_{NO} = 1.6 \cdot 10^{-4}\,m^2\,d^{-1}$
$S_{O2,surface} = 8\,gO_2\,m^{-3}$	$S_{NH,surface} = 3\,gN\,m^{-3}$	$S_{NO,surface} = 2\,gN\,m^{-3}$

Hint for BM:

This problem is best solved if the z-axis starts at the surface of the substratum and time is redefined as z:

RENAME TIME = z

STOPTIME = 0.0005   ; Simulation stops at z = 0.5 mm, units are m

16.21 Bode Diagram

Develop a Bode diagram for a plug-flow reactor and a cascade of 1, 2, or 6 CSTRs. Use $\theta_h = 1\,d$, $k = 5\,d^{-1}$ and the range of frequencies of 0.01 to 100 d^{-1}.

You may extend this exercise to include turbulent flow with the turbulence number N_T as a parameter.

Hints for BM:

BM provides the option *Parameter Plot* where the amplitude of a state variable can be obtained as a function of different parameter values. You may want to use a geometric series and logarithmic scales in your diagram.

Numeric integration can be accurate and fast if the time step DT is adapted to the rate of change of the state variables, which is here related to the frequency f. This can be obtained with init DT = 0.01/f next DT = if 1 then DT else DT (for details see the user manual).

16.22 Dynamic Nitrification

This example is based on realistic data.

Figure 10.5 shows the influent and effluent ammonium concentration of an activated sludge plant operated with an SRT of 5.3 days at approximately 14°C. The aeration tank had a volume of 11 m³, was mixed (surface aerator, CSTR), and the effluent samples were taken from the effluent of the aeration tank.

Table 16.10 Variation of flow rate and ammonium concentration in the pilot test of Fig. 10.5

Time hrs	$S_{NH4,in}$ gN m^{-3}	Flow Q m^3 d^{-1}
0	11.8	65
4	11.8	65
4.001	6.0	65
8	6.0	65
8.001	22.4	90
10	22.4	90
10.001	16.8	90
12	16.8	90
12.001	11.4	90
14	11.4	90
14.001	12.0	90
20	12.0	90
20.001	12.8	65
24	12.8	65

Develop a model for the behavior of this plant, using the kinetic model provided in Table 16.21. The loading conditions are given in Table 16.10.

1. What is the required maximum growth rate μ of the nitrifiers that fits to the
   effluent data? Choose $b_N = 0.1 \cdot \mu_N$.
2. If the secondary clarifier has a volume of $10\,m^3$ and can be modeled as a sequence of two CSTRs, what is the attenuation of the ammonium concentration
   in the effluent?
3. What would be the performance of the plant if the aeration reactor were divided
   into three equal, completely mixed compartments in series?

Hints:

A solids retention time of SRT $= 5.3$ days can be obtained if the excess sludge is
removed directly from the aeration tank and no solids are allowed in the effluent:
$Q_{ExSl} = V/SRT$. The secondary clarifier can be modeled with a bifurcation just
after the aeration tank in which all solids are recycled and soluble materials are
split in the ratio of the flows.

A repeating diurnal variation of ammonium and flow can be obtained by reading the data as: SNH0 = #SNH4(24*mod(time,1)).

A data file is easily generated in EXCEL. The data must be saved as an ASCII
file (*.txt) and may then be read into BM under *Model*. If you save the model in
BM the data will be integrated into the model file.

16.23 Nonstationary Flow in Sewers

The difference in flow velocity v and wave velocity (celerity c) in free surface
flow causes the separation of flow rate and pollutant load (see Sect. 10.5). With

the aid of a series of nonlinear reservoirs this behavior can approximately be modeled. The goal of this exercise is to visualize how nonstationary flow conditions may cause very significant load variations.

A nonlinear reservoir is a completely mixed reactor with a variable effluent Q_i. Where Q_i depends on the instantaneous volume V_i according to:

$$Q_i = k \cdot V_i^n$$

n = exponent which relates the volume or flow depth to the flow rate, for sewers $n = 5/3 = 1.67 \, [-]$

k = a constant that describes the quantitative behavior of the sewer; here we choose $k = 3 \, \mathrm{m^{-2} s^{-1}}$ if individual sewer elements have a length of $\Delta x = 1 \, \mathrm{m}$ and a typical flow is $Q = 0.1 \, \mathrm{m^3 \, s^{-1}}$. (This k value applies approximately to a sewer with a diameter of 0.7 m and a slope of 0.002, which is less than half full.)

The sewer must be modeled in very short sections Δx or else dispersion will soon eliminate all waves.

You are expected to develop a model in BM which can describe the flow as well as the transport of pollutants in a sewer that is 1000 m long.

1. What is the flow velocity which results at steady state? (Hint: you may obtain the total volume of water in the sewer with the command arraysum(V[*])).
2. During a rain event the following characteristics of the influent apply (time in seconds): Qin = 0.1*(1 + squarepulse(0,1000)), Cin = 1 − 0.5*squarepulse(1,1000). What is the amplitude of the pollutant load in the influent and after a flow distance of 1000 m?
3. During the rain event applies $Q = 0.2 \, \mathrm{m^3 \, s^{-1}}$. What is now the maximum flow velocity v and what is the celerity c (wave velocity)?

Hint for program development:

You should first develop the program for only 10 m of sewers (n = 10) and once it is debugged, you can then change the length to 1000 m (n = 1000). It may be easier to follow the mass of pollutant in the individual elements than directly the concentration. In order to reach steady state, you may set the start time way before you use the results (STARTIME = −5000 ; seconds).

16.24 Stochastic Measurement Error

This example is based on realistic data.

In some situations it is possible to separate infiltrating groundwater from polluted wastewater (which originates from drinking water) by using natural tracers such as [18]O isotopes. The concentration of isotopes is expressed as deviation from a standard in ‰($\delta^{18}O$) and can be used in simple mixing computations similar to a concentration. Table 16.11 summarizes the results for some samples taken throughout a 24-h day in a combined sewer.

1. How much water flows through this sewer during the experiment? Derive an expected value and characterize its error. (You can solve this either analytically or by MC simulation).
2. What fraction of wastewater at noon (12:00 h) originates from groundwater infiltration? How accurate can you make this statement? (Again, you can solve this either analytically or by MC simulation).
3. How much infiltrated groundwater is discharged at noon? How accurate is your statement?
4. What is the diurnal pattern of groundwater infiltration? What are the error bounds of this pattern?
5. How much groundwater is discharged throughout the day? How accurate can you answer this question?
6. There is a diurnal pattern in the groundwater flow, whereas we would expect a constant infiltration. Can this pattern be explained by backwaters caused by sediments and uneven joints? If so, what is the volume of this nearly stagnant water? (This last question requires additional modeling and parameter identification).

Table 16.11 Isotopic composition and flow rate of wastewater in a sewer. All measuring errors are normally distributed and consider multiple samples

	Mean μ, ‰(^{18}O)	Stand. dev. σ, ‰(^{18}O)		
Drinking water[1]	−11.30	0.02		
Groundwater[1]	− 9.55	0.04		
Combined wastewater			Flow Q	Stand. dev.σ
Time, hrs	Mean μ, ‰(^{18}O)	Stand. dev. σ, ‰(^{18}O)	$10^{-3}\,m^3\,s^{-1}$	$(10^{-3}m^3\,s^{-1})$
0[2]	−10.68	0.08	17	2
2	−10.40	0.08	11	2
4	−10.17	0.08	10	2
6	−10.32	0.08	14	2
8	−10.60	0.08	28	2
10	−10.76	0.08	26	2
12	−10.70	0.08	26	2
14	−10.81	0.08	26	2
16	−10.70	0.08	23	2
18	−10.63	0.08	24	2
20	−10.75	0.08	26	2
22	−10.62	0.08	22	2
24	−10.61	0.08	17	2

[1] These values apply to the mean of many (n>24) samples
[2] These values apply to the mean of triplicate analysis

Hints:

Be careful when you decide at what time a new stochastic value should be assigned to a specific concentration.

16.25 Systematic Measurement Error

Many elements of wastewater treatment plants are designed for a specific maximum flow rate Q_{max}. If the flow rate is estimated incorrectly, the design will be either too small or not economical. Flow rates are, however, chronically subject to large measurement error (see also Fig. 11.6).

A wastewater treatment plant was expanded with an additional tertiary filter based on the flow measurements of the past (Fig. 16.6). When the filter was placed into operation, it proved to be too small. A recalibration of the flow measurement device indicated an error of 33%.

Figure 16.7 provides the reported average information of this treatment plant for the year 1977.

1. Was the data collected on the plant plausible?
2. What would have been your estimate of the average flow rate?
3. What do you conclude from this exercise?

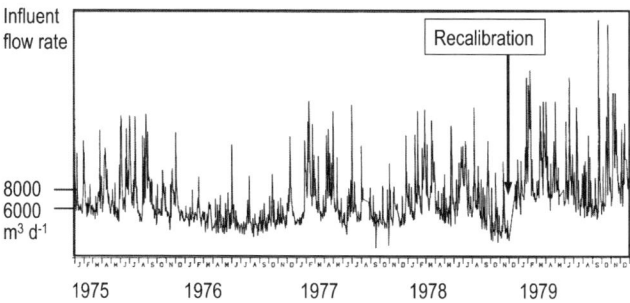

Fig. 16.6 Measured daily flow rate in a wastewater treatment plant. The measuring device was recalibrated in 1978, after an extension of the plant proved to be hydraulically limited

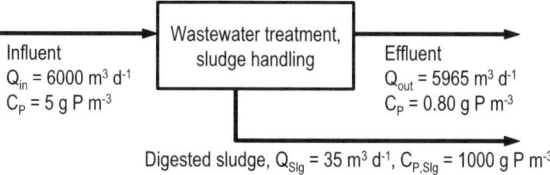

Fig. 16.7 Average operating information for the treatment plant reported in 1977

16.26 Sensitivity and Parameter Identification

You perform an experiment in a batch reactor. Compound A is degraded in a first-order reaction and you measure the concentrations indicated in Table 16.12.

1. Find an analytical solution for the development of the concentration C_A over time and name the parameters of your model.
2. Implement the analytical model in EXCEL and identify the parameters with the aid of the SOLVER routines (these routines must be activated under *Tools – Add-Ins*).
3. Derive the absolute relative sensitivity functions for all parameters. First analytically and then numerically with the aid of EXCEL. Compare the two results.
4. Implement your model in BM, use the mass balance equation not the analytical solution. Import the data and identify the parameters in BM. Are there differences relative to EXCEL?
5. Determine the absolute relative sensitivity functions with the aid of BM. Remember that BM only provides you with partial derivatives as sensitivity functions.
6. Are there structural problems in this model? Would another reaction order provide a better fit to the data?
7. How could you improve your experiment if you wanted to obtain a good estimate of the reaction order?

Table 16.12 Measured concentrations in a batch reactor

Time in min	C_A in g m^{-3}
1	137
5	101
8	81
10	72
16	51
20	43
24	36
30	27

Hints for BM

An ASCII file that contains the data may be written with EXCEL and can then be used to enter the data into BM.

You can obtain the absolute relative sensitivity function as indicated in Example 12.6.

16.27 Sensitivity

Another, rather complex problem relating to sensitivity is provided in Example 12.7: Linear combination of sensitivity functions.

16.28 Error Propagation with Correlated Uncertainty

Two communities have agreed to operate a common wastewater treatment plant and to participate on the cost based on their individual contribution to the pollutants as measured with COD. Two metering stations are built and give the results indicated in Fig. 16.8. Due to the careful calibration of the flow meters you accept their results as accurate and do not want to consider any uncertainty arising from this side. However there is some debate about the pollution (COD) which is measured online with a novel electrode.

You carefully analyze all arising sources of uncertainty for the flow proportional concentration of the pollution. As a major source of error you identify incorrect calibration related to erroneous standard solutions used in the calibration procedure. Since both electrodes are calibrated simultaneously these uncertainties are correlated.

You come to the conclusion that the correlation coefficient between the two uncertainties of the pollutant loads is rather high, with $r_{A,B} = 0.90$. Expected values m_A, m_B and the standard deviations of their uncertainty s_A, s_B are provided in Fig. 16.8.

1. How large is the load of the wastewater treatment plant? Provide an expected value and a characterization of its uncertainty.
2. How large is the expected error in the cost contribution for the operation of the treatment plant that the two communities make, provided the amount will be calculated based on total cost and the fraction of the pollutants based on expected values? Total costs per year are 1 million Euros.

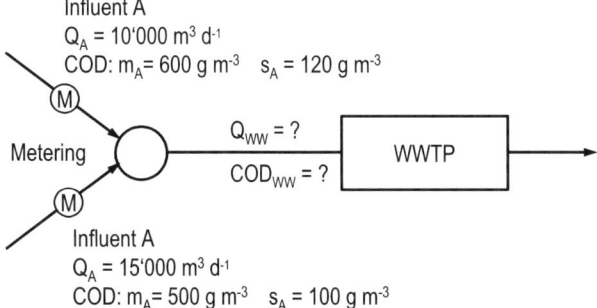

Fig. 16.8 Influent of two wastewater streams to a wastewater treatment plant. The value of the mean, flow weighted COD concentration is uncertain with a coefficient of variation of 20%

16.29 System Identification

This example is based on real (adapted) data.

In order to drain as many nutrients as possible from a small eutrophic lake, the effluent is drawn from the anaerobic depth, which contains a significant ammonium concentration. Nitrification in the effluent brook oxidizes this ammonium to nitrate (Fig. 16.9). For one situation a length profile of the inorganic nitrogen concentrations is available (Fig. 16.10 and Table 16.13).

In a shallow brook, nitrification is due to the microbial activity of the biofilm growing on sediment surfaces. A simple model for the relevant processes is defined in Table 16.14. Nitrification starts only after sufficient reaeration after $x > x_{anaerobic}$. Since BM cannot provide standard errors of parameters and their correlation matrix, the model was implemented in AQUASIM (Reichert, 1998) to obtain this information (see Table 16.15 and Table 16.16)

From an analysis of biofilm kinetics, you derive the (realistic) temperature dependencies of the two rate parameters which are given in Table 16.17.

1. Derive the relevant mass balance equations for the nitrification model.
2. Implement the model and the data in BM and identify the parameters – are they identical to those in Table 16.15? Why?

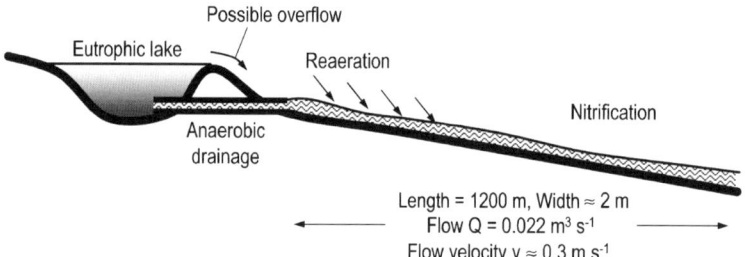

Fig. 16.9 Drainage of anaerobic water from the depth of a small, eutrophic lake

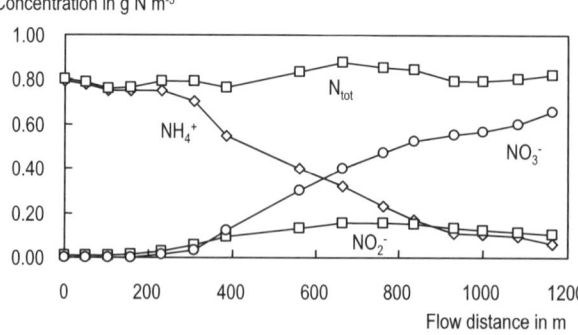

Fig. 16.10 Length profile of ammonium concentrations at $Q = 0.022 \, \text{m}^3 \, \text{s}^{-1}$ and 18°C

3. Identify the absolute relative sensitivity of the two rate constants k_{NH4} and k_{NO2}. Is the lack of correlation between the values identified for these two parameters understandable?
4. How does the length profile change, when the temperature changes? Use 5°C and 25°C.
5. How does the length profile change if the flow rate Q is doubled? Assume a constant flow velocity of $v = 0.3\,m\,s^{-1}$.
6. What is the response if the flow rate is doubled due to an overflow of the lake, when the overflow does not contain any nitrogen?
7. Analyze points 4–6 above, but argue based on absolute relative sensitivity.
8. What is the uncertainty in your predictions? Assume that the temperature coefficients θ in Table 16.17 are only known with a standard error of 10%. Assume the parameters not to be correlated.
9. Repeat point 8 with correlated parameters, using UNCSIM (Reichert, 2002).

Table 16.13 Measured nitrogen concentrations along a small brook with nitrification (T = 18°C, see Fig. 16.10). Total nitrogen is computed from the sum of $NH_4 + NO_2 + NO_3$

Flow distance x in m	NH_4–N g m^{-3}	NO_2–N g m^{-3}	NO_3–N g m^{-3}	N_{tot} g m^{-3}
0	0.79	0.01	0.00	0.80
52	0.78	0.01	0.00	0.79
104	0.75	0.01	0.00	0.76
158	0.75	0.02	0.00	0.76
231	0.75	0.03	0.01	0.79
310	0.70	0.05	0.04	0.79
387	0.55	0.09	0.12	0.76
562	0.40	0.13	0.30	0.83
664	0.32	0.16	0.40	0.88
763	0.23	0.15	0.47	0.85
836	0.17	0.15	0.52	0.84
932	0.11	0.13	0.55	0.79
999	0.10	0.12	0.56	0.79
1084	0.09	0.11	0.60	0.80
1162	0.06	0.10	0.66	0.82

Table 16.14 A simple model for nitrification in a shallow brook with low nitrogen concentrations

Process	S_{NH4} g N m^{-3}	S_{NO2} g N m^{-3}	S_{NO3} g N m^{-3}	ρ
Ammonium oxidation	−1	+1		$k_{NH4} \cdot S_{NH4} / h$
Nitrite oxidation		−1	+1	$k_{NO2} \cdot S_{NO2} / h$

k_{NH4}	=	mass transfer coefficient for ammonium [L T^{-1}]
k_{NO2}	=	mass transfer coefficient for nitrite [L T^{-1}]
h	=	hydraulic radius of the flowing water ≈ Q/(W·v) = cross section/wetted circumference, a measure of biofilm surface [L]

Table 16.15 Identified parameter values with their standard error (AQUASIM, Reichert, 1998)

Parameter	Expected value μ	Stand. dev. σ	Unit	Coeff. of variation cv
k_{NH4}	2.4	0.1	$m\ d^{-1}$	0.04
k_{NO2}	5.0	0.3	$m\ d^{-1}$	0.06
$S_{NH4,0}$	0.76	0.01	$g\ N\ m^{-3}$	0.01
$S_{NO2,0}$	0.01	0.01	$g\ N\ m^{-3}$	0.91
$S_{NO3,0}$	0.04	0.02	$g\ N\ m^{-3}$	0.49
S_{Ntot} (computed)	0.81	0.023	$g\ N\ m^{-3}$	0.03
$x_{anaerobic}$	282	9	m	0.03

Table 16.16 Estimated correlation matrix for the six parameters (AQUASIM, Reichert, 1998)

Parameter	k_{NH4}	k_{NO2}	$S_{NH4,0}$	$S_{NO2,0}$	$S_{NO3,0}$	$x_{anaerobic}$
k_{NH4}	1	0.00	0.02	0.02	−0.17	0.57
k_{NO2}	0.00	1	0.09	0.07	−0.24	0.07
$S_{NH4,0}$	0.02	0.09	1	−0.18	−0.20	−0.35
$S_{NO2,0}$	0.02	0.07	−0.18	1	−0.20	0.24
$S_{NO3,0}$	−0.17	−0.24	−0.20	−0.20	1	0.02
$x_{anaerobic}$	0.57	0.07	−0.35	0.24	0.02	1

Table 16.17 Adaptation of the mass transfer coefficients of the nitrification model to different temperatures, 18°C values are identified from data, 5°C and 25°C values are estimated with a biofilm model. Interpolation is possible with $k(T) = k(T_0) \cdot exp(\theta \cdot (T - T_0))$

Temperature	5°C	18°C	25°C	Units	θ	Units
k_{NH4}	1.0	2.4	3.9	$m\ d^{-1}$	0.07	$°C^{-1}$
k_{NO2}	3.0	5.0	6.6	$m\ d^{-1}$	0.04	$°C^{-1}$

Hints for BM:

Nitrification in this brook is in the steady state, therefore real time is not a relevant system variable. Here we use BM to integrate over the length of the brook. In order to make the code and the graphics more readable we can use the statement:

RENAME TIME = X

Correlated parameters can be generated with *randsamp.exe* of UNCSIM (Reichert, 2004, see handbook). The generated text file can then be edited in EXCEL (add a first column that numbers the rows, 1, 2, 3, etc.) and then read the file as data into BM. Assign each column to a variable as a function of the variable run, then increase run from 1 to n in a *Batch run*.

run = 1
kNH4 = #kNH4(run)

16.30 Uncertainty, Error Propagation

In the example of disinfection with ozone in Sect. 16.9, the result is deterministic. Disinfection kinetics is however rather uncertain. Table 16.18 summarizes your uncertainty about the parameters of the model. Flow rate and ozone concentration are subject to systematic measurement error. Reaction rates suffer from experimental error, the range of n (number of reactors in the cascade of CSTRs) characterizes the uncertainty about the hydraulic conditions, and the range of possible volumes V indicates that there may be some dead space in the reactor.

1. Neglecting the uncertainty stemming from the hydraulic model (n, V), what is the resulting uncertainty of the predicted performance of this disinfection reactor? What fraction of organism remains in the effluent? Obtain a result with Gaussian error propagation and a MC simulation, compare the results.
2. What is the additional effect of the uncertainty about the hydraulic model?
3. What is the performance that you can guarantee to the owner of the plant? How would you explain your result to the owner?

Table 16.18 Uncertainty applying to the disinfection plant. All uncertainties are independent

Parameter	Expected value μ	Distribution	Units	Remark
Flow, Q	10,000	$\sigma = 500$	$m^3 d^{-1}$	Normal distribution
k_{O3}	10	$\sigma = 2$	d^{-1}	Normal distribution
k_D	1500	$\sigma = 150$	$m^3 gO_3 d^{-1}$	Normal distribution
$S_{O3,1}$	0.5	$\sigma = 0.05$	$gO_3 m^{-3}$	Normal distribution
n	6	4–8	–	Range, evenly distr.
V	1000	800–1000	m^3	Range, evenly distr.

Hints:

You may obtain a random distribution of an integer value in the range of 4–8 with $n = int(random(4,9))$

The standard deviation of an evenly distributed variable is $\sigma_x = (x_{max} - x_{min})/\sqrt{3}$. Gaussian error propagation is valid for all continuous distributions with defined variation and not only for normal distributions.

16.31 Process Control, Two-Position Controller

The wastewater from about 1000 inhabitants flows into a pumping well from where it has to be pumped up to the main sewer. The pump has a capacity of $0.012 m^3 s^{-1}$. In order not to overheat, it should not be switched on earlier than

5 min after it has been stopped. The wastewater flow varies between 0 and $0.01 \, \mathrm{m^3 \, s^{-1}}$, and the flow can be simulated with the following code for BM:

STARTTIME = 0
STOPTIME = 1
DT = 0.0001
init Q = 0.005   next Q = 0.99*Q + random(−0.00035,0.00045,1)

1. Qualitatively draw up a system which may fulfill the requirements. If possible, choose a two-position controller. Identify the controlled variable x and the position y.
2. Qualitatively draw the time course of the controlled variable x and the position y and develop the characteristic curve of the controller.
3. Parameterize your controller and your system and simulate its behavior in BM. What guarantees that the pump stops for at least 5 min?

16.32 Process Control, PID Controller

An industry produces wastewater that contains high concentrations of dissolved, readily biodegradable organic material but does not contain any phosphorus. The treatment cost at the public treatment plant is high. You suggest to run the wastewater through an aerated CSTR, in which biomass would grow. The treated wastewater could then be discharged to the sewer and the biomass could settle in the primary clarifier of the public plant; thus, it would not affect expensive biological treatment. However, phosphorus must be added to the wastewater in order to treat it. Here we design the controller for the dosing mechanism.

Table 16.19 defines the simple kinetic model to be used in the design. Table 16.20 summarizes the relevant operating conditions of the plant. The autoanalyzer for phosphorus has a dead time T_t of 5 min and an additional delay of sixth order with a total time of five more minutes.

1. Develop the code in BM to model the steady state of this plant, $Q = 100 \, \mathrm{m^3 \, d^{-1}}$, $S_{COD,in} = 3000 \, \mathrm{gCOD \, m^{-3}}$, $Q_P = 0.08 \, \mathrm{m^3 \, d^{-1}}$.
2. Implement the PID controller (either from Table 13.3 or directly from a file). Identify the optimal parameters of a P, PI, and PID controller either with an unstable controller or a step input for Q_P.
3. Test the controller at steady state.
4. Implement the variable flow rate and COD concentration and test your controller again.
5. Try to identify the best parameters for the controller by minimizing the root-mean-square offset.

Table 16.19 Simple kinetic model for the growth of microorganisms in wastewater

Process	S_P gP	S_{COD} gCOD	X_H gCOD	Process rate ρ
Growth	$-i_P$	$-\dfrac{1}{Y_H}$	1	$\mu \cdot \dfrac{S_{COD}}{K_{COD}+S_{COD}} \cdot \dfrac{S_P}{K_{PO4}+S_P} \cdot X_H$

Model parameters: $i_P = 0.02 \text{ gP g}^{-1}\text{COD}$, $Y_H = 0.6 \text{ gCOD g}^{-1}\text{COD}$,
$\mu = 5 \text{ d}^{-1}$, $K_{COD} = 5 \text{ gCOD m}^{-3}$, $K_{PO4} = 0.01 \text{ gP m}^{-3}$

Table 16.20 Operating conditions of the CSTR treatment plant

Parameter	Value	Units
Volume V	100	m^{-3}
Flow rate Q	Q = if mod(time,1) < 0.3 then 60 else 200*mod(time,1)	$m^3 d^{-1}$
Influent COD $S_{COD,in}$	SCODin = if mod(time,0.5) < 0.25 then 2000 else 3500	$gCOD m^{-3}$
Dosed P conc. $S_{P,in}$	50,000	$gP m^{-3}$
Dosed flow Q_P	0–0.5 (variable)	$m^3 d^{-1}$
Setpoint for S_P	1.0	$gP m^{-3}$

Hints for BM:

Be careful when you combine code from another source (here the PID controller) in order not to use the same variable names twice (here K_P).

Oscillations may be due to large time steps DT as well as instability of the system. The option *Compute/Check DT* makes information on the effect of the size of the time step available.

16.33 Time Series Analysis

Today we have sensors available that allow the collection of data at very high temporal density. If we use standard statistical techniques to identify parameters from such dense time series, we artificially reduce the resulting uncertainty of the parameter values. Time series analysis may be used to check whether the assumptions underlying our statistical techniques are valid.

This example is similar to the one introduced in Sect. 11.5 but relies on real data which has been sampled at high frequency. It requires downloading a file with data from the home page of this book.

Nitrifying organisms are observed in a batch test. The oxygen concentration in the presence of a nonlimiting amount of ammonium and nitrite is recorded in 1.5-s intervals. The goal is to obtain the best estimate for the value of the Monod saturation coefficient K_O of this culture of organisms. The model is simple and has the following form:

$$\frac{dS_O}{dt} = r_{O,max} \cdot \frac{S_O}{K_O + S_O}$$

S_O = oxygen concentration $[gO_2\,m^{-3}]$
K_O = oxygen saturation coefficient $[gO_2\,m^{-3}]$
$r_{O,max}$ = maximum consumption rate of oxygen $[gO_2\,m^{-3}\,d^{-1}]$

1. Develop a model in BM, read in the data from the relevant file and estimate all parameters of the model. What is the remaining root-mean-square of the residuals? Is this value compatible with oxygen measurement with an electrode?
2. Analyze the time series of the residuals. Does visual inspection reveal any problems? How many sign changes are there? Do the residuals show significant autocorrelation?
3. Introduce an error model to eliminate possible systematic errors of the electrode. Does this improve the properties of the residuals? How does it affect the parameter values?
4. Using sensitivity functions, which parameter will be the most uncertain one? Which one do you think can be estimated quite accurately?
5. One possibility to eliminate autocorrelation from data is adding additional noise to the data. Gaussian noise is normally distributed with an expected value of 0. How large is the standard deviation of the noise, which has to be added, in order to obtain either an acceptable number of sign changes in the time series or uncorrelated residuals? Does this noise affect the identification of the most probable parameter values? Why? Does the noise affect the estimated uncertainty of the identified parameters?

Hints:

Unfortunately BM does not provide us with the covariance matrix of estimated parameters. AQUASIM does provide this information but has other limitations. You may want to use BM to obtain the required noise and then AQUASIM to obtain the covariance matrix.

16.34 Design under Uncertainty, Nitrification

You want to design a nitrifying activated sludge plant, but you are insecure about some of the kinetic parameters of the nitrifying organisms. In order to evaluate the consequences of your uncertainty, you perform a MC simulation.

Figure 16.11 shows the flow scheme of your first design of the plant. The excess sludge is removed directly from the activated sludge reactor, which simplifies

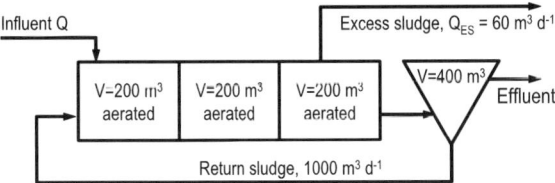

Fig. 16.11 Flow scheme of the planned activated sludge plant

the control of the SRT (sludge age). The volume of the secondary clarifier is 400 m³ and its hydraulic behavior may be simulated with two equal CSTRs in series. You assume that the clarifier does not affect the ammonium concentration in the return sludge. In addition, you assume that all suspended solids are removed in the excess sludge and none in the effluent.

Table 16.21 gives the simplified kinetic model for the nitrifying organisms. It is assumed that only nitrifiers interfere with ammonium. Table 16.22 identifies your uncertainty about the kinetic parameters of the nitrifiers under winter conditions (approximately 10°C) and provides information on the variation of operating conditions.

The discharge requirements for this plant prescribe that the maximum daily ammonium concentration may in winter not exceed $5\,\mathrm{gN\,m^{-3}}$ in at least 80% of the days.

1. What is the effluent ammonium concentration at steady state, based on expected values?
2. What is the maximum, daily ammonium concentration in the effluent of the activated sludge plant, based on expected diurnal variation?
3. For the expected diurnal variation, what is the maximum nitrate concentration in the effluent? (Assume no nitrate in the influent and $r_{NO3} = -r_{NH4}$)
4. How does the secondary clarifier influence the result to the previous two questions?
5. What is the risk that your design will not satisfy the discharge requirements?

Table 16.21 Kinetic model for nitrification

Process	Ammonium	Nitrifiers	
	S_{NH}	X_N	Rate ρ
	gNH_4-N	$gCOD$	
Growth	$-\dfrac{1}{Y_N}$	$+1$	$\mu_N \cdot \dfrac{S_{NH}}{K_{NH}+S_{NH}} \cdot X_N$
Decay		-1	$b_N \cdot X_N$

Table 16.22 Uncertainty and variation in the design of a nitrifying activated sludge plant

Parameter	Units	Expected	Uncertainty	Remarks
μ_N	d^{-1}	0.3	random(0.2, 0.3)	
K_{NH}	$gN\ m^{-3}$	1.0	random(0.5, 2.5)	
b_N	d^{-1}	0.03	random(0.02, 0.06)	
Y_N	$gCOD\ g^{-1}N$	0.24	–	Would not affect result
Parameter	Units	Expected	Variation	Remarks
Q	$m^3 d^{-1}$	1000	normal(1000, 200)	Changes daily
$S_{NH,in}$	$gN\ m^{-3}$	25	$25 + 15 \cdot \sin(2 \cdot \pi \cdot time \cdot f)$	Same every day, $f = 1\ d^{-1}$
Q_{ES}	$m^3 d^{-1}$	60	normal(60, 6)	Changes daily
R	$m^3 d^{-1}$	1000	–	Unchanged

Hints for BM:

An uncertain parameter has to be chosen based on its distribution, but it is kept constant throughout one simulation run. This is obtained with:

init mue = random(0.2,0.3)   next mue = mue

A variable parameter may have to be kept constant for one day but will then be changed to another random value. This can be obtained with:

init QES = normal(60,6)
next QES = IF mod(time,1) < DT then normal(60,6) else QES

The maximum daily value of a state variable is obtained with

DTout = 1 and init Smax = 0
next smax = if mod(time,1) < DT then 0 else max(Smax, Sout)

With a *Batch* run several runs can now be performed with different parameters. The results can then be exported as a two-dimensional table to EXCEL and be analyzed. The problem is that the matrix must be sorted, which may require programming. Be aware that EXCEL cannot accept tables wider than 256 elements.

16.35 Integrated Problem: Nitrification in an RBC

A wastewater treatment plant, which does not nitrify, has been extended with a rotating biological contactor (RBC) for tertiary nitrification. You want to develop a model of this extension in order to analyze some problems in the context of rain events.

The flow scheme of the plant is given in Fig. 16.12. After the secondary clarifier, three reactors, each with $A = 20,000\ m^2$ of biofilm surface and a water volume of $V=100\ m^3$, are followed by a final settling tank with a volume of $1000\ m^3$.

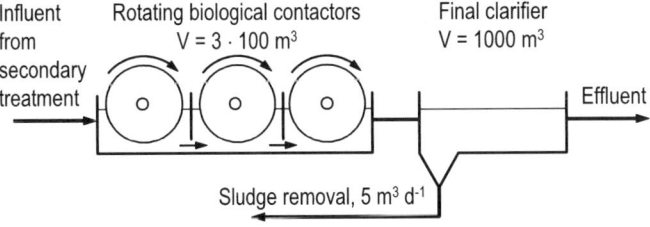

Fig. 16.12 Flow scheme of the RBC plant

Tracer concentration S_T, g m^{-3}

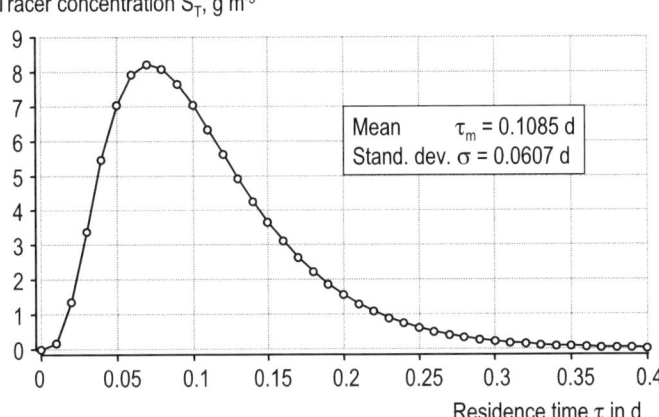

Fig. 16.13 Hydraulic residence time distribution for the entire tertiary plant

You obtained the hydraulic residence time distribution between the influent to the RBC and the effluent of the final clarifier, which is given in Fig. 16.13.

Table 16.23 characterizes the influent to the RBC; Table 16.24 shows the biokinetic model which you want to use for the RBC. You assume that biomass production and erosion from the biofilm are in equilibrium.

1. Is the flow rate in the report of the operator (Table 16.23) compatible with your RTD? If not continue with the correct value. What is it?
2. You want to model the final clarifier with a cascade of CSTRs. How many reactors in series do you choose?
3. Determine the missing composition factors and stoichiometric coefficients in Table 16.24. Indicate their units.
4. If you measure 1 g m^{-3} NO$_2$–N and 10 g m^{-3} NO$_3$–N in the effluent, how large is then the biomass concentration in the sludge removal if you assume perfect separation of the biomass in the settler?
5. What are the steady-state concentrations for ammonium and nitrite in the effluent? Can you estimate the nitrate concentration without including nitrate in the simulation?

Table 16.23 Characterization of the influent to the tertiary plant as reported by the operator (you may assume a steady state, except in the case of a rain event)

Parameter	Symbol	Value	Units
Influent flow rate	Q_{in}	10000	$m^3 d^{-1}$
Sludge removal rate	Q_{SR}	5	$m^3 d^{-1}$
Ammonium in influent	$S_{NH,in}$	15	$gN\ m^{-3}$
Nitrite	$S_{NO2,in}$	0	$gN\ m^{-3}$
Nitrate	$S_{NO3,in}$	0	$gN\ m^{-3}$
Biomass	X_{in}	0	$gCOD\ m^{-3}$

Table 16.24 Microbial processes for nitrification. j_{NH4} is the mean flux of ammonium into the biofilm in $gN\ m^{-2} d^{-1}$ (averaged over one revolution). The ratio A_{RBC}/V_{RBC} transforms the units from area to volume

Process	Oxygen S_O gO_2	Ammonium S_{NH} gN	Nitrite S_{NO2} gN	Nitrate S_{NO3} gN	Biomass X $gCOD$	Process rate ρ
Nitritation	-3.22	?	$+1$		?	ρ_1
Nitratation	-1.11	?	-1	$+1$	?	ρ_2
Conservatives						
gTOD	?	?	?	-4.56	1	
gN	?	?	?	?	0^1	

[1] This is chosen here to simplify the example. In reality this value would be about 0.07 gN g^{-1}TOD. You may want to solve the problem with this value.

$$\rho_1 = j_{NH4} \cdot \frac{S_{NH}}{K_{NH} + S_{NH}} \cdot \frac{A_{RBC}}{V_{RBC}} \quad \text{and} \quad \rho_2 = j_{NO2} \cdot \frac{S_{NO2}}{K_{NO2} + S_{NO2}} \cdot \frac{A_{RBC}}{V_{RBC}}$$

$$j_{NH4} = 4\ gN\ m^{-2} d^{-1}\ (20°C) \qquad K_{NH4} = 2\ gN\ m^{-3}$$

$$j_{NO2} = 5\ gN\ m^{-2} d^{-1}\ (20°C) \qquad K_{NO2} = 1\ gN\ m^{-3}$$

6. Considering uncertainties, is it possible to accurately estimate the parameters of the kinetic model from a length profile of the concentrations for NH_4 and NO_2 in the influent and the effluent of the three RBC reactors at steady state?

7. During a rain event the flow rate is doubled for 2 h, but the ammonium concentration remains unchanged because of the large volume of the primary and secondary treatment. By chance you have observed the ammonium concentration in the effluent of the final clarifier with high temporal resolution (results not shown). Do you think that you can use this data to evaluate the four parameters of the kinetic model (j_{NH4}, K_{NH4}, j_{NO2}, and K_{NO2})?

8. During the rain event under point 7 you observed a maximum ammonium concentration in the effluent of the final clarifier of $9\ gN\ m^{-3}$. Is this concentration compatible with your uncertainty about the values of the kinetic parameters below Table 16.24? Your estimate as an expert is that the coefficient of variation ($cv = \sigma/\mu$) of all four parameters is 15%. The estimates are independent of each other. You assume normal distribution of the uncertainty.

Hints for BM:

You can plot the sensitivity functions for a length profile by choosing [i] as the x axis (*Graph – Choose variables – x Axis*).
You can generate the rain event as characterized with a SQUAREPULSE(t,d) (pulse of height 1 starting at time t with duration d).

16.36 Integrated Problem: Analyzing a Fish Pond

A small brook with a flow of $Q = 0.01 \pm 0.001 \, m^3 \, s^{-1}$ flows through the property of a rich, rather mean old man (Fig. 16.14). On his property he maintains a fish pond with a volume of $2000 \pm 200 \, m^3$. Some water fountains serve for good mixing of the pond. The owner is allowed to divert up to 30% of the flow of the brook into the pond and to discharge the effluent back into the brook. The owner feeds his fish daily with 5 kg of food. He never catches any fish.

The brook is rather clean, and you cannot observe any algae growing above the property, but below you observe green, suspended algae. You suspect that these algae originate from the pond and that they grow on the nutrients released by the fish. On the property, the brook is 500 m long and flows with a velocity of $v = 0.5 \pm 0.05 \, m \, s^{-1}$.

Based on literature information, you estimate the dispersion coefficient in the brook as $D_D = 0.05 \, m^2 s^{-1}$. For the growth of the algae you develop the model in Table 16.25.

All this information is not very accurate and ± indicates the range of a standard deviation. The concentration of the tracer you will be using can only be obtained with a repeatability of ±2%.

You would like to check how much water the owner is using in his pond. This could be simple, but the property is fenced, and the dogs of the owner look vicious.

1. How do you proceed? What will you measure? What do you expect? You should develop a model which you may use to evaluate the results of your experiment.

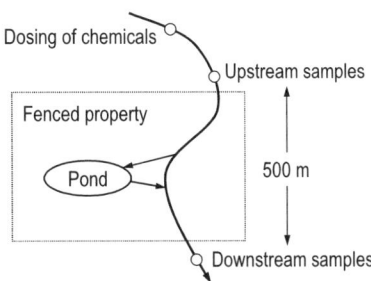

Fig. 16.14   Situation of the property

Table 16.25 Growth of algae

	Phosphorus S_P gP	Algae X gTSS	Fish-food gTSS	ρ
Growth	?	1		$\mu \cdot \dfrac{S_P}{K_P + S_P} \cdot I_S \cdot X$
Respiration	?	−1		$k_R \cdot X$
Feeding fish	?		1	k_{Food}
Composition				
Phosphorus gP	1	0.02	0.05	
TSS gTSS		1	1	

I_S = light intensity of the sun (in BM: IS = if mod(time,1)<0.5 then sin(2*pi*time) else 0)
μ = 2 d^{-1}, max growth rate of algae at full light intensity
K_P = 0.05 gP m^{-3}
k_R = 0.2 d^{-1}

2. You do not want to use a visible tracer in order not to give away the time when you perform the experiment. When do you take samples? You should realize that this tracer is expensive, and its analysis is costly.
3. The results of the experiment indicate that the owner extracts 40% rather than the allowed 30% of the water. Would you dare to go to court based on your results?
4. What concentration of algae would you expect in the effluent of the property? Assume expected values and a use of 30% of the water.
5. You consider adding 1000 gP to the brook upstream of the property and then, recording continuously the turbidity in the effluent of the property, to follow the development of the algae concentration. Could you derive an estimate of the fraction of the water which is extracted to the pond from such an experiment?
6. The owner is quite frustrated about the regular algae blooms in his pond. In order to provide him a service, you want to make some positive suggestions how he could decrease the algae growth. Which parameters are the most sensitive? Could changing the flow pattern in the pond be successful?

Literature

Beck B.M. (1983) A procedure for modeling in Mathematical Modeling of Water Quality, Orlob
G.T. editor, Int. Series on App. Systems Analysis, Vol. 12, Wiley, S. 11–41

Benjamin J.R. and Cornell C.A. (1970) Probability, Statistics and Decision for Civil Engineers,
McGraw-Hill, New York

Bird R.B., Stewart W.E., and Lightfoot E.N. (1960) Transport Phenomena, 2nd Edition, Wiley

Bronstein I.N., Semendjajew K.A., Musiol G. und Mühlig H. (1993) Taschenbuch der Mathe-
matik, Verlag Harri Deutsch, ISBN 3-8171-2001-X

Brun R., Reichert P., and Künsch H.R. (2001) Practical identifiability analysis of large environmental simulation models. Water Resour Res 37, 1015–1030

Coleman, H.W. and Steele, W.G. (1998) Experimentation and uncertainty analysis for engineers,
2nd edn. Wiley, NY. ISBN 0-471-12146-0

Elder J.W. (1959) The dispersion of marked fluid in turbulent shear flow. J Fluid Mech 5(4):
544–560

Fischer H.B., List E.J., Koh R.C.Y., Imberger J., and Brooks N.H. (1979) Mixing in Inland and
Coastal Waters, Academic

Fischer H.B. (1975) Simple method for predicting dispersion in streams. J Environ Eng Div
ASCE 101(3):453–455

Gujer W. and Zehnder A.J.B. (1983) Conversion processes in anaerobic digestion. Water Sci
Technol 15(8–9): 127–167

Gujer W., Henze M., Mino T., and van Loosdrecht M.C.M. (2000), Activated Sludge Model
No.3, Scientific and Technical Report No.9. IWA, London

Gujer W. (2002) Microscopic versus macroscopic biomass models in activated sludge systems.
Water Sci Technol 45(6):1–11

Gujer W. and von Gunten U. (2003) A stochastic model of an ozonation reactor. Water Res
37(7): 1667–1677

Gujer W. (2004) Systems analysis in environmental engineering: how far should we go? Water
Sci Technol 49(8): 37–42

Gujer W. (2006) Siedlungswasserwirtschaft, 3. edn. Springer-Verlag, Berlin Heidelberg, ISBN
3-540-43404-6

Hemmi P. und Profos P. (1997) Grundlagen der Messtechnik. Oldenburg, München

Henze M., Grady C.P.L. Jr, Gujer W., Marais G.v.R., and Matsuo T. (1987) Activated Sludge
Model No. 1, IAWPRC Scientific and Technical Report No. 1. IAWPRC, London

Huisman J. L., Burckhardt S., Larsen T. A., Krebs P., and Gujer W. (2000) Propagation of waves
and dissolved compounds in sewer. J Environ Eng ASCE 126(1): 12–20

Imboden D.M. und Koch S. (2003) Systemanalyse – Einführung in die mathematische Modellie-
rung natürlicher Systeme. Springer Verlag, ISBN 3-540-43935-8

Ives K.J. (1960) Rational design of filters. Proc Inst Civ Eng (UK) 16:189–193

Levenspiel O. (1999) Chemical Reaction Engineering, 3 edn, Wiley

Levenspiel O. and Smith W.K. (1957) Chem Eng Sci, 6:227, cited in Levenspiel (1999)

Linde R.L. (1999) Handbook of Chemistry and Physics, 80th edn. CRC Press

Maniak U. (1997) Hydrologie und Wasserwirtschaft, 4. Aufl. Springer-Verlag, ISBN 3-540-63292-1

Mann H., Schiffelgen H., and Froriep R. (2000) 'Einführung in die Regelungstechnik', 8. Aufl., Hanser Lehrbuch, Carl Hanser Verlag, München Wien

Manser R., Gujer W., and Siegrist H. (2005) Consequences of mass transfer effects on the kinetics of nitrifiers. Water Res 39(19): 4633–4642

McGhee T.J. (1991) Water Supply and Sewerage, 6th edn. McGraw Hill S. 389

Michaelis L. und Menten M. L. (1913). Die Kinetik der Invertinwirkung. Biochem Zeits 49: 333–369

Morgan M.G. and Henrion M. (1990) Uncertainty, Cambridge University Press, Cambridge. ISBN 0-521-42744-4

Murphy K.L. and Boyko B.I. (1970) Longitudinal mixing in spiral flow aeration tanks. J Sanitary Eng Div ASCE 96:211–221

Port, E. (1994) Anforderungen an die Eigenüberwachung bei kommunalen Kläranlagen, WAR, TH Darmstadt. ISBN 3-923419-68-6, 75:353–361

Reichert P. (2004) UNCSIM package, www.uncsim.eawag.ch

Reichert P. (1998) AQUASIM 2.0 – User Manual, Swiss Federal Institute for Environmental Science and Technology, CH 8600 Dübendorf, Switzerland, ISBN: 3-906484-16-5

Reichert P. (1999) Umweltsystemanalyse, Vorlesungsskript, ETH Zürich

Reichert P. (1995) Design techniques of a computer program for the identification of processes and the simulation of water quality in aquatic systems. Environ Softw 10(3):199–210

Reichert P. (1994) AQUASIM – A tool for simulation and data analysis of aquatic systems. Water Sci Technol 30(2):21–30

Reichert P. (1994) Concepts underlying a Computer Program for the Identification and Simulation of Aquatic Systems, Swiss Federal Institute for Environmental Science and Technology, CH 8600 Dübendorf, Switzerland, ISBN: 3-906484-08-4

Rieckermann J., Neumann M., Ort C., Huisman J.L., and Gujer W. (2005) Dispersion coefficients of sewers from tracer experiments. Water Sci Technol 52(5), 123–133.

Ross S. (2003) *Peirce's Criterion for the Elimination of Suspect Experimental Data*, J Eng Technol, Fall, 2003

Stahel W. (2002) Statistische Datenanalyse, 4. Aufl., Friedr. Vieweg, ISBN 3-528-36653-2

Streeter H.W. und E. Phelps (1925) A Study of the Pollution and Natural Purification of the Ohio River, Bulletin 146, U.S. Public Health Service

Thomann M.P. (2002) Datenkontrolle von Abwasserreinigungsanlagen mit Massenbilanzen, Experimenten und statistischen Methoden, Schriftenreihe IHW/ETH Zürich, Band 15, ISBN 3-906445-15-1 sowie Dissertation ETH No. 14824 (available online)

van der Laan E.T. (1958) Notes on the diffusion-type model for the longitudinal mixing in flow. Chem Eng Sci 7:187–191

Ziegler J.G. and Nichols N.B. (1942) Optimum settings for automatic controllers. Trans ASME 64:S759–768 (details also in Mann et al., 2000)

Index